Proton Radiotherapy Accelerators

Proton Radiotherapy Accelerators

Wioletta Wieszczycka
Waldemar H. Scharf
Warsaw University of Technology, Poland

Published by

World Scientific Publishing Co. Pte. Ltd.

P O Box 128, Farrer Road, Singapore 912805

USA office: Suite 1B, 1060 Main Street, River Edge, NJ 07661

UK office: 57 Shelton Street, Covent Garden, London WC2H 9HE

British Library Cataloguing-in-Publication Data
A catalogue record for this book is available from the British Library.

PROTON RADIOTHERAPY ACCELERATORS

ISBN 981-02-4528-9

Printed in Singapore by World Scientific Printers (S) Pte Ltd

This book is dedicated to my Parents and my brother Karol

Wioletta Wieszczycka

PREFACE

Hadronic radiotherapy uses particle beams to treat tumors located near critical body structures and tumors that respond poorly to conventional photon and electron beam radiotherapy. Intital research in hadronic radiotherapy was performed using accelerators built for physics research. The good results of the proton and ion therapy programms have enhanced the tendency to use protontherapy as a routine method. There are about 20 working protontherapy facilities (first, second and third generation) and more than 30 centers is planned.

This book presents the first comprehensive overview of the field with a discussion on the fundamental basis of particle physics and radiobiology, as well as review of clinical and technical specificatioons and designs for proton radiotherapy. In particular, the current designs of proton and heavy ion accelerators, beam delivery systems, gantries, beam monitoring and dosimetry systems, patient positioning and immobilisation devices, control and safety systems, accelerator and whole center shielding, and ancillary treatment facilities are widely discussed. The book contains the review of the global costs and financial analysis of the activities of the proton radiotherapy center. Finally, a proposal of a dedicated hospital-based protontherapy facility is presented.

The book is devoted to engineers and physicians involved in the design and construction of proton radiotherapy accelerators, accelerator and biomedical physics and devoted to radiotherapists, as well as undergraduate and graduate students in high energy physics.

The present book owes a great deal to Dr. André Laisné (Pantechnik, France) who has shown a great interest and offered a great help by reviewing the whole manuscript of *Proton Raditherapy Accelerators*. The authors also want to express their thanks to Prof. Ugo Amaldi from CERN and Prof. Sadayoshi Fukumoto and Kuninori Endo (both from KEK, Japan) for valuable materials and remarks on the book. We also wish to give a special thanks to Dr. S. Ternier (IBA), Dr. I. Akifumi (Hyogo), Dr. A. Maruhashi

(PMRC), Prof. A. Mazal (CPO), Dr. P. Chauvel (CAL), Prof. A.G. Molokanov(JINR), Dr. S. Fukuda (Wakasa Way), Dr. T. Nishio (NCC) - for interesting materials and valuable suggestions.

Wioletta Wieszczycka
Waldemar H. Scharf
Warsaw, Poland
January 2001

CONTENTS

Chapter 1

INTRODUCTION

1.1 Cancer and Radiation Therapy

Cancer can broadly be defined as the uncontrolled growth and proliferation of groups of cells. It is reported that in the year 1982 were 1.2 million new cancer incidents in the countries of European Community [1.1]. 750 000 deaths were attributed to cancer in 1985, which means that one out of five cancer patients dies from this disease [1.3]. An increasing trend of the fatal cancer cases is still observed. In developed countries about 30% of people suffer from cancer and about half of these die from this disease. It corresponds to about 1 million deaths per year because of tumor [1.4, 1.5]. Table 1.1 compares some available statistics and estimations concerning cancer patients worldwide presented in different papers.

Table 1.2 and Fig. 1.1 present situation in cancer treatment. The prognosis in individual cases varies greatly and depends on tumor type, stage and diagnosis, general health of the patient, etc. It is seen, that in Europe 45% of all the patients are "cured", which means that these patients have a symptom-free survival period exceeding five years. One can also read that 90% of the cured patients (i.e. 40% of the total) are cured because of loco-regional control of the primary tumor. Thus among various cancer treatments: surgical removal of the tumor tissue, radiotherapy, chemotherapy and immunotherapy, the first two are today of crucial importance. Surgery and radiotherapy alone are successful in 22% and 12% of cases respectively. When combined they account for another 6% of the cases, so that radiotherapy is involved in almost half of the curative treatments of loco-regional type. For 18% of all cancer patients the local control of the primary tumor without metastases fails. These patients could be cured successfully if improved local or loco-regional treatment techniques were available. Among these improved treatment modalities is the application of proton and ion beams in radiation therapy [1.5].

Country	Year	Country population [mln]	Number of cancer patients per year	Number of deaths per year	Dedicated for radiation therapy	Estimated for Proton-therapy	Ref.
world wide	1975		6 mln.				1.3
USA	1991	~245		225 000		130 000[1]	1.6
	1993		1 170 mln		660 000[2]		1.3
Japan	1991	~124		223 604			1.8, 1.3
	2000			500 000			1.8, 1.3
India	1989	~778	1,5 mln.				1.3
	2000		3 mln.[3]				1.3
Europe	1982		1,2 mln.				1.1
	1985			750 000			1.9, 1.3
	1992					111 000[4]	1.10
	1997	~320			600 000	60 000	1.12
						280 000 (25 000)[5]	1.12
	2000			1 000 000			1.9, 1.3
Germany	1993		400 000		250		1.13, 1.3
	1996	~82		50-70 000			1.14
	1997					8 700	1.5
Austria	1994	~8				2 500 (4 500)[6]	1.10
Italy	1996	~57			120 000	12 000	1.4, 1.15
UK	1994	~56		40 000			1.16
Czech	1997	~10				700-4500	1.17
Poland	1989			70 000			1.3, 1.18
	1998	~39,6		112 200[7]		5925–6600[8]	1.19, 1.26

Table 1.1. Statistics and estimations concerning cancer patients worldwide.

Table1.2 shows that, contrary to widespread belief; all the other systematic treatments account for only 5% of the cured patients. There is ample space for improvements here, because 37% of the tumors are metastasized at the moment of the diagnosis and cannot be cured with loco-regional treatments alone [1.4].

[1] Number of patients dedicated for protons and Three Dimensional Conformal Radiation Therapy 3D CRT.
[2] There are 230 patients per one radiotherapeutical accelerator.
[3] According estimations of author: 0.5 mln. of new cancer cases are added every year.
[4] Estimations made by EULIMA 1992.
[5] First priority cases.
[6] Estimated patients from Austria, northern Italy and southern Germany.
[7] According calculations concerning whole Poland based on data that there are following numbers of patients per 10 000 inhabitants: Warsaw – 32, Płock – 28, Katowice – 25.
[8] Estimations made by W. Wieszczycka in [1.26]

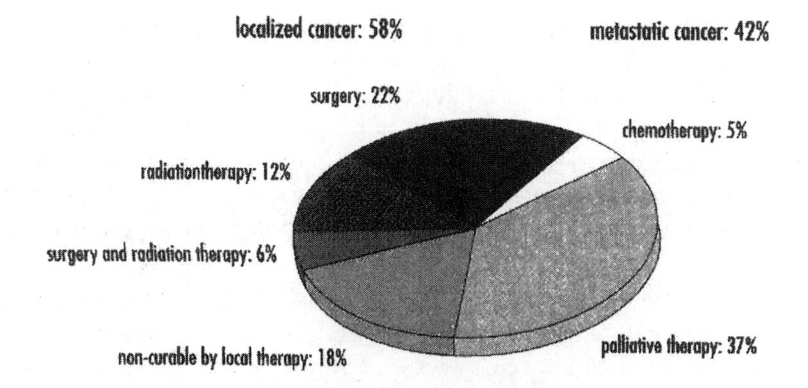

Figure 1.1. Present situation in cancer treatment; after [1.24].

	At the moment of diagnosis		
	Primary tumor	Metastasized tumor	
Treatments used:			
surgery alone	22%		
radiotherapy alone	12%		
both combined	6%		
all other treatments and combinations including chemotherapy		5%	
Patients curable now	40%	5%	45%
No cure available	18%	37%	55%
	58%	42%	

Table 1.2. Types of therapies used in cancer treatment; after U. Amaldi et al. [1.4]

1.2 Historical Development of Radiation Therapy

Within a few months of Roentgen's discovery of x-rays in 1895, investigators realized that X rays could be used for both diagnostic and therapeutic purposes. Table 1.3 made on the basis of the annex of the report *Towards Co-ordination of Cancer Research in Europe* describes the progress of radiotherapy in about hundred years.

The earliest radiation sources were provided by *gas-filled x-ray tubes*. It was W.D. Coolidge who, in 1913, did the groundwork for the present-day x-ray techniques by developing a *vacuum tube* with a hot tungsten cathode.

The earliest tubes of this type, manufactured for general use, operated at a peak voltage of 140 kV with a 5 mA current. By 1937, x-ray tubes used in radiation therapy operated at 400 kV with 5 mA of current and tubes operating at 200 kV had currents of up to 30 mA. x-rays generated by these tubes were, unfortunately, fairly soft and from the medical point of view, the depth-dose curves were particularly disadvantageous since the maximum dose would be delivered at the skin surface and then would rapidly fall off with depth in tissue. For this reason, during the early experience with vacuum x-ray tubes, first attempts were also made at searching for sources of a more penetrating radiation with attention focused on gamma emitting radionuclides. The first device using harder radiation was probably a *radium cannon* developed by Koenig at a Women's Hospital in Breslau (now Wroclaw) in 1912 for the treatment of pelvic cancer. The first radium cannons, utilizing a few grams of radium-226, a radionuclide that emits gamma rays with energies ranging between 0.24 and 2.20 MeV, were installed at only some dozen or so places. The main obstacle to a wider use of these devices was their low availability and the high cost of radium. Therefore, as early as 1935, when it was generally accepted that the future developments in radiotherapy would involve *megavoltage therapy*.

In the early half of the 1950's the era of *cobalt sources* for radiotherapy, containing several kilocuries of ^{60}Co, began. Cobalt-60, emitting gamma rays with energies of 1.17 MeV and 1.33 MeV, made it possible to obtain far better depth-dose curves than those for the x-radiation emitted by an x-ray tube. The ^{60}Co depth dose curve had its maximum at about 5mm below the skin surface, thus markedly decreasing the dose to the skin when deeper layers had to he irradiated. The first three ^{60}Co sources were produced in 1952 and installed in Canada and the USA.

In the early 1930's several attempts were made to use *resonance transformers* in order to attain peak voltages of 1 to 2 MV. However, accelerators equipped with transformers of this kind did not find wide application in radiotherapy as greater attention was being focused at that time on a high voltage accelerator developed in 1932 by R. Van de Graaff and later named the *Van de Graaff machine* after its inventor. The first hospital-based accelerator of this type was a 1 MeV air-insulated machine installed in Boston in 1937. The second machine, a pressure-insulated 1.25 MeV version, was installed in 1940 at the Massachusetts General Hospital also in Boston. The real era of Van de Graaff accelerators began in 1946, when the High Voltage Engineering Corporation was formed by R.

Van de Graaff and began commercial production of 2 and 2.5 MeV machines. A total of 40 accelerators were built, but their production was discontinued in 1959. They were of quite great size and rather unmaneuverable, occupied a large hospital area, and involved high irradiation costs. In spite of that, they turned out to be very reliable and at least 12 of them were still in operation as late as 1983.

Dramatic increase in the photon radiation energy was made possible by the development of a new accelerator, the *betatron*, by D.W Kerst, who, in 1943, suggested its use as a radiation therapy tool. The first patient was irradiated in 1949 with x-rays generated by 20 MeV electrons from a Kerst betatron installed in Urbana (USA). Later, the Allis-Chalmers company started commercial production of 20 - 25 MV betatrons for radiotherapy. In 1977, 45 betatrons were in operation in the USA alone.

In Europe, the earliest work on the medical application of betatrons was started in 1948 in Gottingen (6 MV Gund betatron) and the first 31 MeV betatron, produced by the Brown-Boveri company and adapted for radiation therapy, was put in operation in 1951 at the Cantonal Hospital in Zurich (Switzerland). Through 1966, this company manufactured more than 40 betatrons. The Siemens company (Germany) produced more than 70 betatrons in the same period and in the early 1970's about 200 betatrons were in medical use the world over.

Betatrons have played a significant role in the development of radiation therapy since they deliver x-rays with far better properties, such as skin sparing, higher depth dose and less side scatter than those of the radiation generated by x-ray tubes and radioisotope sources. The betatron's major disadvantages, such as its weight which makes this machine unwieldy, the relatively low intensity of the x-ray beam, as well as the small treatment field area have contributed to the discontinuation of its production in the mid-1970's. Since then, the use of betatrons has shown a gradual decline. In 1977, in the USA there were 45 betatrons in use, while in 1982 their number dropped to 32.

The advances made during World War II in the construction of magnetrons for radar techniques have made it possible to use microwave generators for electron acceleration in radiation therapy. In England, D.W. Fry demonstrated the first *traveling-wave linear accelerator* in 1946. On this basis, C.W. Miller, in cooperation with Metropolitan Vickers Ltd. and the Atomic Energy Research Establishment at Harwell (England), designed the first British stationary rf linear accelerator, which was subsequently installed at Hammersmith Hospital in 1952. The first patient was treated on this machine in 1953, and the last one in 1969.

50% of cancer patients are treated by radiotherapy 18% of cancer patients are treated by radiotherapy either alone (12%) or in combination (6%)			
DATE	**PROCESS**	**ENERGY IN MeV**	**COMMENT**
X-RAY RADIOTHERAPY			
1905	x-ray tube	0.05 –0.45	Superseded
1947	Van de Graaff	3	Superseded
1948	Betatron	25	Superseded
1953	Linear accelerator	4–25	More effective for deep seated tumors
GAMMA-RAY RADIOTHERAPY			
1910	Radium needles	1-3	Superseded
1951	Cobalt "bomb"	1.17, 1.33	Still used
ELECTRON RADIOTHERAPY			Useful for superficial cancers
1947	Van de Graaff	3	Superseded
1948	Betatron	25	Superseded
1963	Linear accelerator	4-25	Still used
NEUTRON THERAPY			Treatment of tumors insensitive to gamma therapy thanks to a better bio-logical effect (for instance prostate, he-ad and neck cancers)
1969	Cyclotron	30	
1975	Deuterium-tritium accelera-tors	14	
1968	Californium-252		
PROTONTHERAPY			
1955	Cyclotrons, synchrotrons, li-near accelerators	60-250	Treatment of tumors with very high ac-curacy, required when the tumor is clo-se to a vital structure (spinal cord, bra-in, etc.).
NEW THERAPIES UNDER DEVELOPMENT			
Boron neutron capture radiotherapy (BNCT)		High accuracy treatment of tumors, which take up boron, for instance gliomas.	
Light ion radiothe-rapy	energies 400 MeV / u	Combines the advantages of neutron and proton therapies, i.e. treatment of tumors insensitive to gamma therapy thanks to greater biological effect plus high accuracy to spare healthy tis-sues.	

Table 1.3. Outline of the progress of radiotherapy; on the basis of U. Amaldi et al. [1.4].

In the same period, intensive work began in the USA where H.S. Kaplan at the Stanford Medical Center, in cooperation with a team from the Stanford High Energy Physics Laboratory built a prototype linear accelerator machine for x-ray therapy at an energy of 5 MeV. It was installed at Stanford in 1955, and treatment was started in early 1956. Soon after, a clinical linear accelerator construction program was implemented and Varian Associates designed and built a prototype isocentric accelerator, which permitted a full rotation around the patient. This machine was installed at the UCLA Medical Center in 1962, and the first commercial machine was set up at the Stanford University School of Medicine (Palo Alto).

Thus, in radiation therapy the mid-1960s saw the beginning of the era of rf linear accelerators, which were soon to take up a dominant place on the world market of medical accelerators. At that time, the limitations of betatron technology became more evident. These were, above all, the low dose rate, 40 cGy / min (40 rad / min) at 100 cm treatment distance, and energy of 22 MeV as well as small treatment field areas which, at 100 cm, were 12.5×12.5 cm^2 and 8 cm max diameter for x-ray and electron therapy, respectively. The principal disadvantage of betatrons was that they could not be rotated a full 360° and thus were limited to arc therapy only. It is generally agreed that it was this fact, together with the availability of high dose rates, that contributed to the rapid growth in the number of rf linear accelerators and their dominant role among world's radiotherapy machines. In fact, linear accelerators can deliver dose rates of 200 - 500 cGy/min (200 to 500 rad / min), with field capabilities of up to 40×40 cm^2.

The subsequent development of radiation therapy linacs is shown in Fig. 1.2. The design of the first generation of linacs in the mid-1950s was based on the so-called in-line position of the accelerating wave-guide (Fig. 1.2a). The linear energy gradient was 4 MeV m^{-1}, which made it impossible to be rotated round the table, with partial rotation axis placed rather low, at 112 cm above the floor (Fig. 1.2a). In the 1960s, machines appeared featuring a horizontally mounted fairly long accelerating waveguide that generated high energy beams. This structure made use of a magnet bending the beam through a 90° angle and thus making the vertical dimension of the waveguide considerably shorter.

In this design, the treatment head became small enough to pass under the table during a full 360° rotation of the gantry with the isocenter placed relatively low (Fig. 1.2b).

In the early 1970s, a new generation of compact accelerators was developed, which used *achromatic magnets*, emitting a beam through an angle of about 270° (Fig. 1.2c). On substituting achromatic magnets for *non-achromatic magnets* the beam parameters and the stability of the treatment fields were greatly improved.

What was important in the above design was that a waveguide with a *traveling wave* was replaced by a structure using a *standing wave*. This made it possible to increase the linear gradients up to the currently obtainable 12 to

18 MeV m^{-1}, which, in turn, led to considerably shorter accelerating structures with comparable energies. In this way, the accelerator became smaller, and the rectilinear waveguide could again be employed (Fig. 1.2d), and therefore bending magnets could be eliminated. Full rotation of the treatment head was ascertained, though the position of the isocenter had to be increased. Such a design is being currently employed in accelerators with lower energies of 4 to 6 MV.

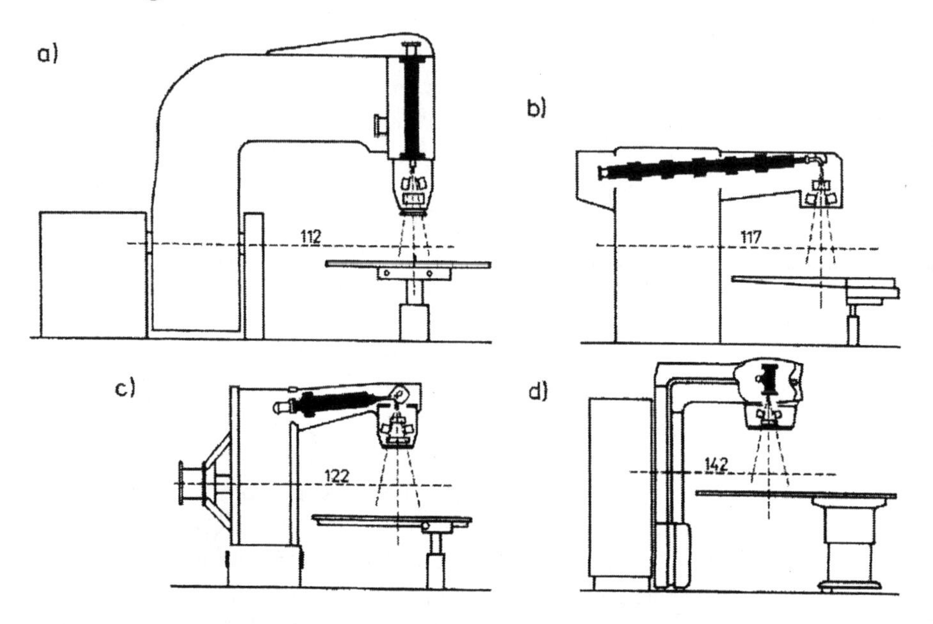

Figure 1.2. Development of radiation therapy linear accelerators (shaded parts represent accelerating structures drawn to the same scale; the distance of the iso-center from floor is given in cm): (a) – 4 MeV AEI linac with an in-line treatment head, mfd. in 1955; (b) - the SL75 Philips linac with a horizontal accelerating structure and a 90° bending magnet, suitable for full rotation therapy, mfd. in 1964; (c) - the Dynaray 4 linac with a 270° achromatic magnet, mfd. in 1971; (d) - a present-day system with an in-line treatment head, without beam bending, e.g. 4 MEV LMR-4 Toshiba Linac with 35 cm long accelerating structure.

In consequence of the above development, the maximum treatment fields were considerably increased, up to the presently available 40 × 40 cm^2 or more, at the rotation axis plane.

The photon dose rates have been increased from the initial values of 100 – 200 cGy min^{-1} to 400 – 500 cGy min^{-1}. In this way, the exposure times became much shorter even for wedge filters. In order to achieve optimum treatment of both thin and thick layers, the initial energy range of 4 – 5 MV has been extended to 4 – 25 MV. At the same time, multiple-energy mode of operation has been implemented, that is the possibility of using two photon energies and a wide electron energy range has been made use of.

The reliability of accelerator operation has been greatly improved. For some critical parts such as microwave tubes or electron guns, it is up to several years. The guaranteed lifetime of electron guns is, at present, as long as five years. The previously used vacuum oil pumps have been replaced with ion pumps. This, together with the development in the design and construction of the accelerating structure, has led to the improvement in the quality of the vacuum, in turn, eliminated sparking and breakdowns, previously common especially at higher energy gradients.

Actually there is a strong trend of increased use of accelerators compared with Co-60 units [1.22].

1.3 Modern Tools for Conventional Radiotherapy

1.3.1 Main Parameters of Conventional Therapy Beams

Today, as sources of radiation for modern radiotherapy with collimated beams, the *electron linear accelerators* are used. In general, they have energy range between 3 and 25 MeV. Such accelerators, generally of the standing wave type, where the required energy for the acceleration of the electrons is supplied by microwaves generated by a klystron (at a frequency of a few GHz), are capable of producing:

- beams of practically *monoenergetic electrons*, with energies varying between 3–4 MeV and 20-25 MeV and of cross-section. in the range between a few cm^2 and few tens of cm^2 at the treatment distance (of the order of 1 m);
- *photon beams*, obtained by slowing down the accelerated electrons in a heavy target (bremsstrahlung); these beams are characterized by a continuous energy spectrum, with maximum energy corresponding to the electron energy, and cross-section. equal to the ones achievable with electrons [1.4].

Figs. 1.3 and 1.4 present the monodimensional curves of the in-depth transmission of absorbed dose (in water), for electron and photon beams produced by a linear accelerator.

For electrons the maximum range, expressed in centimeters, is almost equal to half the initial energy of the beam, expressed in MeV. Due to these characteristics, electron beams are suitable for the treatment of superficial or semi-deep tumors (a few centimeters, starting from the skin surface).

Photon beams are characterized by the absorption of exponential type, after a maximum, which is reached for example at 3.5 cm for beams of 25 MeV maximum energy. As a consequence of this *build-up effect* in a high-energy photon irradiation the skin dose is relatively low. The photon beams from a linear accelerator, equally characterized by a low local diffusion, are suitable for a very efficient treatment of deep-seated tumors situated at a depth of many centimeters inside the body with respect to the skin surface. In order to irradiate selectively such targets sophisticated irradiation techniques have been developed, which entail the need of using multiple beam entry ports onto a point usually coinciding with the geometrical center of the target (technique of multiply converging beams). In order to apply these irradiation techniques, it is necessary for the whole accelerating structure to rotate around a fixed point in space (*isocenter*) so as to utilize any pre-established beam entry point with respect to the patient.

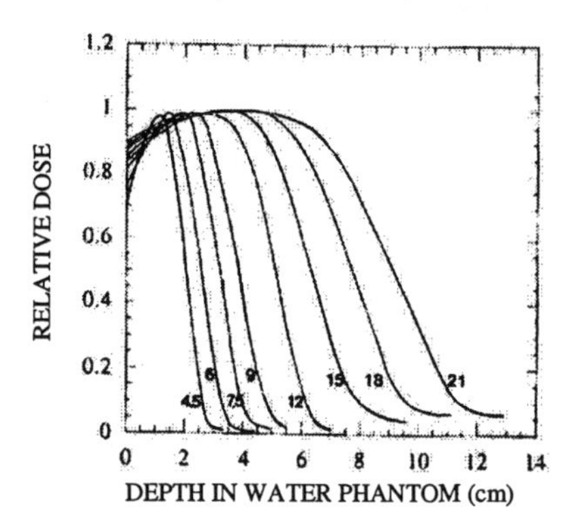

Figure 1.3. Depth–dose curves in water for electron beams in the energy range between 4.5 and 21 MeV. Measurement conditions: field size $10 \times 10\ cm^2$, SSD = 100 cm; after U. Amaldi et al. [1.4].

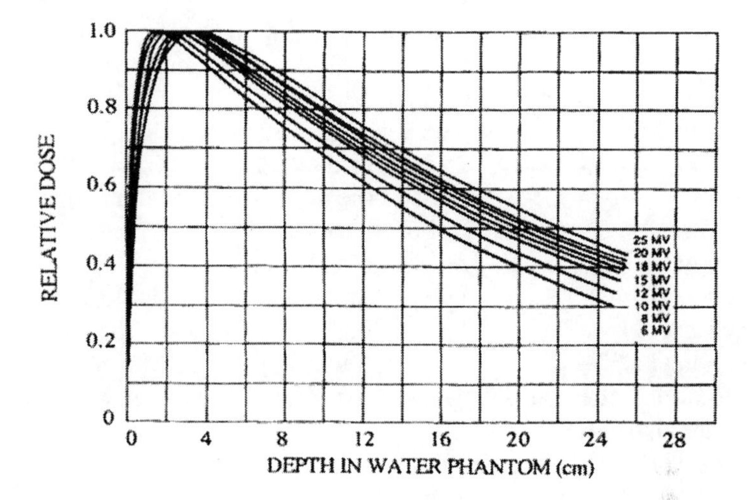

Figure 1.4. Depth–dose curves in water for photon beams in the energy range 6 - 25 MV. Measurement conditions: field size 10×10 cm^2, SSD=100 cm; after U. Amaldi et al. [1.4].

1.3.2 Elements of Conventional Accelerator System

In general, typical accelerator system for conventional radiotherapy consists of: (1) rf generator equipped with magnetron or klystron, (2) accelerating structure, (3) treatment head.

In accelerators for routine radiation therapy the design currently used is that shown in Fig. 1.5. If the magnetron is employed as a source of rf energy, the whole electronic system, together with the accelerating structure and the treatment head is contained in a gantry which can rotate through a full 360° angle around the rotation axis which passes through the rotation point X called *isocenter* (in the right and center part of Fig. 1.5). In some designs, the head can rotate through an angle of 365° to 370°. The rotation speed is controlled typically in the range between 0.1 and 1.0 rev min^{-1}.

The possible axes of rotation and directions of linear movement are identified in Fig. 1.5 (after IEC Documents 601-2-1 and 976). The movements shall be designated as follows:

axis (1) - rotation of the gantry,
axis (2) - roll of the radiation head,

axis (3) - pitch of the radiation head,
axis (4) - rotation of the diaphragm system (beam limiting devices),
axis (5) - isocentric rotation of the table,
axis (6) - rotation of the table top,
axis (7) - pitch of the table (treatment coach),
axis (8) - roll of the table,
direction (9) - height of the table,
direction (10) - lateral displacement of the table,
direction (11) - longitudinal displacement of the table,
direction (12) - axis 1 to radiation source.

In addition, the following geometric parameters (not indicated in Fig. 1.5) are also specified:

direction (13) - height of the radiation source, from the isocenter,
dimension (14) - dimension X of the radiation field at a specified distance,
dimension (15) - dimension Y of the radiation field at a specified distance.

Axes (1) to (2), direction (9), (12) and (13) and dimensions (14) and (15) shall be provided with scales. Rotation scales shall be graduated in units of degrees, using positive numbers, for example ...358°, 359°, 0°, 10, 2°... irrespective of movement available. Linear scales shall be graduated in units of centimeters with subdivision at intervals of 5 mm or less.

In the system shown in Fig. 1.5, the patient lies on the treatment couch, which has a wide range of linear and rotational movements. For most treatment techniques, the patient lies prone or supine and is immobilized. The tumor center must be located at the isocenter, a point defined by the orthogonal intersection. of the axis of rotation of the head and the therapy beam axis. Patient positioning is usually performed using three orthogonal light (laser) beams projected vertically from the ceiling of the therapy room and from its two sides. Due to these very small aperture, beams and skin markers the irradiated field can be very well localized. The patient is subsequently positioned and the gantry can be fixed in a predetermined position (field therapy) or can be rotated (rotational or arc therapy).

Thanks to their compact construction, the present-day radiation therapy accelerators can be installed in rooms whose height is a mere 2.5m [1.22].

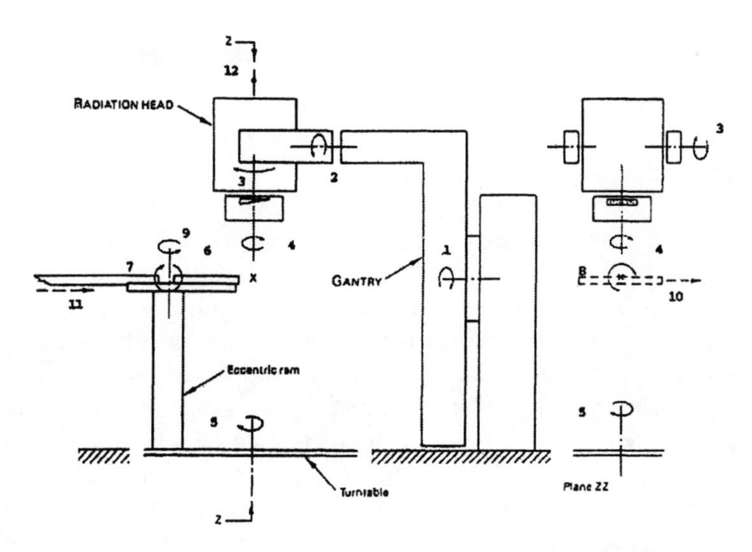

Figure 1.5. The rotary gantry (schematic); after IEC Document 976.

The example of the solution of a modern treatment place equipped in the linear accelerator is presented on the Fig. 1.6.

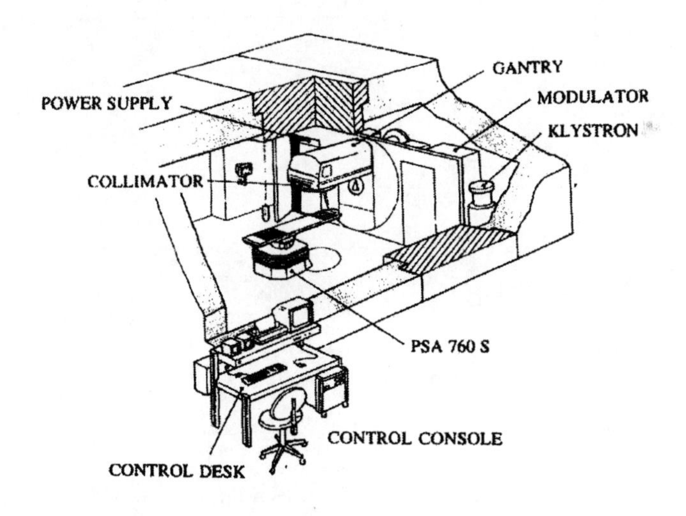

Figure 1.6. A modern treatment place equipped in the linear accelerator; after U. Amaldi et al. [1.4].

1.3.3 Reasons for Development of Non-Conventional Radiotherapies

Despite the advantages of modern electron linacs, extensive clinical experience with photon therapy has shown that some tumors, called *radioresistant tumors*, respond poorly to photon therapy. Sometimes even non-radioresistant tumors cannot be given a tumoricidal dose because of the unavoidable associated dose to neighboring healthy tissue. *Hadronic radiotherapy* uses particles such as neutrons, protons, pions, helium, or heavier ions to treat radioresistant tumors and tumors located near critical body structures such as the spinal cord. Initial research in hadronic radiotherapy was performed using accelerators built for basic physics research. At present, there are about ten hospital-based accelerators built for and dedicated to neutron and protontherapy, but much of the hadrontherapy research continues to take place in physics laboratories. In this respect, the development of hadrontherapy closely parallels the development of photon therapy. It specially concerns the patients with tumors in the brain and base of the skull region, sarcoma and prostate carcinoma [1.4].

Comparison of depth-dose distribution of modalities related to radiation therapy is presented on the Fig. 1.7.

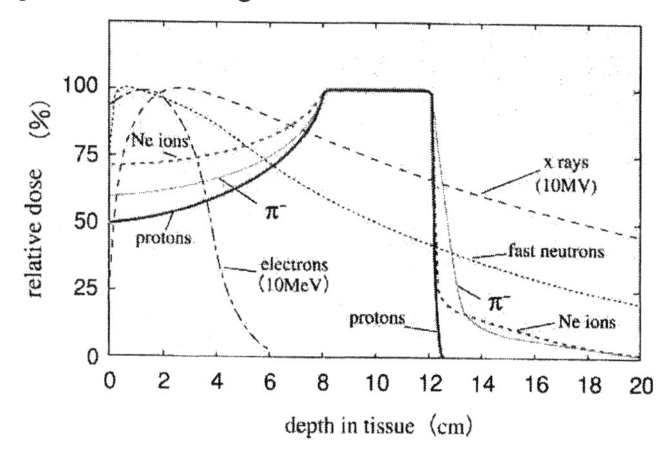

Figure 1.7. Depth-dose distribution of modalities related to radiation therapy; after [1.23].

Three approaches appear very promising for a further increase in selectivity of radiotherapy: protontherapy, radiotherapy with ions of light atoms and BNCT.

1.3.3.1. Protontherapy

Protons have a finite range and that range is energy dependent. By appropriate selection of distribution of proton energies, a depth dose curve can be flat at ≈100% over the depth of interest and then the dose falls precipitously at the end of the range to virtually zero. The dose decreases at the end of range from 90% to 10% over a distance of ≈ 0,6 mm, depending upon the range of the beam in tissue. Similarly, lateral fall-off dose is quite steep [1.20].

1.3.3.2. Radiotherapy with Ions of Light Atoms (Light Ions)

Beams of light ions, such as those of carbon, oxygen and neon, travel in virtually straight lines with negligible sideways spreading and they deposit a large fraction of their energy at the end of range. This allows a well-defined distribution of the dose in depth, better even then protons. In addition, because of their intense local ionization, some authors think that light ions would be more effective than protons against radioresistant tumors [1.4].

In a light ion biomedical facility a wide variety of ion beams can be made available. This include beams of both stable ions and radioactive species as indicated in Fig. 1.8. This Figure shows a portion of the *Chart of the Nuclides* that is relevant to light ion therapy. The black squares represent the stable isotopes that are available for elements from protons to neon (proton number 1 – 10). The open squares represent selected unstable isotopes that can be produced in a light-ion facility and adapted for clinical use. Beams of these species can be formed by accelerating a stable parent beam, and impinging it on a conversion target to form secondary beams with half lives ranging from a few seconds to a few hours. A number of primary beams can be accelerated, including hydrogen (protons), helium, carbon, nitrogen, oxygen and neon. In order to accomodate a versatile program for research and clinical medicine, enough flexibility in the accelerator and the beamline operation needs to be provided to ensure that they can be quickly retuned to change from one ion species to another, from one energy to another, and to provide a quick interchange between stable and radioactive beams. This requirement for flexibility impacts significantly on the choices of accelerator technology and control system design.

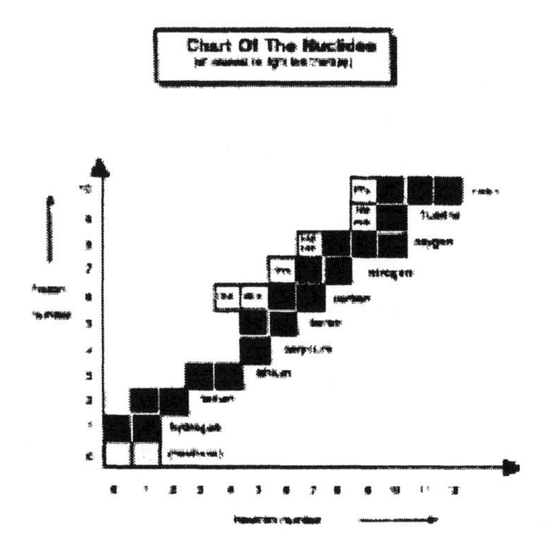

Figure 1.8 Portion of the chart of the nuclides of interest to light ion radiotherapy; after R.A. Gough [1.27].

The second major requirement is that there should be adequate intensities of each ion available to carry out the clinical program. It should be noted that the requirements are different for different ions. In Tab. 1.4, this is illustrated by listing the intensities required for selected ions to deliver a 600 rad dose to a standard volume in a one minute exposure. The intensities required to meet this reqirement vary from $8*10^7$ for argon to $1*10^{10}$ for protons.

Dose Target volume Time	6 Gray (600 rad) 1 litre (200 cm^2 * 5 cm) 1 minute
Hydrogen (1)	$1*10^{10}$ ion/sec
Helium (–4)	$2.5*10^9$
Carbon (–12)	$4.5*10^8$
Oxygen (–16)	$3.5*10^8$
Neon (–20)	$2.3*10^8$
Silicon (–28)	$1.2*10^8$
Argon (-40)	$8*10^7$

Table 1.4 Dose considerations for light ion therapy; after R.A. Gough [1.27].

1.3.3.3. Boron Neutron Capture Therapy (BNCT)

For various reasons some boron compounds may accumulate in certain tumors, particularly in brain tumors where a reduction in the tumor blood-brain barrier encourages their selectivity. Upon irradiation with neutrons, the boron atoms capture the neutrons and give raise to alpha and lithium particles. These have a high energy and their track is less than a millimeter in tissue, so that they do not leave the tumor. It is estimated that in this way, with the same dose to surrounding normal tissue, the dose to the tumor itself can be increased by one-third [1.4].

1.4 Glossary

1.4.1 Beam Intensity

The accelerator physicists and medical physicists to mean quite different things use the term beam intensity, and we wish to clarify its use in this book.

The rigorous definitions of terms involving the number of accelerated particles are as follows:

Number of particles, N

Fluence, ϕ: particles / unit area, $\phi = dN/da$

Flux, F: particles / unit time, $F = dN/dt$

Intensity, I: energy / unit area / unit time

$$I = Ed^2N/dtda = Ed\phi/dt = EdF/da \qquad (1.1)$$

where E is the energy of the particles.

For our purpose of describing mono-energetic particles, it is convenient to define:

Intensity, I: particles / unit area / unit time

$$I = d^2N/dtda = d\phi/dt = dF/da \qquad (1.2)$$

Accelerator physicists, however, uses the term beam intensity to mean the number of particles per unit time, i.e., the same as the flux defined above.

Intensity, I particles / unit time

$$I = dN/dt = F \qquad (1.3)$$

and $qe \cdot I = qedN/dt =$ beam current (1.4)

With this definition, the beam intensity does not vary when the beam cross-section changes. Also the beam intensity modulation means the beam flux modulation. In this book, we will use the beam intensity to mean always particles per unit time (Eq. 1.3). For protons $qe = +1$, and the beam intensity and beam current are used interchangeably.

In the literature, to describe the process of obtaining uniform dose distributions by accumulating raster-scanned beam spots with Gaussian-like intensity distribution, the term intensity is used to mean $d^2N/dtda$. To further confuse the issue, in the description of voxel scanning the term intensity is often used to mean dN/da, which is the fluence.

1.4.2 Beam Range in an Absorbing Medium

The *range of the beam* is defined as the mean depth of penetration measured along a straight line parallel to the original direction of motion of the protons from the point where they enter the medium to the point at which additional displacement is no longer detectable. If a depth dose curve is considered, the range corresponds to the distance between the entrance surface of the beam and the distal point of 80% dose (Fig. 1.9). For clinical applications the range is usually expressed in g/cm^2, in order to be independent of specific cases where inhomogeneities can be present. If inhomogeneities are present on the beam path (for example bone or lung), the range will be slightly different from the value relative to water.

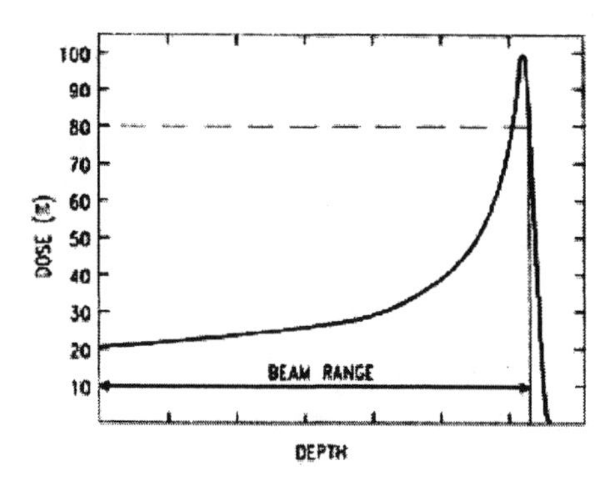

Figure 1.9. Beam range in medium.

1.4.3 Bragg Peak Modulation

This expression refers to the possibility of adjusting the SOBP extension. This is defined on a depth dose curve as the distance between the proximal and the distal points corresponding to the 90% of the peak height. The 100% are associated to the maximum point of the SOBP (Fig. 1.10).

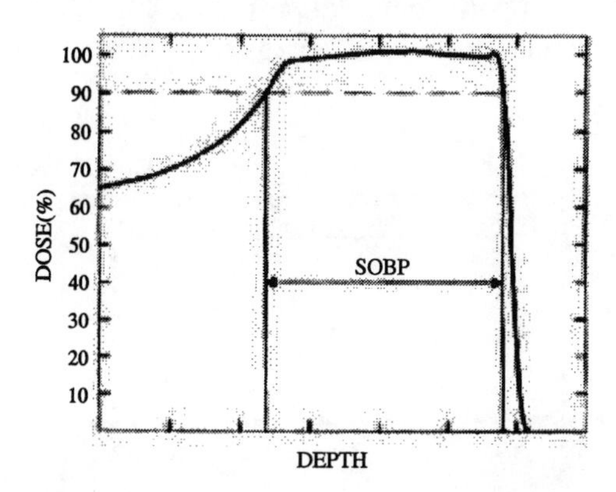

Figure 1.10. Definition of the Spread Out Bragg Peak (SOBP).

1.4.4 Range Adjustment

This expression refers to the possibility of translating the SOBP in depth.

1.4.5 Radiation Field

If the intersection of a beam with a surface perpendicular to the beam axis is considered, the radiation field is defined as the locus of the points where the absorbed dose exceeds 50% of the dose absorbed at the point where the beam axis intercepts the surface (Fig. 1.11).

1.4.6 Field Homogeneity on a Transverse Section

The homogeneity is defined as the ratio:

$$R_t = (P_{max} / P_{min}) * 100\% \qquad (1.5)$$

where:

P_{max} is the maximum absorbed dose, at any point in the irradiation field, averaged over a surface not larger than 1 cm^2 for a beam spread by a passive system and not larger than 0.25 cm^2 for a pencil beam used for magnetic scanning;
P_{min} is the minimum ubiquitous absorbed dose in the homogeneity region, averaged over a surface not larger than 1 cm^2 for a beam spread by a passive system and not larger than 0.25 cm^2 for a pencil beam used for magnetic scanning.

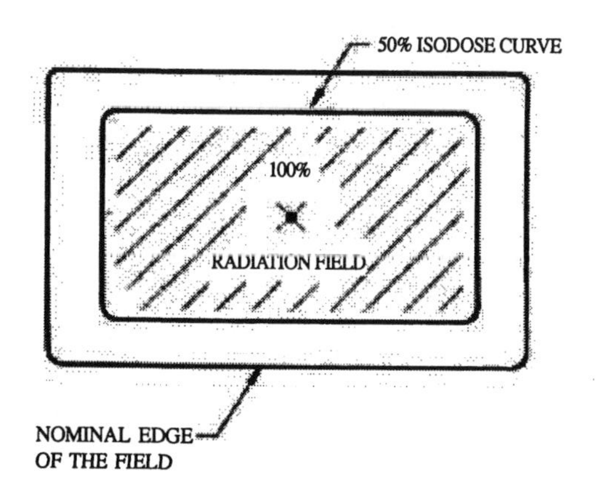

Figure 1.11 Definition of radiation field.

The measurement should be performed at the surface of a water-equivalent phantom. This surface, for a rotating gantry, should be at the isocentric and should he perpendicular to the beam axis. The phantom must extend to at least 5 cm beyond the geometrical edge of the radiation field; the phantom thickness must he at least 5 cm greater than the depth of the Bragg peak. For fixed beams the phantom surface should he at the normal treatment distance.

1.4.7 Homogeneity Region in a Plane Orthogonal to the Beam Direction

The homogeneity region is defined by the lines connecting the points on the principal axes and on the diagonals of a square field as shown in Fig. 1.12. The dimensions d_m and d_e are defining the homogeneity region.

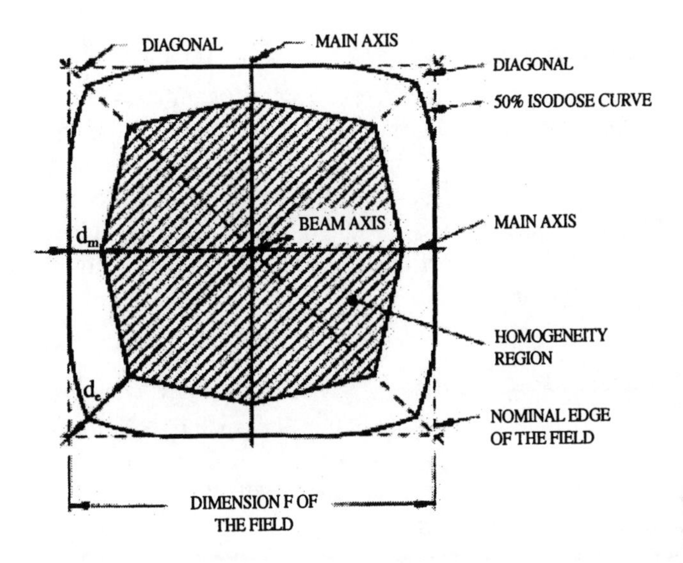

Figure 1.12. Homogeneity region in a plane orthogonal to the beam direction.

1.4.8 Field Homogeneity along the Beam Axis

With proton beams it is also necessary to define the homogeneity along the beam direction in correspondence to the SOBP region. In this case the region of homogeneity is defined as the region between two distal fall-off widths (below) inside the 50% dose point distally and one distal fall-off width inside the 90% dose point proximally. It is assumed that 100% corresponds to the point of maximum dose in the SOBP (Fig. 1.10).

The homogeneity is defined as the ratio:

$$R_1 = (P_{\max} / P_{\min}) * 100\% \qquad (1.6)$$

where:

P_{max} is the maximum dose absorbed everywhere in the irradiation field, averaged on a surface not larger than 1 cm^2;

P_{min} is the minimum ubiquitous absorbed dose in the homogeneity region, averaged over a surface not larger than 1 cm^2.

1.4.9 Field Symmetry

The symmetry, on a plane transverse to the beam axis, is defined as the maximum value S of the ratio between the highest and the lowest absorbed dose, averaged on a surface not larger then 1 cm^2, for every pair of symmetrical positions with respect to the beam axis inside the homogeneity region. The maximum ratio S should be expressed as a percentage. The measurement should be performed in the same conditions as for the determination of the homogeneity.

1.4.10 Lateral Penumbra

If a dose profile measured at the entrance surface of a water-equivalent phantom along one of the main axis of the field is considered, the lateral penumbra is defined as the distance between the points of 80% and 20% absorbed dose. It is assumed that 100% corresponds to the beam axis (Fig. 1.13).

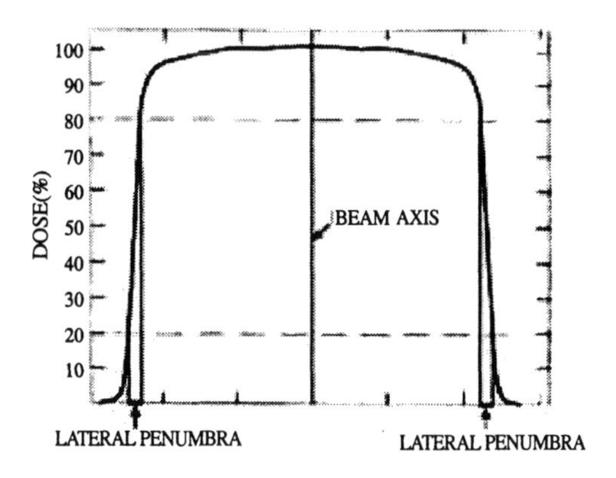

Figure 1.13. Definition of the lateral penumbra.

1.4.11 Distal Dose Fall-off

If a depth dose curve along the beam axis is considered, the distal dose fall-off is defined as the distance between the points of 80% and 20%

absorbed dose along the beam axis beyond thc Bragg peak. It is assumed that 100% corresponds to the maximum in the SOBP (Fig. 1.10).

Chapter 2

PHYSICAL AND RADIOBIOLOGICAL PROPERTIES OF HADRONS

2.1 Basic Facts in Radiotherapy

The main objective of radiotherapy is the loco-regional control of the tumor and, in some situations, of surrounding diffusion paths. In order to reach this objective, one must deliver to the tumor – which may be considered in physical terms as the *target* – a sufficiently high *dose*. So as to destroy it, at the same time maintaining the dose to the surrounding healthy tissues; inevitably involved in the irradiation, within such limits so that they do not undergo serious or even irreversible damage or complications [1.4].

By definition the *absorbed dose* (symbol D), sometimes called *dose*, is the ratio between the energy E_D imparted by the irradiation to a small volume of material (tissue) and the mass m. of this volume of material. The ratio $D=E_D/m.$ is measured in gray (Gy). 1 Gy equals 1J per kg. In conventional radiotherapy with photons and electrons doses of order of 60-70 Gy are deposited in the tumor tissues in amounts of about 2 Gy per session over about 30 days.

2.1.1 Dose-Effects Curves

In the hypothesis of a fairly accurate identification of the target, it is possible to evaluate the probability of obtaining local control of the tumor through the analysis of the so-called *dose-effect curves*. They represent:

- for tumor tissues, the possibility of obtaining the desired effect as a function of the dose delivered,
- for healthy tissues, the probability of provoking serious or irreversible damage, always as a function of the dose absorbed by the same tissue.

On the Fig. 2.1 the thick line represents, as a function of dose absorbed, a hypothetical dose-effect curve for a generic tumor tissue. The dashed line represents a dose-effect curve for the healthy tissue involved in the irradiation. As can be seen, to the absorbed dose necessary to achieve the probability close to 100% of obtaining local control of the tumor corresponds also a very high probability, in most cases to high to be acceptable of producing serious complications in the healthy tissue, when this receives the same dose.

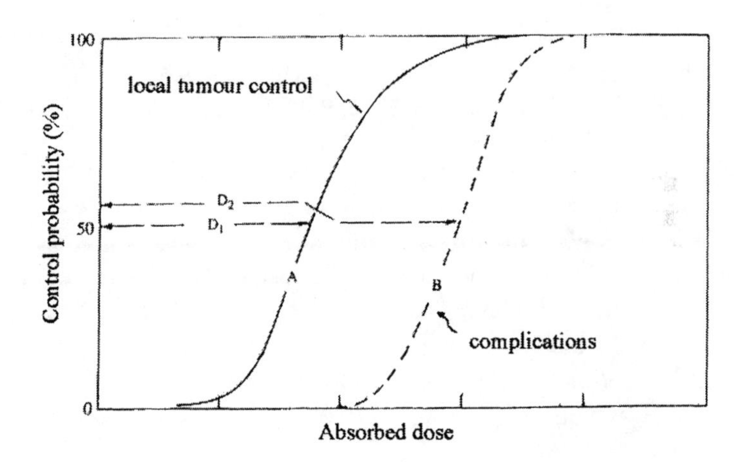

Figure 2.1. Dose-effect curves for neoplastic (A) and normal (B) tissues; after U. Amaldi [1.4].

In the daily practice, the radiotherapist has to find the compromise between the local control of the tumor and the possible emergence of complications. The possibility of finding such a compromise can be expressed quantitatively by the *therapeutic ratio*. It is defined as the ratio D_1/D_2 between the dose corresponding to a 50% probability of producing complications D_2 and the dose corresponding to a 50% probability of obtaining the local control of the tumor D_1 [1.4].

On the basis of these considerations it is clear that the probability of curing the tumor without unwanted side effect increases in line with the *ballistic selectivity* or *conformity* of the irradiation delivered. The term "conformity" means in this case the difference between the dose to the target and the dose to the healthy tissue, which is irradiated. With the beam of charged hadrons one can increase the probability of curing the tumor because the absorbed dose is more concentrated in the tumor tissues than with electrons and photons.

2.1.2 RBE Dependence on LET and OER

The effect of irradiating a tumor does not only depend on the absorbed dose, which expresses at the macroscopic level the local deposition of energy, but at least on two other parameters:

- the modalities of energy transfer from the radiation to the tissues often expressed, at the microscopic level, by the Linear Energy Transfer (***LET***), that is the density of energy deposition along the track of the particles in the tissues.

The Linear Energy Transfer (symbol L_∞) is defined as the ratio between the energy ΔE deposited by a charged particle in a very short track element, and its length Δx. In formula $L_\infty = \Delta E/\Delta x$, the index ∞ indicates that there is no limitation on the amount of the energy ~ released in any single collision of the particle with an atom or a molecule of the traversed medium. Physicists call L_∞ the *unrestricted energv loss* or the *unrestricted stopping power-* Other quantities often used are 'restricted' by requiring, for instance, that $\Delta < 150$ keV; the symbol used in this case is $L_\Delta = L_{150}$. Radiobiologists call it 'LET' or better *restricted LET.*

- the oxygen content of the irradiated tissues. In fact the oxygen content is generally low in the scarcely vascularized tumor tissues and the biological effects usually decrease when the oxygen content reduces. The ratio between the doses required producing a given effect in the absence and presence of oxygen is termed *Oxygen Enhancement Ratio* (***OER***) [1.4].

In formula the definition is $OER = D/D_0$, where D is the dose needed to produce the effect in the actual tissue and D_0 is the dose which would be needed if the tissue were fully oxygenated in an air atmosphere under normal pressure.

Table 2.1 presents the LET values (expressed in keV/μm in water) of the particles and radiations of interest in radiotherapy: carbon ions have a LET in the range 15-200 keV/μm, i.e. a hundred times larger than the LET of conventional photon beams.

The destruction of hypoxic tissues caused by irradiation with x-rays, gamma rays or ions is several times less severe that than of anoxic tissues with a normal amount of oxygen. As a result of the so-called *oxygen enhancement effect*, the unoxygenated central part of a tumor is more resistant to irradiation than the adjacent normal anoxic tissues. The *oxygen enhancement ratio* (OER) is the ratio of the doses of radiation required to produce equal biological effects in the absence of oxygen and in its presence. This ratio depends to a large measure on LET. Fig. 2.2 illustrates the relationship of OER and RBE as a function of LET.

PARTICLE	CHARGE	MASS	ENERGY (MeV)	L_∞ (keV/μm)
Electron	-1e	1 m_e	0,01 0,1 1,0	2,30 0,42 0,25
Photon	0		Co-60 γ-rays 4-25 photons	02-2
Neutron	0	1839 m_e=1.0009u	14	3-30
Proton	+1e	1836 m_e=1.0008u	2 5 10 200	16 8 4 0,4
α particle=He^{+2}	+2e	2u	5	95
Carbon ion=C^{+6}	+6e	6u	10-250 MeV/u[9]	170-14
^{20}Ne	+10e	20u	525 MeV/u[10]	50
^{40}Ar	+18e	40u	730 MeV/u[10]	150
BN ^{4}He	+1e	4u	7 μm[10]	150
CT ^{7}Li	+1e	7u	5 μm[10]	150

Table 2.1. LET (L_∞) of ionizing particles and radiations of interest in radiobiology and radiotherapy, after U. Amaldi [1.4] and D.T. Jones [1.25].

It has been recognized since long ago that the absorbed dose is not a good predictor of biological effects. The differences in the effect of radiations of various types for the same physical dose are accounted for by introducing a *coefficient of relative biological effectiveness* (*RBE*), sometimes also called in radiation protection a *quality factor Q*. The RBE of a given radiation is defined, as the ratio between the absorbed dose of a reference radiation and that of the test radiation, required to produce the same biological effect. For LET lower than 3.5 keV/μm Q=1, for LET=(3.5…7) keV/μm Q=1…2, whereas LET=(53…175) keV/μm Q=10…20 [1.22].

[9] The kinetic energy is usually expressed by dividing the total kinetic energy in MeV by the mass of the atom measured in atomic mass unit. Its symbol is MeV/u. In practice this ratio E_N equals the kinetic energy carried by each proton and neutron of the ion and is usually called *energy in MeV per nucleon*.

[10] Energy for range ~26 cm in water.

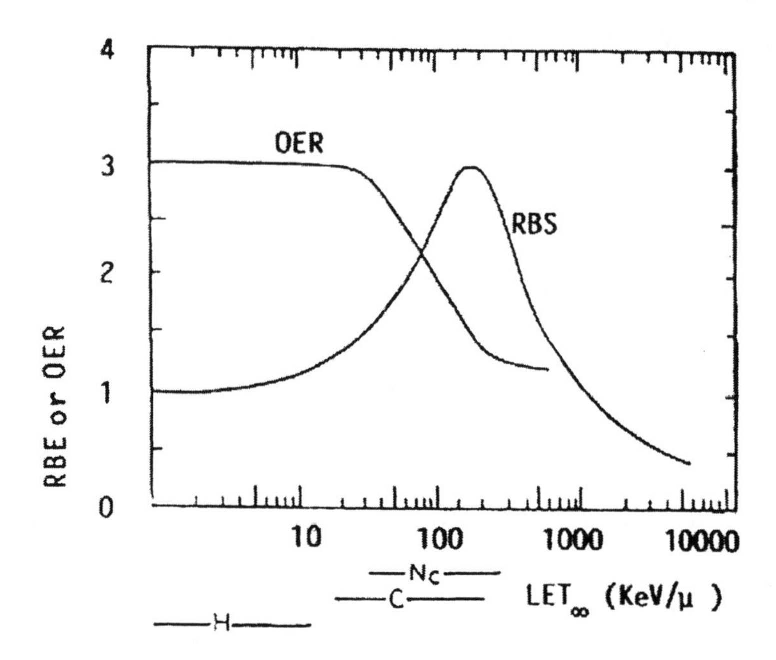

Figure 2.2. Dependence of Q-factor and OER on LET; after W. Scharf [1.22].

For low LET up to about 10 keV / μm both RBE and OER are nearly constant. RBE reaches maximum in the range of 100 120 keV/m, whereas OER falls rapidly. One can also conclude from Fig. 2.2 that in order to diminish the oxygen enhancement effect and to increase the biological effectiveness, high LET radiations should be used [1.22].

From a global point of view, to measure quantitatively the killing effect of radiations on the populations of irradiated cultured cells, the determination of the clonogenic ability of the irradiated cells is by far the most widely used method. The results are presented by means of ***survival curves***, which are obtained by counting the number of clones of a specific type of cells irradiated with a well-defined beam. The relation between dose and cell damage is subject to extensive experiments with cell cultures. The results are usually displayed on a semi-logarithmic scale (fraction of cells surviving the irradiation vs. dose).

One of the methods of determining optimum methods of fractionation is one that involves mathematical modeling of processing that take place in irradiated cell colonies. When ionization radiation interacts with matter some of its energy, defined by the absorbed dose, causes excitation and ionization of atoms. As a result, chemical and biological changes occur in living cells, which lead to transient or permanent transformations in the whole body. One of the models, which describe the effect of radiation on living cells, is the so-called multiple-target model. It represents a cell as a group of active centers to be hit by the radiation. For small doses only some targets become damaged. This leads to sublethal lesions in the cell, which are then capable of regeneration with time. However, when a given threshold dose is exceeded lethal lesions occur, cells are unable to regenerate, and, as a result, they die. Apart from the absorbed dose, other factors, such as the type of the irradiated cells, the degree of their oxygenation, temperature, phase of the cell cycle and the type and spectrum of radiation, all this affect the type of injury.

For the multi-target model it has been assumed that the survival rate of a cell colony which is homogeneous from the point of view of its radiation sensitivity can well be described by:

$$S/S_0 = 1-(1-e^{-d/D_0})^n \qquad (2.1)$$

where: S_0 is the initial number of cells, S number of surviving cells, d radiation dose, D_0 and n are parameters depending on the above factors and the degree of well-oxygenated cells.

The $log(S/S_0)$ vs. radiation dose is referred to as a survival curve. Fig. 2.3a is an example of such a curve expressed by Equation (2.1). It also provides explanation for the parameter n, which is a number taken from extrapolations, and D_q, which is called the *quasi-lethal dose*. D_0 determines the slope of the rectilinear part of the survival curve. The section AB, or, in other words, the shoulder of the survival curve, represents the cumulating of sublethal injuries. The cells are able to repair the injuries with time. If cells are exposed to another irradiation right after the first one, this effect is represented by the section CD on the curve. However, if adequate time is left between irradiations for the cells to recuperate from sublethal injuries a new shoulder of the curve (DCE) is obtained.

Fig. 2.3b shows two examples of cell survival curves. One for low LET irradiation (sparsely ionizing) and one for high LET radiation (densely ionizing) of the same cell line. The shapes of the survival curves depend on two main factors, one of physical and the other of biological nature, i.e. the kind of the radiation used and the type of cell. In many cellular systems an initial "shoulder" is observed in response to sparsely ionizing radiation (Fig. 2.3b). The shoulder progressively reduces and eventually disappears with increasing LET. The large shoulder of the photon/proton curve indicates that a lot of repair occurs in the cells after the irradiation. Almost no repair is possible in the cells irradiated with the same dose delivered by neutrons. Repair is certainly not desired in tumor cells, but it is crucial to avoid relevant damage to normal tissue as much as possible. A fractionation of the whole treatment dose into small daily fractions allows all cells to recover. Small differences in the dose given to the tumor and healthy tissue respectively or small differences in the radiosensitivity of different tissues are amplified by this procedure. Except for a few specialized treatments radiation therapy is always delivered as a fractionated treatment. The shape of the cell survival curve is therefore a very important factor when comparing different radiation qualities. Protons are low LET radiation. Therefore the biological effect of protons is similar to photon and electron beams [2.13].

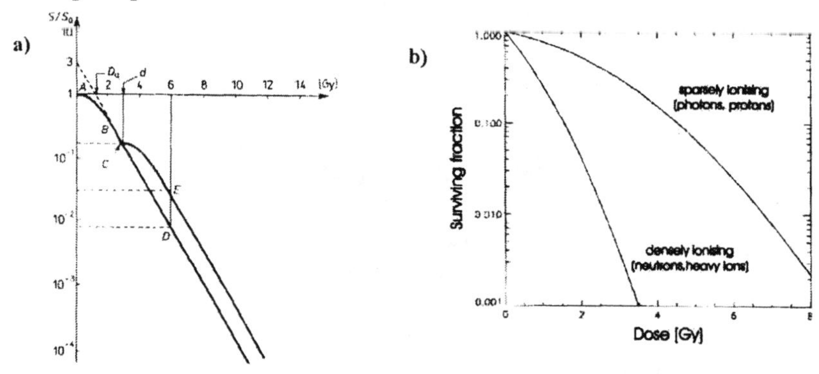

Figure 2.3 a) The cell survival curve (D_0=100 Gy, n=3 (the number of extrapolations) and the quasi-lethal dose D_q=D_0 ln n); after W. Scharf [1.22]. b) Typical cell survival curves for cells irradiated with sparsely and densely ionizing radiation respectively, after B. Schaffner [2.13].

By comparing survival curves obtained with high energy photons and protons it has been concluded that protons in the *plateau* are as lethal as photons, while protons in the Bragg peak show a higher effectiveness (Sec. 2.1.2). This indicates the importance of gaining a better insight in the radiobiological properties of low energy protons, i.e. those having energies in the range of interest of the *Bragg peak* used for therapeutical purposes [2.1].

2.2 An Introduction to Hadrontherapy

Hadrons are compound subatomic particles made of quarks and antiquarks, bound together either in triplets or in doublets by strong force. The hadrons, which are today employed for radiotherapeutic purposes are neutrons, protons and light ions (such as helium, carbon, oxygen and neon). Among the used hadrons, neutrons and light ions are high-LET particles. Protons show almost the same LET as conventional photons and electron beams when they enter the treated body, but have an increasing LET at the end of their path in tissues (Table 2.1, Fig. 1.7) [1.4].

The therapeutical advantages of hadron beams when compared to electron and photon beams are due to:

- the possibility of delivering the dose to the tumor in a physically selective way at the ***macroscopic scale*** (conformal therapy) while sparing the tissues which are traversed by the beam and surround the irradiated tumor,
- the possibility of varying the radiobiological effects of the radiation by influencing the pattern of the energy deposition at the ***microscopic scale***.

These novel therapeutical possibilities follow from the different dosimetric and radiobiological properties of hadron beams with respect to photon and electron beams (Sec. 2.4).

The possibility of varying the radiobiological properties is due to the fact that the biological effects are correlated with the microscopic distribution of the energy deposited. This distribution is influenced by the energy and atomic number of the impinging hadrons (Z=1 for protons, Z=0 for neutrons, Z=6 for carbon and Z=8 for oxygen). For this reason one often speaks of hadrons *densely ionizing* radiation in contrast to the *sparsely ionizing* radiation such as *x*-rays and γ-rays, although it should be noted that this classification depends not only on the nature of the radiation considered, but also on its energy [1.4].

In Table 2.2 the basic physical and radiobiological properties of photons and ions are compared. All of them will be discussed in the following Sections.

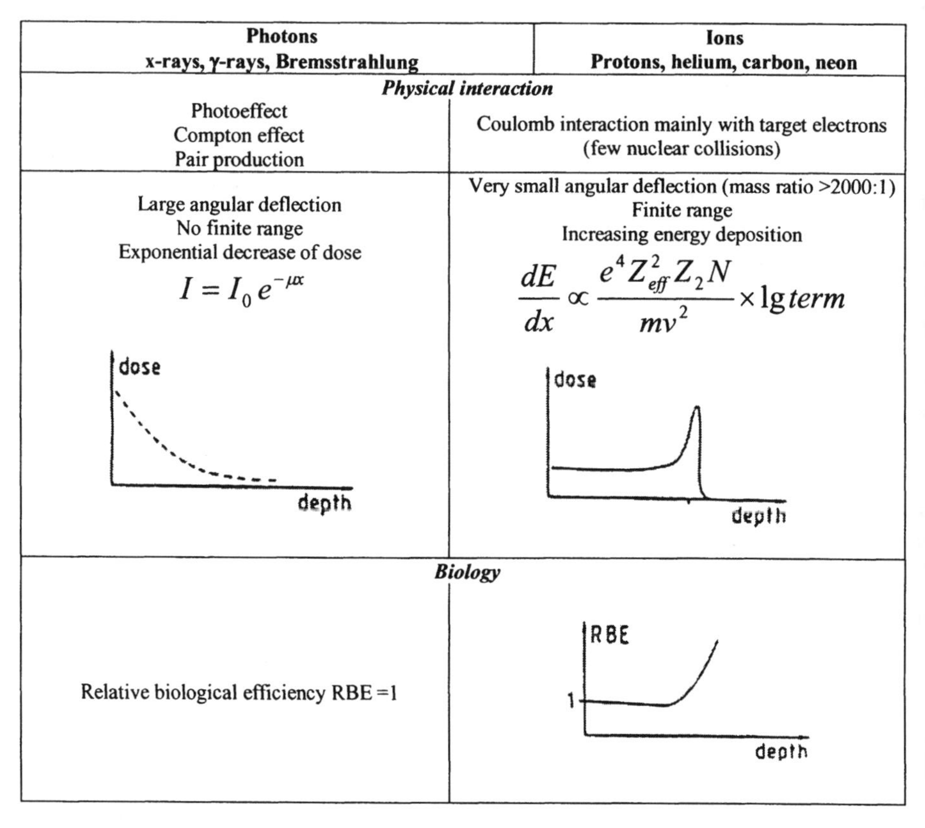

Photons **x-rays, γ-rays, Bremsstrahlung**	**Ions** **Protons, helium, carbon, neon**
Physical interaction	
Photoeffect Compton effect Pair production	Coulomb interaction mainly with target electrons (few nuclear collisions)
Large angular deflection No finite range Exponential decrease of dose $I = I_0 e^{-\mu x}$ dose depth	Very small angular deflection (mass ratio >2000:1) Finite range Increasing energy deposition $\frac{dE}{dx} \propto \frac{e^4 Z_{eff}^2 Z_2 N}{mv^2} \times \lg term$ dose depth
Biology	
Relative biological efficiency RBE = 1	RBE 1 depth

Table 2.2. Differences between photons and ions, after G. Kraft [1.5]

2.3 Energy Loss

For sparsely ionizing radiation like X- or γ-rays, the energy is transferred from the electromagnetic radiation via Compton, photo-processes or pair production into more or less energetic electrons. It is not only this processes but also the secondary collision processes of the liberated electrons that result in ionization and consequently in biochemical and biological damage. With electrons as the primary radiation, these effects are produced by the ionization processes of these electrons and their subsequent electron cascade.

Compton process, photoeffect and pair production are rare and stochastic events. Therefore, the primary intensity of the beam decreases exponentially with the penetration depths and the beam does not have a finite range. In addition the lateral scattering of the secondary electrons is large.

In general, the radiobiological effect of all types of sparsely ionizing radiation are caused by the action of liberated electrons and the primary ionization events. With the exception of very low energy electrons produced for example by soft x-rays, the liberated electrons experience the same slowing down process in the target. In consequence, the biological efficiency differs only some percent between electromagnetic radiation produced from different sources like Co-gamma-rays or the bremsstrahlung spectrum of a x-ray tube [2.1].

For particle radiation like protons, α-particles or heavier ions a much larger variation in the relative biological efficiency is observed. Their biological effect is caused by primary ionizations occurring along the particle trajectory and the action of the liberated electrons and their secondaries. Heavy charged particle when passing through the matter, dissipate their energy mainly via interaction with the electrons of the target material. Differences in the biological efficiency of these particles have therefore to be attributed to the spatial and time correlations of the ionization events caused by the electrons and primary particle [2.1].

The energy loss is caused by the Coulomb interaction of the ions either with the target electrons (electronic stopping) or with the nuclear potential of the target atoms (nuclear stopping).

2.3.1 Nuclear Energy Loss

The most important energy loss mechanism for radiobiology is the interaction of the effective charge with the target electrons but only at very small energies ($E \leq 10$ keV/u) the nuclear energy loss dominates. Nuclear energy loss is caused by the scattering of the projectile at the screened nuclear potential of the target atom (Rutherford scattering). Because the recoil energy transferred to the target atom is frequently much higher than the binding energy, as these atoms are released from their chemical environments can trigger of a reaction chain if kinetic energy is big enough. In consequence nuclear energy loss is biologically very efficient when the biological target molecules like DNA are hit directly. However, the dominance of nuclear collision process is restricted to very low energies and refers also in its spatial extension to very small volumes. Therefore, for the radiobiological and therapeutical applications nuclear stopping is not important and the observed effects are caused predominantly by the interactions with the electrons (Fig. 2.4) [2.1].

2.3.2 Electronic Energy Loss

Swift heavy charged particles when passing through the matter dissipate their energy mainly via interaction with the electrons of the target material. Because of the large difference between the electron mass compared with the mass of the atomic nuclei, the deflection of the projectile ions is very small and only multiple collision process can cause a net deflection of the particle beam. Due to the reaction kinematics, the mean deflection of these electron collisions becomes ever smaller for heavy ions.

The energy dissipating electronic collisions is very frequent. Therefore, the energy loss of the projectile is nearly continuous and depends only on effective charge, i.e. atomic number and the velocity of the projectile and the density of electrons in target material, as given by the Beth-Bloch formula (Eq. 2.2) [2.2]. At relativistic velocities, the ion electron interaction is small but increases with decreasing velocity to the 1/v squared term in Bethe-Bloch formula:

$$\frac{dE}{dx} = \frac{4\pi e^4 Z_{eff}^4 N Z_2}{m_e v^2} \times \ln \frac{mv^2}{I_0 Z_2} \tag{2.2}$$

where
- e electronic charge
- N the density of target atoms
- Z_2 the atomic number of target atoms
- m_e the electron mass
- v the particle velocity
- Z_{eff} the effective charge of the projectile
- I_0 a mean ionization potential of the projectile

The energetical region of electronic energy loss is presented in Fig. 2.4.

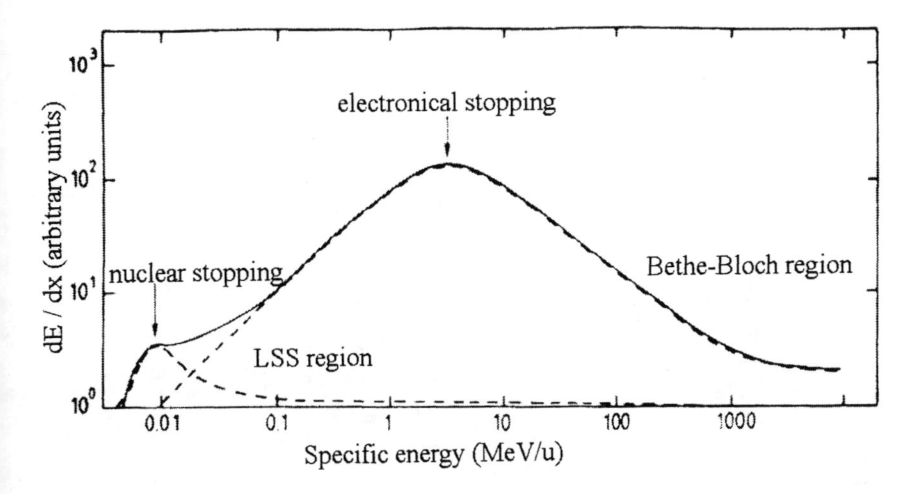

Figure 2.4. Schematic representation of the two different processes of energy loss, after G. Kraft [2.1].

2.3.3 Range and Bragg Curves

From the energy loss curves the range of proton can be calculated by numerical integration:

$$R = \int \left(\frac{dE}{dx}\right) dE \tag{2.3}$$

Ranges calculated in this way are given in the dE/dx tables by many authors. These range values refer only to the primary ions. Secondary ions resulting from nuclear fragmentation are not included.

Frequently in radiobiology the ionization is plotted as function of particle range instead of energy. At high particle energies these curves show a low *plateau* value in the entrance and an increase of ionization towards the end of the particle range to a sharp maximum, the *Bragg peak*. The general structure of the Bragg curves as shown in Fig. 2.5 can be understood very easily on the basis of the energy loss curve from Fig. 2.6.

At high energies, the energy loss is small and the particles travel along the trajectory producing a low and nearly constant ionization density (plateau region). In the lower energies, the energy transfer increases rapidly, and therefore the sharp Bragg peak appears on the depth dose characteristic.

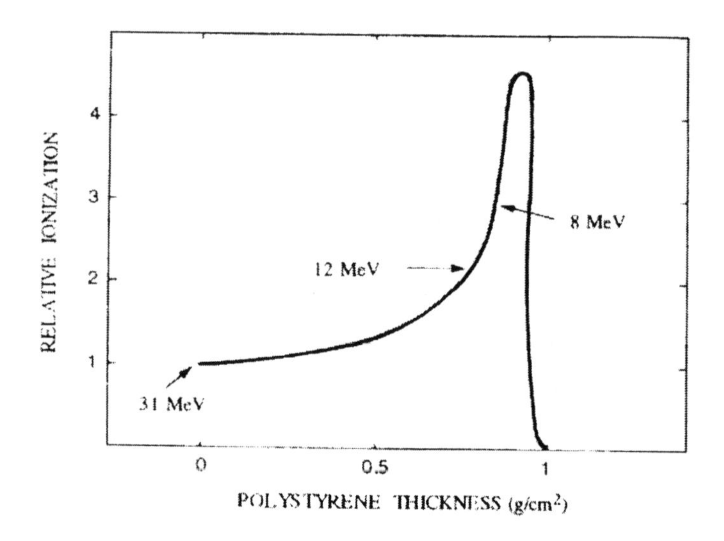

Figure 2.5. Bragg curve for the proton beam used for inactivation of human EUE cells, after U. Amaldi and M. Siliari [1.4].

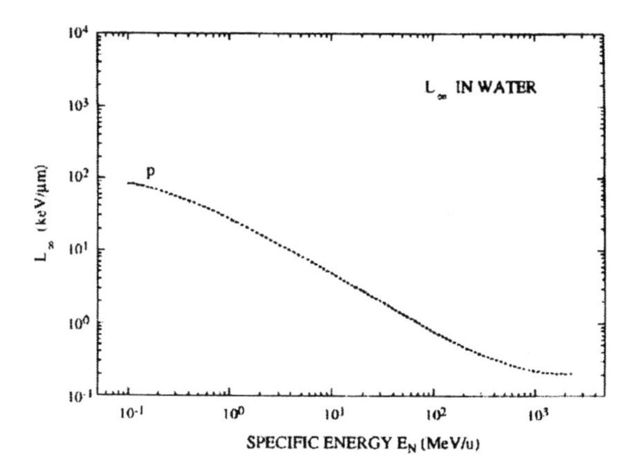

Figure 2.6. Stopping power (or unrestricted LET value) in water for protons as a function of their specific energy E_N, after U. Amaldi and M. Siliari [1.4].

2.3.4 Track Formation and Radial Dose Distribution

In the regime of electronic energy loss, the energy lost from the primary ions is mostly transferred to the electrons. Depending on the initial energy, 65% to 75% of the energy loss are transformed into kinetic energy of the electrons. 15% to 25% are necessary to overcome the binding energy and only 5% to 15% are consumed for electronic excitation. According to this distribution, most of the electrons are liberated from their atomic binding. These electrons dissipate their energy in some distance from the primary collision forming a track of ionization around the projectile trajectory.

In principle it should be possible to predict the biological damage by an comparison of this track structure with the geometry of the biological relevant molecule DNA [2.1].

2.4 Interaction of Hadrons with Biological Matter

2.4.1 Cellular Organization and the Target of Radiation Action

All biological organisms consist of cells as the basic units. For the higher organism like yeast, mammals including men these cells are *e*ukaryotic i.e. they have a cell nucleus containing DNA molecules that carry all the genetic information. Because of the complexity of the cell as an interacting system of thousands of different molecules that can be damaged by the ionizing irradiation, it is difficult to imagine that general rules or reaction mechanisms can be find. However, many experiments have confirmed that the DNA inside the cell nucleus represents the important target for radiation damage. This does not mean that the other cellular molecules are not damaged, but those molecules can be replaced by others of the same species or can be produced newly using the DNA information as template. Because DNA is unique, large and essential the DNA damage that cannot be repaired affects the function and the reproductivity of the cell lethally.

The DNA inside the cell is highly organized. The basic structure is Helix molecule of the sugar-phosphate backbone. The bases are attached to the sugar and connected pairwise with the corresponding base of opposite strand. The four bases, adenine A, cytosine C, guanine G and thymin T combine as AT and CG pairs and form the genetic code for all protein. The double stranded DNA is wound over histon-protein complexes with approx. two rounds over the histon forming a string of pearls, that are organized again in some higher order. In cell division in the DNA protein complex again condenses to chromosomes that are visible size under the microscope. In the time between two mitosis, in the interphase, the chromatin is decondensed and the DNA-histon complex is contained within the nuclear membrane. Because of the importance of the molecules and of the high venerability of the cells, especially all DNA lesions are correctly repaired and non-repairable DNA damage is the exception [2.1].

As an immediate consequence of radiation energy deposition in a biological system physical effects occur in the time scale of up to 10^{-13} s (Fig. 2.7), followed within less than 10^{-3} s by chemical reactions caused by primary and secondary free radicals as well as by excited species. The final biochemical effects of metastable products take hours to appear, and finally, biological effects, such as cancer, may be observed within the time scale of up to several tens of years [1.22].

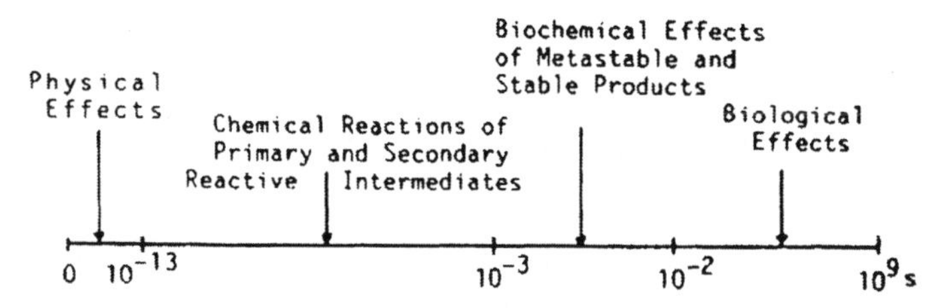

Figure 2.7. Time sequence of radiobiological effects resulting from radiation deposition; after W. Scharf [1.22].

Radiation causes damage at the cellular and subcellular level in a biological system. This damage, which is time dependent, is largely affected by the so-called cell cycle, which is the time sequence of events occurring in a cell in the period between cell divisions. The beginning of the cycle is determined by the moment when the cell is created and the end of the cycle coincides with the moment when the cell divides to form daughter cells. According to the diagram shown in Fig. 2.8a, the cell cycle can be divided into four main phases: (1) the S phase of the DNA synthesis, (2) the G1 phase of the pre-DNA synthesis, (3) the G2 phase of the post-DNA synthesis, and (4) the M phase of mitosis. For most cell systems, mitosis lasts about 30 min. During the 5 phase, the cell synthesizes DNA, whereas no DNA synthesis occurs in the G1 and G2 phases, although other synthetic processes may take place. The G phase lasts between 2 and 4 hours, while the G1 phase varies from cell line to cell line. That is why cells show a timedependent variation in their radiosensitivity, as depicted in the example in Fig. 2.8b. A normal cellpopulation includes, of course, cells in all phases of the cell cycle.

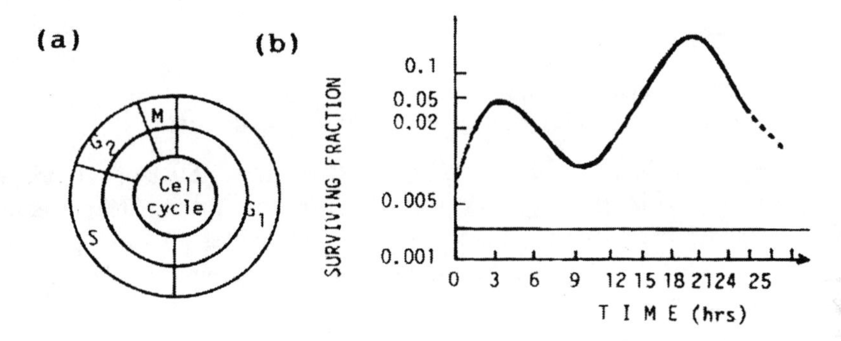

Figure 2.8. Cell cycle: a) cycle phases, b) variation of radiation radiosensivity of HeLa cells (dose 5 Gy); after W. Scharf [1.22].

Radiobiological DNA experiments are either carried out in the lowest order of DNA organization i.e. with pure DNA in a buffer solution or at the chromosomal level or at cellular level in the so called survival tests. Because it is difficult to elutriated intact DNA molecules from mammalian cells without introducing DNA damage, for direct measurement of the induction of single and double strand breaks, small DNA molecules of viruses and phage are frequently used. When a single strand break (SSB) is introduced in such a system, the supercoiled molecule relaxes to an open circle (knocked circle), a double strand break (DSB) produces a linear form. These three DNA, shown schematically in Fig. 2.9, configurations can be separated and the amount SSB and DSB induction as a function of dose can be determined.

Using a proper calibration the number of DSB can be calculated and compared to the local dose, i.e., the Bragg curve (Fig. 2.10a). The ratio between the plateau and "biological" Bragg maximum is smaller than 1:2, while the physical Bragg curve has a ratio of 1:6.

The region of the elevated of the elevated LET in the Bragg maximum shows a large amount of non-rejoinable breaks while breaks in the plateau are rejoined to a large extent. The ratio between plateau and Bragg maximum becomes comparable to the energy deposition but it should be noted that the number of non-rejoinable breaks does not exceed the level of physical dose deposition.[2.7].

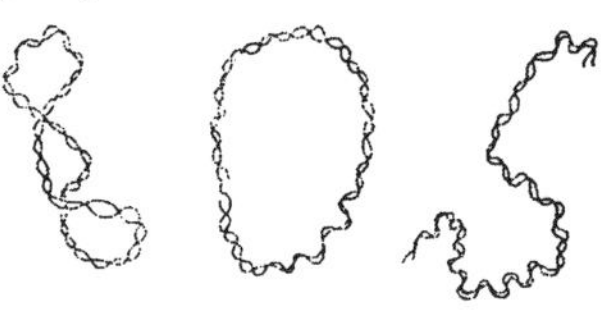

Figure 2.9. The formation of strand breaks in DNA after high –LET irradiation – a comparison of data from in vitro cellular system; after G. Kraft [2.1].

For the carbon beam, as shown in Figure 2.10b, about 80% of the initial lesions are rejoined in the Bragg maximum. For heavier ions in the entrance channel more irreparable lesions are produced. For lighter ions like protons the entrance channel stays the same but the efficiency in the Bragg maximum is reduced.

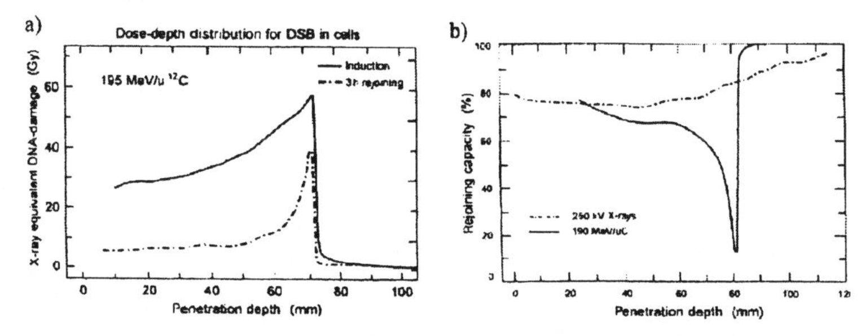

Figure 2.10. a) The amount of double strand breaks immediately after exposure to 200 MeV/carbon ion and after additional incubation time of 3h for DNA repair shown as a function of penetration depth. b) Comparison of rejoining after x-ray and particle exposure as function of penetration; after G. Kraft [2.7]

2.4.2 Hadrons' Track in Biological Matter

Biological effects of ionizing radiation result mainly from ionizations and excitations of atoms and molecules in the matter traversed by the radiation. In a living cell these primary processes initiate a complex chain of events that eventually lead to chemical changes in some important biomolecules (among them the DNA is a most important target) and to biofunctional changes such as mutation, transformation and cell death. To relate the physics information with the cellular structure, and in particular with the double helix of the DNA molecule, a fruitful approach is based on the track structure and in particular on its radial profile.

In this approach, the "tracks" are described in terms of their average profiles of energy density perpendicular to the trajectory of the charged particle. The maximum radial width is determined by the ranges of the most energetic secondary electrons (γ-rays) produced that increase with increasing velocity of the projectile. The track is schematically divided into a *core* (where energy transfer is dominated by the primary particle) and a *penumbra* (that has a radius equal to the maximum range of γ-rays). Both core and penumbra radii are functions of the projectile velocity. Therefore they are also dependent on particles energy and beam range. R. Oozer and co-workers measured the dependence of the penumbra for clinical proton beam. There are differences in the influence of the range of 'entrance' and 'distal' penumbra [2.13]. The entrance penumbra is practically independent on the range of the beam, while the distal penumbra increases linearly with range (Fig. 2.11). It means that in the treatment of the deep-sided tumor, the value of the beam penumbra is an important factor and has to be considering in the calculations during the treatment planning.

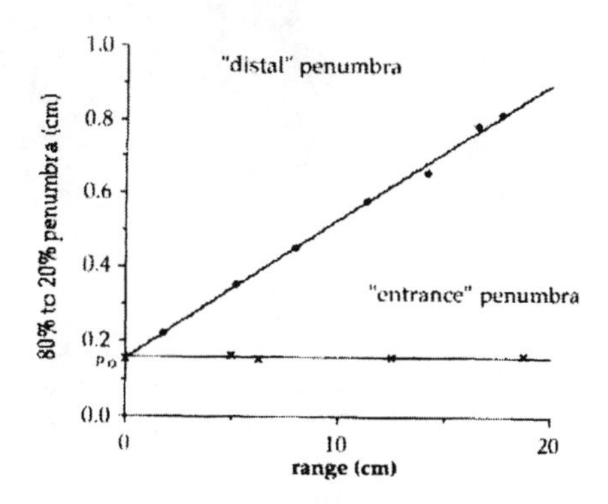

Figure 2.11. Influence of the beam range on the 'entrance' and 'distal' 80% to 20% penumbra; after R. Oozer et al. [2.13]

Fig. 2.12 shows how the detailed spatial structure of energy deposition calculated for various particles compares with the structure of biological relevant targets, such as chromatin, nucleosomes and DNA. It appears that ions have a much larger probability of producing clusters of breaks in both strands of the cellular DNA, while one does not expect such a phenomenon to be important in the case of low LET radiations. The figure shows what was said at the beginning of this Section: DNA, the biologically meaningful target for radiations, has nanometre dimensions. Thus *nanodosimetry* would be needed to understand the effect of the energy depositions due to ionizing radiation and their fluctuations.

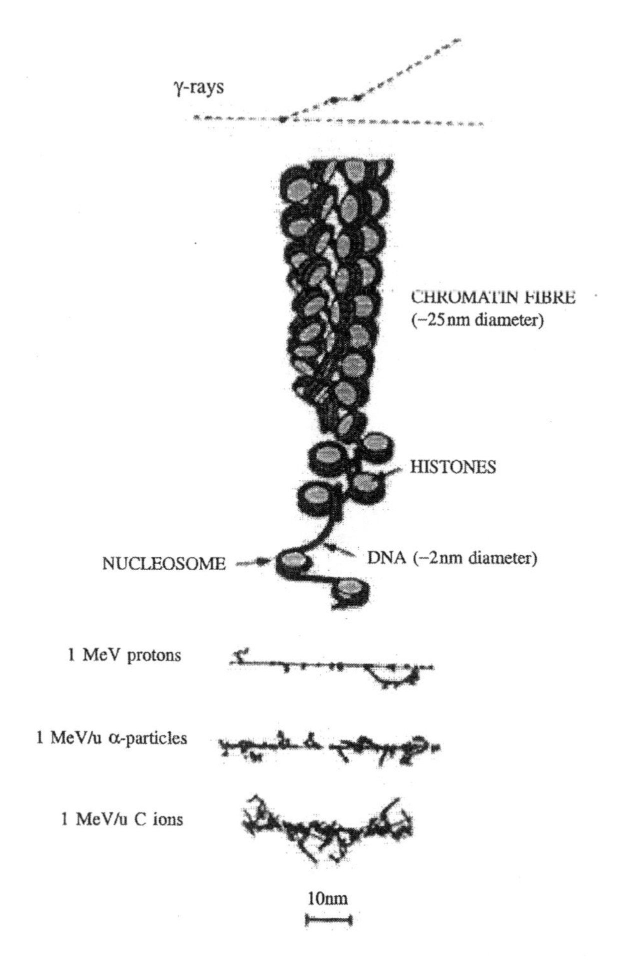

Figure 2.12. Schematic representation of tracks of sparsely and densely ionizing radiations compared with the relevant biological targets, after U. Amaldi and M. Siliari [1.4].

2.4.3 RBE of Hadrons

In order to compare the different biological responses produced by radiation of different quality the term *Relative Biological Efficiency RBE* was introduced. (Sec. 2.1.2). Also in case of a given ion species, the relative biological effectiveness depends on particle energy. Table 2.3 lists the representative data obtained with unspread and spread out Bragg peak [1.4]. The detailed table of RBE values presents L.D. Skarsgard in [2.6]. It is very important to concentrate on determining the RBE under conditions equal or similar to those used for therapy. The apparent discrepancy between the proton RBE value of 1.1 currently used in clinical practice (mainly based on experiments with spread out Bragg peak), and the data obtained from monoenergetic beams needs to be clarified. In fact, the RBE tends to increase as the energy decreases. One of the most interesting conclusions is that protons in plateau are as lethal as γ-rays (or perhaps less effective), while protons in the Bragg peak show a higher effectiveness.

Initial beam energy [MeV]	Beam energy at the cells	Cell line	RBE	Reference
100	plateau, mean energy 90 MeV (LET 0.79 keV/μm) Bragg peak, mean energy 8 MeV (LET 5.5 keV/μm)	CHO (rodent)	0.68±0.09[11] 0.98±0.10[12]	[2.8]
160	spread out Bragg peak, (mid and descending part)	H4 (rodent)	1.0-1.4[2]	[2.9]
160	plateau spread out Bragg peak	V79 (rodent)	1.2[1], 1.41[2], 3.6[13] 1.2[1], 1.46[2], 4.3[3]	[2.10]
31	31, (LET 5.5 keV/μm) mean energy 12, (LET 3.9±0.5 keV/μm) mean energy 8, (LET 5.5±1.2 keV/μm)	EUE (human)	1.0±0.1[2] 1.4+0.2[2] 1.5±0.2[2]	[2.11]
67	spread out Bragg peak, (LET ~5 keV/μm)	V79 (rodent) LS174T (human)	1.28±0.25 [3] 1.08-0.16[3]	[2.12]

Table 2.3. RBE for inactivation of cultured mammalian cells irradiated with proton beams of high initial energies; after U Amaldi and M. Siliari [1.4].

An increasing effectiveness corresponds to an increasing steepness of the survival curves. The energy at the maximum effectiveness corresponds to a

[11] Evaluated from the final slope of the survival curves, i.e. at D≥3 Gy.
[12] Evaluated at survival level at D=2 Gy.
[13] Calculated for D~0

residual range of approximately 0.4-1.4 mm; i.e. the maximum effectiveness is reached close to the end of particle range. At very end, the RBE drops due to saturation effects caused by the very high LET. This systematic can be also illustrated by plotting the RBE at different positions of a monoenergetic Bragg peak (Fig. 2.13a). When extended volumes are irradiated, the sharp peak of RBE is smeared out and the average RBE in extended peak is reduced compared to single peak at comparable dose levels (Fig. 2.13b) [1.5].

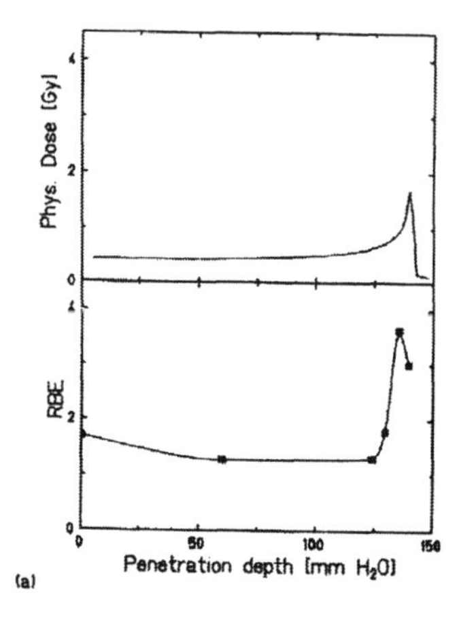

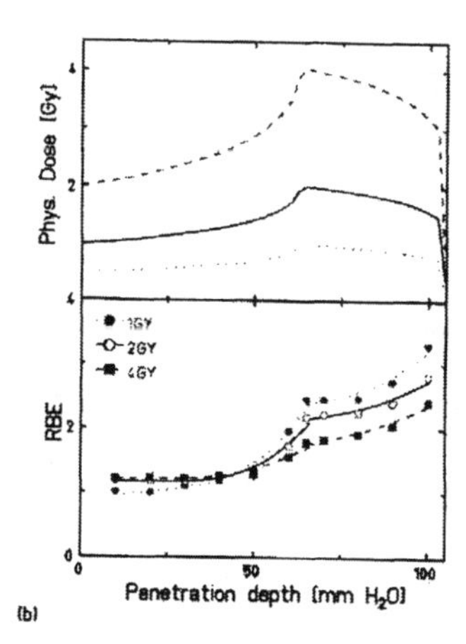

Figure 2.13. Physical dose and RBE as a function of the penetration depth in water for: a) monoenergetic 270 MeV/u carbon ions; b) simulating a tumor irradiation in 6-10 cm depth. The points are measured data and the lines represent model calculations; after J. Debus [1.5].

Because of the shoulder in the x-ray effect curve RBE strongly depends on the effect level for high survival values, like 80%, the RBE is much bigger than for those of low survival (Fig. 2.14). However, the location of RBE maximum is not shifted in the LET range of the height of RBE maximum changes. That means that for low doses corresponding to high survival levels the height of RBE maximum is larger than for low survival caused by high doses [2.1, 2.7].

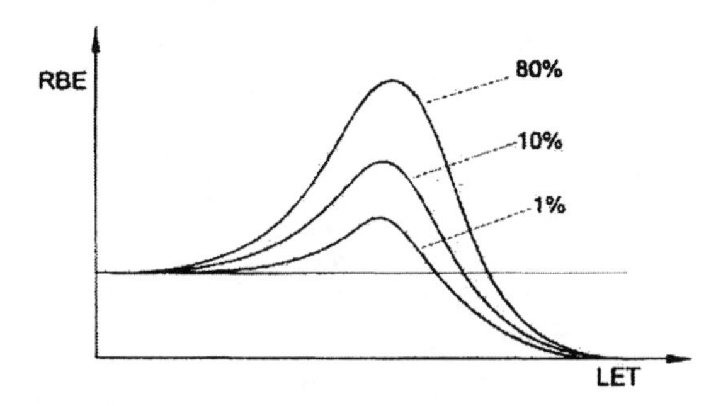

Figure 2.14. A schematic representation of RBE vs. LET; after G. Kraft [2.7]

The RBE is shifted to different LET values, however, if the atomic number of the projectile is changed (Fig. 2.15). For protons the RBE maximum is found at a LET value of 25 keV/μm, for He ions at 100 keV/μm, for carbon 200VkeV/μm and for heavier ions at higher LET values [2.1, 2.7]. The common interpretation of the variation of the RBE with LET is very mechanistic. If the extension of the tracks is neglected the LET can be correlated to a local dose. Therefore, a higher LET is correlated to the induction of a larger amount of local damage.

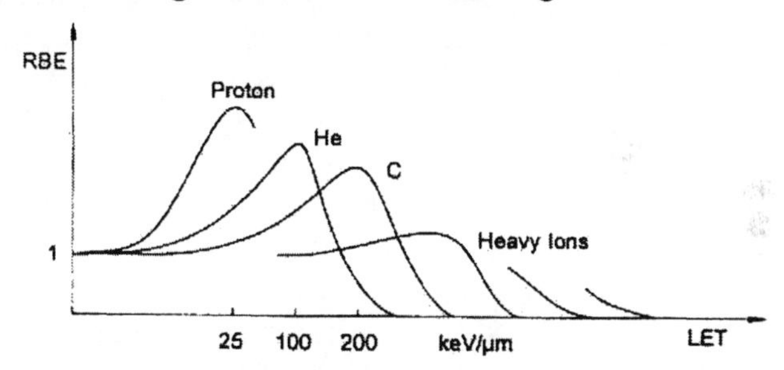

Figure 2.15. The RBE-LET relationship for particles of different atomic numbers; after G. Kraft [2.7].

2.4.4 Fractionation and Oxygen Effect

Fractionation in conventional therapy is performed in order to spare the normal tissue surrounding the tumor volume. In contrast to photon therapy, for charged particle therapy whereas in the entrance channel sparing of the normal tissue can be expected as irradiation. However, two factors allow for a different, i.e. shorter, fractionation scheme. First, the inverse dose profile leads to a lower effective dose in the normal tissue, so that the dose per fraction in the tumor can be increased and fewer fractions are necessary. Second even for high-energy carbon ions in the entrance channel the shoulder of the curve will be slightly reduced compared to photon irradiation. Thus, saturation of the sparing effect will be reached already for a smaller fraction number as compared to photons [1.5].

In the high LET regime, a drastically diminished fractionation effect has been observed. In a few cases a potentiation of the radiation damage has been found for fractionated treatment. At present time, it is not clear whether the potentiation effect is useful for therapy. However, the diminished repair in the tumor volume combined with full repair in the entrance channel will be of clinical relevance.

Many tumors appear to have an inadequate blood supply and may contain a proportion of cells deficient in oxygen: these cells are known as *hypoxic cells*. Hypoxic cells are approximately three times more resistant to x-rays compared with oxygenated cells. The presence of even a small proportion of hypoxic but viable cells in the tumor requires the increase in dose for tumor control. Normal tissues in the radiation field may not tolerate such an increase in dose. However, an increasing proportion of hypoxic cells may become oxygenated during the course of fractionated treatment; this process is known as reoxygenation. The hypoxic cells that become reoxygenated during treatment are not resistant to subsequent fractions of radiotherapy. However, reoxygenation may not take place in tumors, which are advanced or of certain histological types. There is considerable interest in the use of compounds known as *hypoxic-cell sensitizers*, which selectively sensitize hypoxic cells to radiation. High-LET radiations considerably reduce the radiation sensitivity differences between oxygenated and hvpoxic cells; this is one reason for the interest in the use of high-LET particles in radiotherapy [2.12]. For LET values around 100 keV/μm, both types of cells, oxic and hypoxic, exhibit the same radiation sensitivity [2.2].

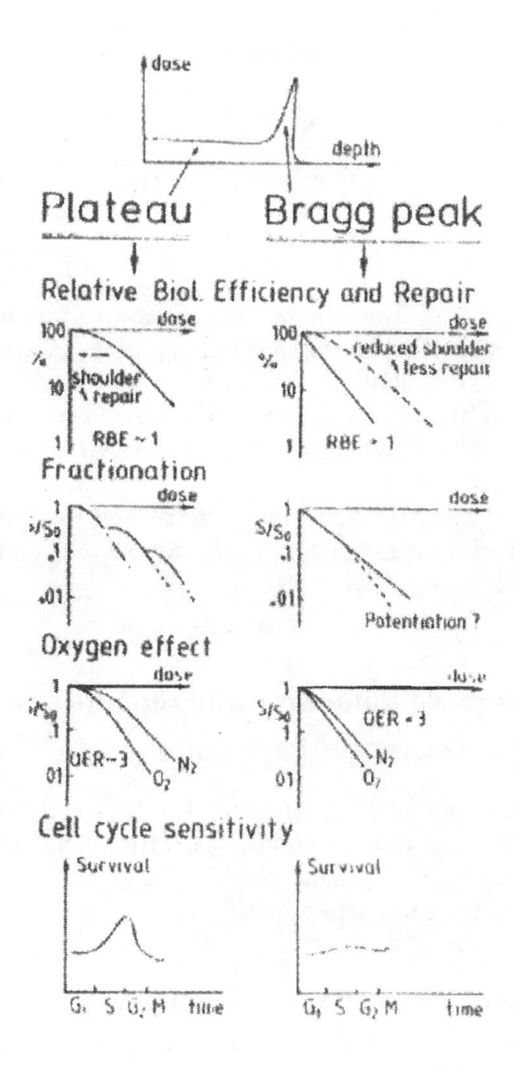

Figure 2.16. Comparison of the radiobiological effects in the plateau and in the Bragg peak; after G. Kraft [2.15].

Chapter 3

STATUS OF CLINICAL RESEARCH IN PROTONTHERAPY

The physics characteristics of a beam of accelerated protons allow a very favourable distribution of the dose in depth, represented by spread out Bragg peak of Fig. 1.10. The continuous variation of the beam and an appropriate system of spreading the protons in space allows a fine control of the dose distribution among the target volume and the surrounding areas (Sec. 2.1.1). These irradiation conditions are much more favorable than those obtained with electrons or photons. The therapeutical results can therefore be improved because of the increased ratio of the doses absorbed by the tumor volume and the healthy tissues.

3.1 Precision in Radiotherapy and the Role of Protons in Improving Treatment Precision

The goal in planning radiotherapy (Sec. 10.2) is to select a set of radiation beams, which combined with an optimal patient immobilization procedure and position confirmation system, yields a treatment volume that most closely conforms to the defined target volume.

3.1.1 Precision in Conventional Radiotherapy

Progress in clinical radiotherapy has been strongly related to technical developments allowing for a more precise irradiation of the tumor. About fifty years ago megavoltage photon beam equipments for external treatment were introduced (i.e. cobalt units, betatrons, and linear accelerators). This greatly improved the ability to reach deeply located targets (increase in depth dose) with a simultaneous decrease of the *skin dose* (*entrance dose*) and a steeper lateral dose fall-off (reduced beam penumbra).

Isocentric gantries, treatment simulators, portal radiographs, secondary collimation and patient immobilization devices are some of the presently existing standard tools helping to improve precision and treatment reproducibility with megavoltage equipment. The more recent progress in medical imaging: computerized tomography (CT), magnetic resonance imaging (MRI), and positron emission tomography (PET), allows for a precise definition of the target and critical structures. Additional technical innovations in computer technology is helping to optimize treatment planning by displaying *beam's eye views* of the anatomical structures through which the beam passes and by calculating the isodose contours around the target volume three-dimensionally (Sec. 10.2). Dynamic photontherapy and *multileaf collimator systems* for linear accelerators are being introduced in clinical radiotherapy and are expected to further increase precision and to shorten the treatment time by a rapid and automatic change of secondary collimation during irradiation.

Precision in radiotherapy is limited by irradiation of the patient or of a tumor bearing-organ during an individual treatment. Plastic masks and bite blocks achieve a reduction of the head movements to 1-2 mm. A totally rigid immobilization system, such as a *head-holding frame* screwed to the skull for radiosurgical procedures, can reduce mobility to 0.1-0.5 mm. Although fitting casts have been used for pelvic radiotherapy, significant reduction of the pelvic motion has not always been observed. Using on-line visual monitoring systems as mentioned earlier could further reduce the effect of motion. Because of the range of motion during inspiration, mobility of the chest may be as much as several centimeters. The latter effect can be reduced by gating the treatment with the respiratory cycle [3.1].

3.1.2 Role of Proton Beams in Improving Treatment Precision

Protons are similar to photons from a biological point of view. However, unlike photons, they have a finite range in tissue and a rapid distal and lateral (energy-dependent) fall-off in dose, potentially providing a superior dose distribution in several anatomic sites.

In case of conventional radiotherapy, the dose delivered to the structures located proximally to the target is equal to, or greater than the dose received by the target volume. A monoenergetic beam of protons, however, has a well-defined range of penetration, depending on the initial beam energy and the electron density of the absorbing material. This is because the primary energy loss mechanism for protons is via frequent collisions with atomic electrons, with each collision causing only a small energy loss to the incident proton (Sec. 2.3.3). As the particle proceeds through the medium losing energy and velocity the rate of energy loss increases sharply near the end of its range (Bragg peak region, Fig. 3.1). The high-dose region is a narrow peak with limited use in the treatment of most commonly occurring human tumors [3.1, 3.16].

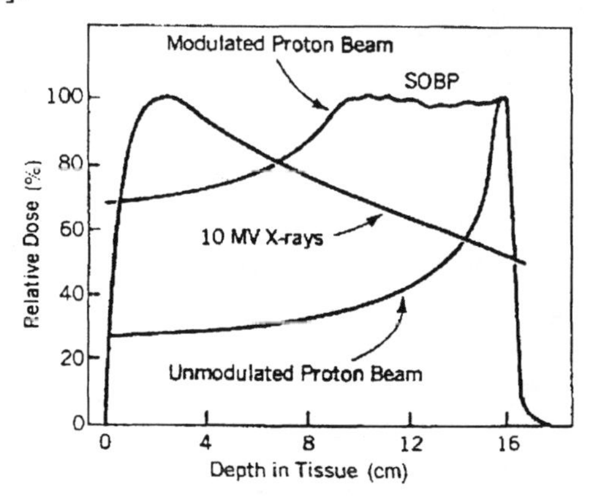

Figuure 3.1. The depth dose on tissue for a 160 MeV nonmodulated and modulated proton beam compared with that for a 10 MV x-ray beam; after R. Miralbell [3.1].

Techniques have been developed for producing a spread-out Bragg peak (SOBP), which is an area of uniform ionization at a depth sufficient to cover the intended target volume thickness (Fig. 3.1). A *range-shifter wheel* rotating in the beam can provide the *range modulation*. The 100% dose level region is deposited with constant thickness in depth (*fixed range modulation*). A *compensator bolus* is used to adjust the distal edge of the dose distribution to exactly correspond to the distal edge of the target volume. The dose can therefore be accurately conformed to the distal part of the target surface. Some unwanted dose is, however, deposited on the proximal site of the target volume. This can be minimized with active magnetic scanning of the beam (*variable range modulation*). The techniques of spreading out the Bragg peak will be discussed in Sec. 7.5.2.

In order to clinically use proton beams it is necessary to know the 3D distribution of tissue densities and stopping powers for the entrance region and the target volume with the patient in the treatment position. A CT-scan-based 3D treatment planning system is essential for the design of the treatment technique since the range of protons is determined by the electron density of the materials, traversed by the beam. The target volume and adjacent normal structures are viewed in sequential CT transverse slices as well as in reconstructed sagittal and coronal Secs. through the area of interest. The defined structures can be viewed on the display screen from any beam direction (*beam's eye view*). A treatment plan, which spares the normal structures while optimally encompassing the target volume is obtained. Dose-volume histograms (DVHs) display the uniformity of the dose throughout the treatment volume of a structure (as a percentage of the whole) that has received at least a given dose of radiation. DVHs allow quantification of the differences in dose distribution between different treatment plans which otherwise might be difficult to assess by using only isodose contours viewed over multiple slices [3.1]. Therapy planning will be disccussed in Secs. 10.1.2.1 and 10.2.

3.1.3 Range Uncertainties of the Protons in the Patients

The Bragg peak is the great advantage of protons, but can also cause major problems if applied improperly. The small high dose area and the steep dose gradients of a deposited proton pencil beam require a high precision in the control of the range of the protons. An accurate range control requires a high precision of the application technique (steering of the beam, patient setup and repositioning) and an accurate modeling of the beam [3.2]. There are two main points, which have to be addressed:

3.1.3.1 Position of the Bragg Peak in Depth

The depth of the Bragg peak in a homogeneous medium is defined by the initial energy of the beam, and the stopping power of the medium. The energy of the beam can be measured rather accurately. Therefore, the precision of modeling the range of the beam depends mainly on the precision of the knowledge of the stopping power properties of the human tissues. A direct measurement of proton stopping powers in the human body is theoretically possible using energy loss proton tomography. Proton tomography images have already been taken in 1981 [3.3]. Due to the success of computer tomographic (CT) methods using photon attenuation measurements, this technique has not been developed any further.

Proton stopping power and photon attenuation are both, a function of the electron density. In high-density tissue and at low photon energies (used in CT imaging), however, a considerable portion of the photon attenuation is due to the photoelectric effect. Hence, a calibration curved is needed to convert CT images of the patient into stopping power images [2.1].

3.1.3.2 Shape of the Bragg Peak

The precise of the protons in the human body could be obtained by integrating the relative stopping power values from the calibrated CT images along proton trajectories, if their paths would be completely known in advance. Coulomb scattering however causes the path of the protons to deviate randomly from straight lines. A proton detected first at a point A (Fig. 3.2b), and then at a point B can have taken a multitude of paths between the two points. The energy loss suffered by a proton following the different trajectories depends on the stopping power of the traversed materials. Thus the energy and therefore the residual range of a proton following trajectory p_1 is different from one following p_2 [2.1].

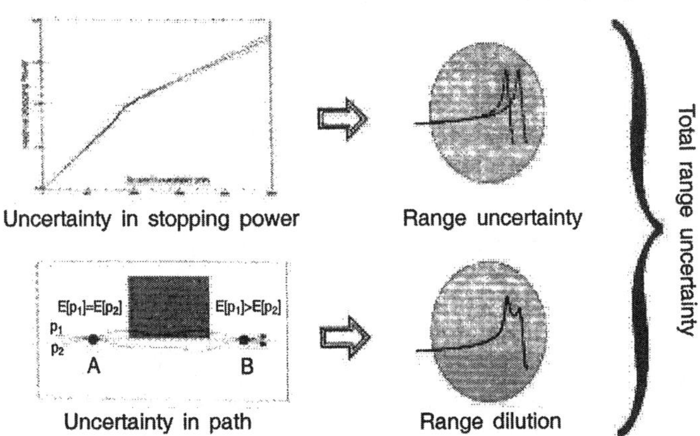

Figure 3.2. The two main sources of uncertainties in proton radiotherapy treatment planning and their effects on the Bragg curve. The uncertainty in the determination of the stopping power leads to an uncertainty in the position of the Bragg peak. The uncertainty in the path of the protons causes a dilution of the range and the degradation of the peak; after B. Schaffner et al. [3.2].

3.2 Clinical Experience in Protontherapy. Patient Statistics

About 31 838 patients were treated with all particles by July 2000, at 30 particle radiotherapy facilities around the world [3.6]. Table 3.1 gives an overview of charged particle therapy centers and the number of patients treated world-wide. In 23 Protontherapy facilities they were treated 27 434 patients, 3 304 patients were trated with heavy ions and 1 100 with pions.

Treatment schemes using proton beams are frequently described as being in one of three major categories:

- **Single fraction** - the dose is delivered in a single session to the target volume though several portals may be used in that session. The target is usually intracranial (pituitary tumors, arteriovenous malformations, other vascular lesions).
- **Eye treatments** - uveal melanomas accounted for ~49% of all protontherapy treatments world wide in 1994, making the treatment of this tumor the most common application of Protontherapy. At HCL, the current treatment plan uses 5 fractions, each of 14 CGE (*Cobalt Gray Equivalent*[14], RBE = 1.1).
- **Fractionated therapy** - protons alone, or in conjunction with other modalities, are used to treat a variety of sites, both benign and malignant. The number of proton fractions delivered depends on the site in question. For example, at HCL in 1994, over 2700 fractions were delivered to 113 patients, ranging from 2 to 40 per patient.

3.3 Clinical Indications for Protontherapy

After the experience of a few decades, the main clinical indications for ion beam therapy remain proximity of the target area to critical structures (where maximum selectivity of dose distributions is of paramount importance), low tumor radiosensitivity necessitating high doses, and a high benefit-to-cost ratio. For a well defined number of tumor types which meet these criteria, such as uveal melanomas and chondrosarcomas, and chordomas of the base of skull and spinal region, protontherapy may be superior to photon beam therapy. Health care providers acknowledge the cost for the treatment of patients with these indications in particle therapy centers abroad. A typical

[14] The CGE is the proton dose in Gy multiplied by the proton RBE (1,1).

example of the improved cure rates after ion beam therapy is presented in Fig. 3.3.

Laboratory	Particle type	Accelerator type	Energy [MeV]	Date of first – last RX	Recent patient total	Date of total
Donner Lab., Berkeley, USA	p	synchrocyclotron	340	1954-57	30	
Donner Lab., Berkeley, USA	He	synchrocyclotron		1957 1992	2054	June 91
Werner Inst. Uppsala, Sweden	p	synchrocyclotron	85	1957-76	73	
Harvard Cycl. Lab. , USA	p	synchrocyclotron	160	1961	8558	July 00
Dubna, Russia	p	synchrocyclotron	90-200	1964-74	84	
ITEP, Moscow, Russia	p	synchrotron	70-200	1969	3268	June 00
LIJF, St. Petersburg, Russia	p	synchrocyclotron	70-1000	1975	1029	June 98
Donner Lab., Berkeley, USA	heavy ion	synchrocyclotron		1975-1992	433	June 91
Los Alamos, NM. USA	π^-			1974-1982	230	
NINP, Chiba, Japan	p	synchrocyclotron	70	1979	133	April 00
PMRC, Tskuba, Japan	p	synchrotron	250	1983	629	July 99
TRUMF, Canada	π^-			1979-1994	367	December 93
PSI, SIN, Switzerland	π^-			1980-1993	503	
PSI, Villigen, Switzerland	p	cyclotron isochronous	70-590	1984	3014	December 99
Dubna, Russia	p	synchrocyclotron		1987	58	June 00
Werner Inst. Uppsala, Sweden	p	synchrocyclotron		1989	236	June 00
Clatterbridge, UK	p	cyclotron isochronous	62	1989	999	June 00
LLUMC, Loma Linda, USA	p	synchrotron	70-250	1990	5262	July 00
Louvain-la-Neuve, Belgium	p	cyclotron	90	1991-1993	21	
Nice, France	p	cyclotron	65	1991	1590	June 00
CPO Orsay, France	p	synchrocyclotron	70-200	1991	1527	December 99
NAC, South Africa	p	cyclotron	200	1993	367	June 00
IUCF, Davis, IN USA	p	cyclotron	200	1993	34	December 99
UCF-CNL,CA USA	p	cyclotron	200	1994	284	June 00
TRIUMF, Canada	p	cyclotron	180-520	1995	57	June 00
PSI, Villigen, Switzerland	p	cyclotron	200	1996	41	December 99
HIMAC, Chiba	heavy ion	synchrotron		1994	745	December 99
GSI, Darmstadt, Germany	heavy ion			1997	72	June 00
Berlin, Germany	p	cyclotron	72	1998	105	December 99
NCC, Kashiwa, Japan	p	cyclotron	235	1998	35	June 00
					1100 pions 3304 ions 27434 protons	
				TOTAL	31 838	all particles

Table 3.1. World wide charged particle patient total, July 2000; after J. Sisterson [3.6]

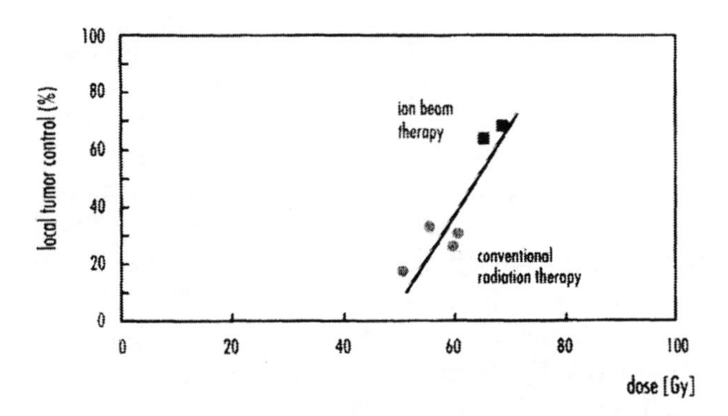

Figure 3.3. Local tumor control as a function of delivered dose; after J. Debus et al. [1.5].

For other malignancies including tumors of the brain, head and neck, esophagus, prostate, rectum, female reproductive system, as well as soft tissue sarcomas, where improved local control is likely to result in higher rates of definitive cure, Protontherapy has claimed to have an advantage over conventional photon radiotherapy. However, all of these were comparative treatment studies in small series of patients with an often too short follow-up period and clearly require verification in larger studies. Moreover, some non-malignant lesions such as cerebral arteriovenous malformations and pituitary adenomas can also successfully be treated by Protontherapy.

3.3.1 Ocular Melanomas

Intraocular tumors treated with enucleation are ideal targets for proton beams, which are capable of concentrating large radiation doses in the tumor while sparing the normal uninvolved intraocular structures. Uveal melanoma is the most common primary intraocular tumor in adults. It is the tumor arising from the middle coat of the eye comprising the choroid, the ciliary body and the iris. The common treatment for this tumor is the enucleation of the eve. At the beginning local irradiation using ^{60}Co plaques sutured by the choroid was used for destroy this kind of tumor. However, it was associated with morbidity and loss of visual function. ^{125}I plaques are now being used because they give less radiation dose to uninvolved normal tissues [3.8].

Patient treatments for choroidal melanoma using protons from the Harvard cyclotron started in 1975. Treatments were given in five fractions over 8-9 days. The total doses were increased in steps gradually from 47,3 to 85,7 CGE. Local control was obtained for all nine patients. The Harvard group has treated more than 2360 patients (1997) [3.9]. For most of the patients, a total dose of 70 CGE was delivered in five fractions over 8 or 9 days. An impressive 5-year local control of 96% was achieved, within the irradiated eye. The goal of the treatment is to preserve the visual function of the treated eye. In a small, but significant number of patients, enucleations had to be performed, mainly because of late effects of the radiation. The visual outcome was found to be dependent on the proximity of the tumor to the fovea and optic disc on the size of the tumor. Unfortunately 50% of uveal melanomas occur <3mm from the fovea or optic nerve.

From 1984 until the end of 1999, 3000 patients suffering from tumors of the eye were treated with protons at the OPTIS facility at PSI. This technique was introduced in Europe by PSI in 1984, and since then applied in collaboration with the Hôpital Opthalamique, Lausanne. Patients received a total dose of 60 CGE in four fractions over 5 days. The 5-year local control rate was ~95. The 5-year survival rate was 85%. The most important parameters determining survival after multivariate analysis were: the largest tumor diameter (60% 5-year survival for lesions >20 mm); age (worse for >45 year-old patients); and local status after treatment (60% 5-year survival in local failures). Useful vision was preserved in 50% of the treated eyes. Cataracts were observed in 19% of the patients; rubeosis iridis and neovascular glaucoma in 20%; macular vasculopathy in 9%; and optic atrophy and radiation papillitis in 10%. Check-ups over 10 years show that 98% of the tumors have disappeared, and that the preservation of the irradiated eye and the survival of the patient depend upon the size of the tumor [3.1, 3.19].

In conclusion, the HCL and PSI experiences have shown that a high rate of eye retention can be achieved while at the same time maintaining or improving the useful vision and without jeopardising the potential for cure compared with enucleation.

Eye tumors have been treating using protons at ITEP since 1976. A total of about 400 patients were treated with good results. Uveal melanoma treatments using protons were also started at MRC in Clatterbridge, UK (463 patients , Jan. 1994), GWI in Uppsala, Sweden (20 patients, May 1993), Centre do Prothontherapie d'Orsay (235 patients, May 1993) and at the Centre Antoine-Lacassagne in Nice, France (216 patients, April 1993) [3.8].

3.3.2 Skull Base and Cervical Chordomas and Chondrosarcomas

Chordomas and chondrosarcomas of the skull base and cervical spine are rare tumors. They represent a difficult treatment planning problem because of their proximity to important normal structures, such as the brain stem and spinal cord, the optic nerves, and the optic chiasm. Most patients treated with conventional radiotherapy (50-55 Gy) after surgery die of locally progressive disease (35% 5-year local control rate).

Between February 1972 and September 1993, 354 such patients were treated at the HCL with tumor doses ranging from 56.8 to 80 CGE (median 68.4 CGE). Almost all the patients received one fraction of 4 or 10 MV x-rays for each 160 MeV proton treatments to reduce the skin dose (too high with exclusive protons). An analysis of the outcome of the first 254 patients treated until December 1991 has been published [3.10]. The estimated 5-year local recurrence free survival (LRFS) was 77±5% for the skull base tumors patients and 58 ± 14%, for cervical spine tumors. LRFS was significantly better for chondrosarcoma patients than for chordoma patients (only for skull base tumors): 95 ± 4% vs. 62 ± 9%. Female chordoma patients fared significantly less well than males. Overall survival was also better for patients with skull base tumors (worse for cervical spine tumors) and with chondrosarcoma (worse for chordoma). Considering the high doses delivered the complication rate has been relatively modest. Radionecrosis with significant brain injury has occurred in approximately 6% and severe visual complications in 2% of the skull base patients. Pituitary insufficiency after treatment occurred in 10 - 15%, being partial in most [3.1].

The apparently higher cure rates with proton beams are most likely due to higher doses delivered and / or to the improved 3D planning techniques employed. A clinical trial is now in progress randomizing tumor doses of 66 versus 72 CGE by Harvard and Loma Linda groups. However, because of a large and significantly lower control rate in female patients with chordomas of the skull base and in both female and male patients with cervical spine sarcomas are now randomized to 72 or 79 CGE. The chondrosarcomas of the skull base and all malae patients with skull base sarcomas will continue to be randomized to 66 or 72 CGE [3.8].

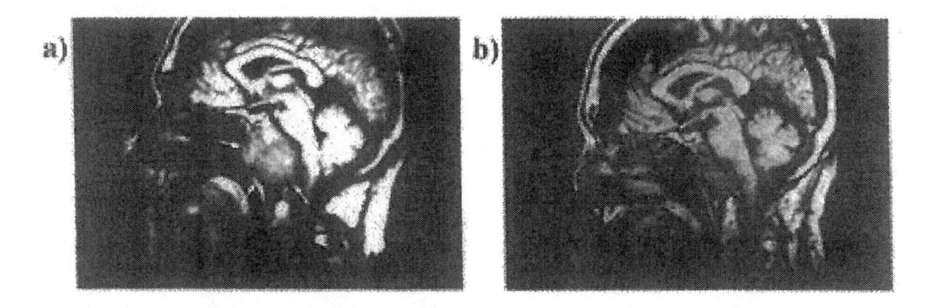

Figure 3.4. Skull base tumor: a) chordoma; unresectable because of the adjacent brain stem; b) after proton irradiation shrank remarkably; after [1.23].

3.3.3 Intracranial Tumors

Other targets, which have been treated with fractionated protontherapy at HCL, include primary intracranial tumors such as gliomas. meningiomas and craniopharyngiomas. Early experience with high-grade gliomas was rather disappointing. In an analysis of 36 patients with benign meningioma treated between 1968 and 1986 with radiotherapy at MGH, 25 patients were treated with conventional radiotherapy, while 11 patients were treated with combined photon and proton beams (dose ranging from 55,8 to 71,6 CGE). There was a trend toward improved local control in patients treated with combined photons and protons. None of these patients relapsed after a median follow-up of 53 months.

Trial has been recently started to prospectively evaluate local control and morbidity after combined photon and proton treatment of incompletely excised or recurrent benign meningiomas to doses of 55,8 or 63 CGE. Twelve patients with craniopharyngiomas have been treated with doses ranging from 52,7 to 63,4 CGE. After more than 5 years, median follow-up 10 patients remain alive without tumor progression or complication. It is widely accepted that a moderate dose such as 50,4 Gy in 28 fractions is sufficient to successfully treat craniopharyngiomas. Although this dose level can easily be delivered with conventional radiotherapy, proton techniques can achieve a significant reduction of the integral dose to the non-target normal brain [3.1].

3.3.4 Pituitary Tumors

The majority of pituitarity tumors do not contain hormone-producing cells, and are generally treated by surgery, with or without conventional high-voltage radiotherapy given postoperatively. In contrast with endocrine-active tumors, these tumors are relatively radiosensitive and may be cured with with a very high percentage success rate by radiation doses below the tolerance level of the surrounding normal tissues.

There are a number of pituitary tumors that are associated with an overproduction of hormones such as growth hormones in acromegaly and adenocorticotropic hormones and cortisone in Cushing's disease. The typical symptoms of patients with acromegaly are enlargement of the skull, hand, and foot bones and an increased risk of early stroke. The corresponding symptoms of Cushing's disease are increased production of cortisone from the adrenal gland, leading to obesity, high blood pressure, weakness and porosity of the bone [3.8].

3.3.4.1 Acromegaly

Kjellberg et al. [3.11] obtained issoeffect data for brain necrosis as a function of proton beam diameter. They found that the threshold for brain necrosis was 50 Gy for for the 7 mm beam. A 1% risk for brain necrosis was established between 50 Gy for the 7 mm beam and 10,5 Gy for the 50 mm beam, and a 99% risk line between 150 Gy for the 7 mm beam and 100 Gy for the 50 mm beam. As of January 1992, they had used protons to treat 582 acromegaly patients. The treatments are given in a single fraction. Six portals are used on each side of the head for a total of 12 portal entry sites, and a dose of up to 120 Gy in the central region of the target is delivered. The dose to the optic nerve is kept <6 Gy by choosing the location of the beam spot. They found intrasellar adenomas to be well suited for proton Bragg peak therapy because they were able to deliver an adequate dose to destroy the lesion without risking optic nerve damage. Because of the risk of optic nerve damage or oculomotor disturbance. adequate doses cannot be delivered when the adenomas were extended to the suprasellar or lateral regions. Large isoeffect doses were found to be required for both acromegaly and Cushing's disease. Preservation of normal pituitary function is also an important consideration for the treatment of. acromegaly . This is achieved by selecting the dose to the central region to produce ablative effects, but not to the outer isodose zone within the sella. A 20-year experience of treating acromegaly patients with the proton Bragg-peak was reviewed by Kliman et al. [3.12]. They concluded that proton Bragg-peak therapy offers an effective therapy without mortality risk, and the potential effectiveness of the treatment remains for life. They recommended that Bragg-peak Protontherapy be given as a first choice for patients with active acromegaly. They also stated that the prospects for improvement or a cure for acromegaly can he enhanced by early diagnosis and treatment with protons.

In Russia, there are two centers using protons for the treatment of acromegaly. Minakova et al. [3.13] using the Protontherapy facility at ITEP, have treated about 160 patients. A total dose of 50-65 Gy is delivered in 2-3 fractions over 1 week. Konnov et al. [3.14] have treated 160 patients at synchrocyclotron at Gatchina. They used the high-energy region of 1000 MeV protons, delivering a dose of 100-125 Gy in one fraction. The results are also reported to be excellent.

3.3.4.2 Cushing's Disease

The primary objective in proton Bragg-peak therapy of Cushing's disease is to correct the excessive adenocorticortopic hormones as well as to restore the normal adrenocortical function. As of January 1992, 182 patients with Cushin 's disease were treated using Harvard cyclotron. A single dose of 120-140 Gy was used. The results were found to be comparable with the encouraging results seen in acromegaly. A high remission rate of 85% was obtained with low incidence of side effects. Protontherapy was found to be a useful alternative to transsphenoidal surgery.

A number of 320 patients with Cushing's disease were treated with ITEP cyclotron in Moscow. A total dose of 70-90 Gy was delivered in one to four fractions over 1-2 weeks. They also reported excellent results: a complete, or near complete, clinical and biochemical remission in 96% of the patients. Also in Gatchina about 60 patients were treated with a single dose of 100-125 Gy, using the 1000 MeV proton plateau region, and reported a remission rate of about 75%. Routine treatment of Cushing's patients with protons is being continued at both ITEP and Gatchina [3.8].

3.3.5 Arteriovenous Malformations (AVM)

Arteriovenous malformations (AVMs) of the brain are congenital lesions. They manifest themselves clinically through intracranial bleeding, seizures, headaches progressive neurologic defects, etc. Surgical excision through craniotomy is a craniotomy is a common procedure. However, AVMs are often inaccessible for surgery because of their size or location in critical regions of the brain.

The first to use protons for treating AVMs in 1965 was the Harvard group. They have treated mostly inoperable cases. As in radiotherapy, the dose selected should be high enough to be effective on abnormal blood vessels, and at the same time, low enough to spare normal brain tissue surrounding the AVM. Beam diameters varying from 7 to 50 mm, modified to conform to the size and shape of the AVM were used. No immediate effect was observed during the first 12 months after the treatment, while the clinical improvement was evident between 12 and 24 months after treatment. The results of treatment of 709 AVM patients varied from total obliteration of the AVM to no evident change. Totally obliterated in 20% of patients, reduced by 50 % in 56% of patients and no change in 13% of patients. By January 1992, thy had treated 1351 AVM patients. Proton Bragg-peak therapy is a useful technique for treatment of intracranial arteriovenous malformations unsuitable for surgical excision or embolization.

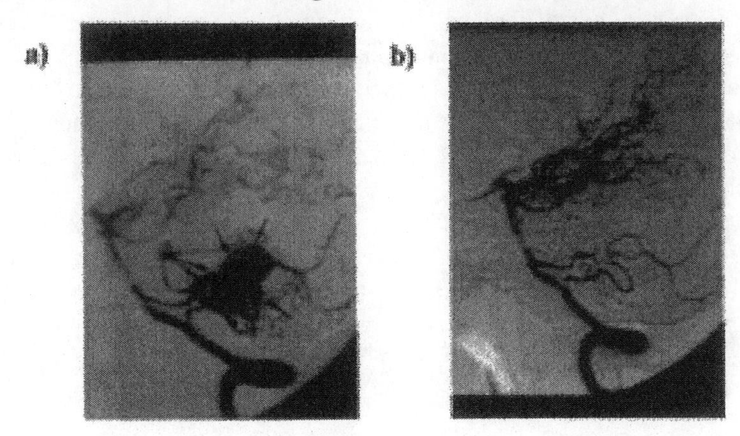

Figure 3.5. Arterovenous malformation: a) abnormal vascular net between artery and vein; cerebellar hemorrhage might occur when untreated; b) after proton beam irradiation, abnormal vascular net disappeared; after [1.23].

Protons were also used for treatment 91 AVM patients in ITEP. Doses in the range 40 - 60 Gy were delivered in one or two fractions, depending on the size and location of the AVM. The mean obliteration rate for all volumes was found to be approximately 60%, with a partial obliteration in 25% of the cases. At Gatchina 187 patients were treated using the high-energy plateau region of the 1000 MeV proton beam.

3.3.6 Adenocarcinoma of the Prostate

Between 1976 and 1979 a pilot study of a perineal proton boost following large-field photon pelvic irradiation was undertaken at HCL. Sixty-five patients with prostate cancer (T2, T3 and T4) were included in this study. All patients received 50,4 Gy to the pelvis, and the prostate was boosted to 73,5 CGE (8-10% greater than the usual dose given with conventional megavoltage x-ray techniques). Late treatment sequelae, however, were not different from those seen after conventional x-ray treatments. A randomized trial was opened in 1981 to determine whether an increase in total dose from 67,2 Gy photons to 75,6 CGE using a proton boost would provide significantly improved clinical local control of T3 and T4 tumors, relative to that seen with photon techniques, without increased rectal or bladder toxicity. The trial was closed in 1992 after 203 patients had been enrolled. At 7 years no significant differences in local control, disease-free survival, or overall survival have been observed between the two arms. However, the local recurrence-free survival was improved in the high dose arm for poorly differentiated tumors (i.e., 85 vs. 37%). Posttreatment rectal bleeding from telangicctatic rectal mucosal vessels (none requiring surgery or hospitalization) occurred in 34% of patients boosted to the higher dose with protons and in 16% of patients treated to the lower dose with x-rays only. The volume of anterior rectum receiving >75 CGE correlated with the probability of remaining free of rectal bleeding. This probability was 80% in 27 patients in whom <40% of the anterior rectum received > 75 CGE but less then 40% in patients in whom >40% of the anterior rectum received >75 CGE [3.1].

3.4 Hadrontherapy Diseases Categories

Clinical indications for hadrontherapy have been divided into categories according to a decreasing priority based on the experience obtained in a limited set of pathologies. In preparing the list have been excluded highly radiosensitive tumors, such as seminomas and lymphomas (where the low dose level required allows the use of any megavoltage photon beam). The tumors situated in regions where the dose distribution obtainable with conventional radiation is already optimal and cannot be further improved (skin, breast) make this group larger [1.4].

The main criteria used to define a priority scale of the clinical indications for protontherapy, are the following:

- the proximity of the target volume to very critical structures whose damage could involve serious life risk or a severe permanent infirmity (need of maximum selectivity in the dose distribution),
- low radiosensitivity (need of high doses),
- the possible impact of the improvement of the local results on the long term survival or on the life quality (high ratio benefit / cost).

Besides these main criteria, other clinical characteristics, related to the whole exploitation of the proton dose distribution, have been taken into consideration. In particular:

- the possibility of easily identifying, by means of modem imaging techniques, the site and the contours of the tumor,
- the limited possibility of a modification of these elements, due to the physiological movements,
- the ratio between the volume of the tumor mass and the Sec. of the body where it is situated.

The possibility was also considered of employing proton beams not for the whole treatment, but as a *boost* on a reduced volume after the irradiation with proton beams over a larger volume.

By using the previously described criteria, the pathologies characterized by a potential indication for proton radiotherapy have been subdivided into four categories, in order of decreasing priorities (after U. Amaldi et al. [1.4]).

Category A includes pathologies (such as uveal melanomas, tumors of the base of the skull and of the spinal cord, etc.) which are characterized by the closeness to highly critical structures. In these conditions the use of proton beams is the only way to administer a radical dose without producing serious side effects. For this category, the advantage of proton over photon beams does not need to be demonstrated, because significant clinical data are already available in the literature. The main diseases in this cathegory are as follow: uveal melanoma, parasellar meningnoma, paraspinal tumors, acoustic nerve schwannoma, base of the skull chordoma, base of the skull chondrosarcoma, optical nerve glioma, hypophysis adenoma, arteriovenous malformations (AVM).

Tumors belonging to *category B* comprise pathologies characterized by a prevalently local evolution (so that the local control will lead, with great probability, to definitive cure) and by low radiosensivity. For these pathologies conventional radiotherapy already gives good or partial results, which could be improved by increasing the dose delivered. The advantage of the proton beam for these indications has still to be demonstrated clinically, but in any case the potential interest is great. The main diseases in this cathegories are cancer of prostate, rectum, salivary glands, portio, retrioperitoneal sarcomas, undifferentiated thyroid gland tumor.

Tumors belonging to *category C* are advanced head and neck cancers, biliray tract, tumors of esophagus, lung, thymona. It includes the pathologies for which the main indication for the use of protons is the boost on a restricted volume. The use of protons for treatment is less interesting in these pathologies for several reasons.

The list of pathologies of *category D* includes those with locally advanced conditions of very unfavorable prognosis but which may allow quite a long survival with heavy symptomatology. The main diseases of this category are: pancreas cancer, local recurrence of pelvic tumors after surgery, isolated cerebral metastasis, patparaorthis metastatic adenopathies, pelvic recurrences in previously treated areas, high grade glioma, rhinopharyngeal recurrences in previously irradiated areas. In these cases palliation with a conventional beam is often unsatisfactory and may be followed by side effects that can worsen the quality of the patient's life, while protons allow a more complete and protracted palliation with fewer side effects. Category D in Poland numbers some 500 patients per year.

As the exemple, Table 3.2 presents overall number of potential users for protontherapy in Poland.

Category of pathology	Number of cases expected for protontherapy
Category A	1020 – 1630
Category B	2445
Category C	2130
Category D	500
Total	6095 – 6705

Table 3.2. Overall number of potential users for protontherapy in Poland. The estimations made according to the data of the Italian TERA project; after W. Wieszczycka [1.26].

3.5 Number of Hadrontherapy Patients World-Wide

It is already established that proton beam therapy is effective for uveal melanoma, intracranial tumor (chordoma from skull base, meningioma, arterio-venous malformation, pituitary tumor), paraspinal tumor and advanced diseases in the pelvis (uterus and prostate cancer). It is difficult to treat these with conventional modalities because critical organs are close to these tumors. Site distribution of cancer patients treated with the beam from Harvard cyclotron is used, are shown in Table 3.3.

Site	Number of patients
Uveal melanoma	2242
Skull base / cervical spine sarcoma	479
Prostate	166
Soft tissue sarcoma	108
Other bone sarcoma	96
Head / neck	89
Glioma	66
Meningioma, cranio, etc.	96
Other	207
Neurosurgery	2929
Star	148
Total	6626

Table 3.3. Site distribution of cancer patients treated by MGH / MEEI / HCL (1974-1995); after W. Krengli et al. [3.15].

The number of patients so far treated with proton beam there is the highest among the institutions in the world. The results are better than that of conventional therapy for all presented kinds of tumors.

Table 3.4 lists tumor sites and patient number world-wide for the year 1993. The actual number of patients treated with proton beam recently is presented in Table 3.1.

Site/Tumor	Berkeley, USA 1954-57	GWI, Uppsala, Sweden 1957-93.5	Boston, USA 1963-92.12	Moscow, Russia 1969-92.1	Dubna, Russia 1964-92.8	Gatchina, Russia 1975-91.6	Chiba, Japan 1979-93.1	Tsukuba, Japan 1983-93.7	Villigen, Switzerland 1983-92.12	Clatterbridge, UK 1989-93.5	Loma Linda, USA 1990-93.5	Louvain-la-Neuve, Belgium 1991-92.6	Orsay, France 1991-93.5	Nice, France 1991-93.4	Total (%)
Uveal melanoma			1 670	355			34		1 250	369	25	14	235	209	4 165 (34)
Other							10		108		3			7	128 (1,0)
Pituitary tumors	26	26	510			287					30				864 (7,1)
Brain tumors		8	55	1 297							84				1 473 (12)
AVM		7	439			94									556 (4,5)
Skull base and cervical spinal			179												222 (1,8)
Other			1 768								41				1 768 (14,5)
Head and neck tumors		20	42					21			37				120 (1,0)
Thoracic tumors								23			2				25 (0,2)
Gastro-intenstinal tumors			23					106							129 (1,1)
Urogenital tumors			147					32			292				553 (4,5)
Gynecological tumors		41		82				126							227 (1,9)
Bone and soft tissue tumors				160				1			7				8 (0,1)
Cutaneous tumors							34	14							48 (0,4)
Other	4	4		77			8	72	8		14				187 (1,5)
Unknown		27	709	597	108	338									1 761 (14,4)
Total	30	107	5 542	2 550	108	719	86	353	1 370	369	535	14	235	216	12 234 (100)

Table 3.4. Tumor sites and patient numbers for proton beam therapy; after [1.23].

Chapter 4

HADRONTHERAPY FACILITIES WORLD-WIDE

4.1 Hadrons Acceleration and Beam Delivery Technologies for Radiotherapy

To test the therapeutic potential adequately, the physical and biological advantages of heavy charged particle beams must be fully exploited for clinical application. One must develop technologies to deliver optimum radiation dose distributions, i.e., delivering a maximum dose to the tumor, and at the same time, minimizing the radiation dose delivered to surrounding sensitive, normal structures of the body. This short introduction concentrates on the presentation of typical instrumentation developed for the production, modification, control, and monitoring of the clinical beams [4.1].

Fig. 4.1 shows a preliminary layout of a portion the beam hardware designed for the NPTC. The names of subsystems are located near the appropriate equipment.

Proton beam is accelerated to the required energies by an **accelerator**; in the case of NPTC it is a fixed energy 230 MeV cyclotron. Then the beam is transported through an *energy selection system.* It contains a degrader, which is a material of variable thickness intercepting the proton beam used to degrade the energy of the beam. Such a system is necessary in case of fixed energy accelerators. *Beam lines* provide the required focussing and achromatic bends to deliver a round beam with the appropriate trajectory for transmission through the *gantries* or the *fixed beam lines*. The beam delivery system includes a *passive scattering system* and *active scanning system.* They will be widely discussed in Secs. 4.1.1 – 4.1.5.

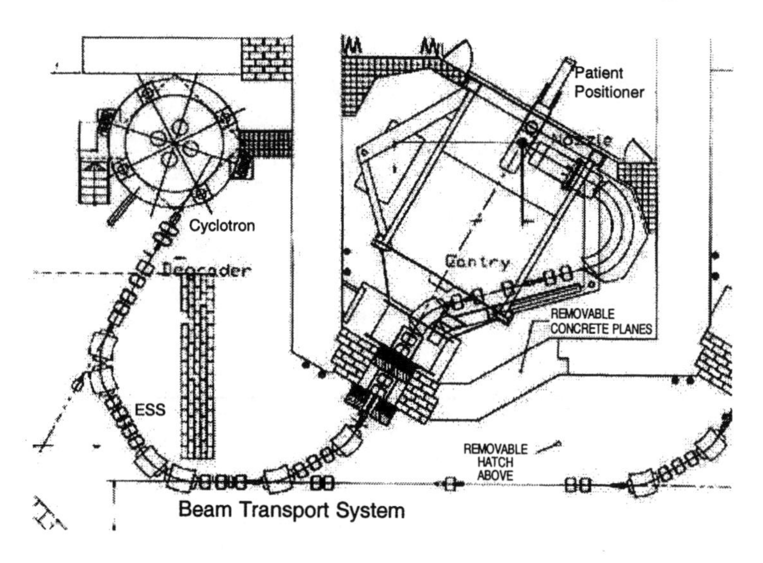

Figure 4.1. Layout of the beam line equipment for the NPTC, after J. Flanz [4.2].

4.1.1 Accelerator for Radiotherapy with Hadrons

As it is summarized in Sec. 6.1, a hospital accelerator intended for protontherapy should deliver beam of accelerated protons with the maximum energy of at least 250 MeV. What is particularly important is that this energy should be easily controlled over a wide range of values because of the need to have the depth distribution quickly modified. For the dose rate of several $Gy\,min^{-1}$ to be produced in the volume of $1000\,cm^3$ the beam intensity should be about $10^{10}\,p \times s^{-1}$. Modern accelerator technology exists today to meet all of the clinical requirements within the reasonable budget for hospital-based hadrontherapy facilities. There is currently many debates concerning the optimum specifications and choice of accelerator for use in high-energy protontherapy. A hospital-based machine is rather different from a physics machine. The accelerator should run in a hospital environment without technical support services found in nuclear physics laboratories. Therefore, technological choices are particularly relevant.

First generation of accelerators used so far for proton radiotherapy was designed for high energy physics studies. They have to be specially adapted to protontherapy, which often required quite complex technical procedures.

Accelerators of second generation employed for protontherapy are those added to medical cyclotrons originally intended for the production of radionuclides and neutron therapy.

The facilities based on existing accelerators built previously for physics research purposes will be presented in Sec. 4.2.

In addition to the existing accelerator facilities, a number of design studies have been undertaken for a dedicated medical accelerator. The three principal options for the accelerator designed specifically to operate in the hospital settings, are the *cyclotron*, the *synchrotron* and the *linear accelerator*. The recent increase of in the number of patients treated with this kind of radiotherapy reflects the increase in the number of operating dedicated facilities.

The main protontherapy facilities will be presented in Secs. 4.3, 4.4 and Chapter 6. The projects of dedicated proton accelerators will be presented in Sec. 6.3.

The *cyclotron* is fixed-energy accelerator capable of high beam currents. These beam currents call be a thousand times greater than that required for protontherapy, but may be useful for additional applications such as radionuclide production. Cyclotrons have been used for medical applications for many years and produce what is essentially a continuous beam. It has been claimed that a cyclotron with a conventional magnet is the machine that will give the greatest reliability in clinical use. One disadvantage of the conventional design of cyclotrons is the weight of the magnet system, which ranges, depending on the design of cyclotron from 80 tons to 190 tons.

The *synchrotron* has the advantages of being able to produce proton beams of variable energy, and having low weight and power consumption but requires a complex ion injector system. The big possible disadvantage is the beam output, which is limited by the design of the injector and the pulse repetition rate[15]. Even with the extremely high extraction efficiency the maximum beam current that can be obtained is quite near the lower limit required for radiotherapy. The synchrotron can be designed to produce either a long beam pulse of between 200 ms and 1 s (slow extraction) or a very short beam pulse of 25 ns (fast cycling).

The *linear accelerator* like the cyclotron has a more than adequate beam output but with the disadvantage of requiring a large space allocation, although this may not be a limitation with careful site planning. It can, in

[15] The long term cutting of the synchrotron beam makes this generator inappropriate for the voxel scanning technics, which is probably the most promissing technics in the near future.

theory, deliver beams of variable energy but this requires extraction of the beam after different stages of acceleration. The linear accelerator produces a pulsed proton beam [4.57].

The general performances for dedicated Protontherapy accelerators are discussed in Sec. 6.1, the examples of designs are widely presented in Sec. 6.3.

4.1.2 Energy Selection System

The *energy selection subsystem* (*ESS*) consists of a variable thickness degrader, followed by slits to limit beam emittance, and a momentum analyzing, selection and recombination system all of which produce a near monoenergetic variable energy beam of suitable quality for radiotherapy from a fixed energy source. The ESS is the price one has to pay if one uses a fixed energy cyclotron, for example, and requires the sharpest possible distal beam fall-off. In return one gets simple and highly reliable performance and at least in some situations, reduced cost [4.3].

4.1.3 Beam Transport

The extracted beam is transported from the accelerator to the treatment room by the beam transport system, a series of dipole and quadrupole magnets. The arrangement of magnets, vacuum chambers, and diagnostic instrumentation is called a beam line.

A stable and efficient transport of the beam from the accelerator to the treatment room is required for reproducible dosimetry and thereby reliable patient treatments. The stability of the centroid of the beam position must typically be better than 1.0 mm. This requirement places constraints on the stability of the bending and focusing magnets needed to control the beam position and profile. The ease of adjusting the beam through the beam line, in what is called *tuning a beam line*, and the *reproducibility of the beam-line tune* are critical for efficient and reliable clinical operation. *Beam tuning* means adjusting the beam optics to transport the given beam to the desired location with the desired parameters at that location [4.1].

4.1.4 Gantry

In order to share the dose into the surrounding healthy tissue between several entrance ports, a standard technique conventionally used in radiotherapy is to treat the tumor several times with various incidences. Since most patients are lying during the treatment, this implies the need of a device providing variable incidence angle between the horizontal plane and the beam. The physicians call this device *rotating gantry*. The beam is coming in along the axis, guided away from the center and bent back towards the axis to cross it orthogonally. If the beam line ends at the axis (i.e. the patient is located on the axis) the gantry is called *isocentric*.

The gantry is considered by the radiotherapists as a major component of the facility: all technical studies of critical technology areas for future protontherapy projects include a gantry design optimization. Moreover this component contributes a significant fraction to the total costs of the facility. The rotating gantry can be the part either of the beam transport system or the beam delivery system.

It requires major mechanical components and tends to be costly. The corresponding beam optics should produce a non-dispersed image in the target plane and leave room for the appropriate range modulation and beam spreading after the final bend. Some alternative designs are shown in Fig. 4.2. In all cases, the equipment rotates about the original beam direction as axis. The *classic gantry* (Fig. 4.2(1)) has the advantage of simplicity, but has been criticized for its long length which implies a large shielded volume. In the *corkscrew gantry* (Fig. 4.2(2)) the beam is first bent through 90° by B1 and B2, then through 270° in the plane perpendicular to the axis by the coplanar bends B3 and B4.before arriving at the isocentre. 1. A variant of the corkscrew which gives more room for the patient is achieved by spiralling the beam along the main axis. This leads to the *supertwist gantry* shown in Fig. 4.2(3). The spiral structure introduces some coupling between the transverse planes, but this can be corrected by adding small skew quadrupoles, leading to a good final image with no dispersion in position or direction in both planes. Fig. 4.2(4) shows another design put forward by IBA, with a wedge at the dispersed intermediate image to compress the momentum spread in the beam and thus give a more precise end point in range [4.4].

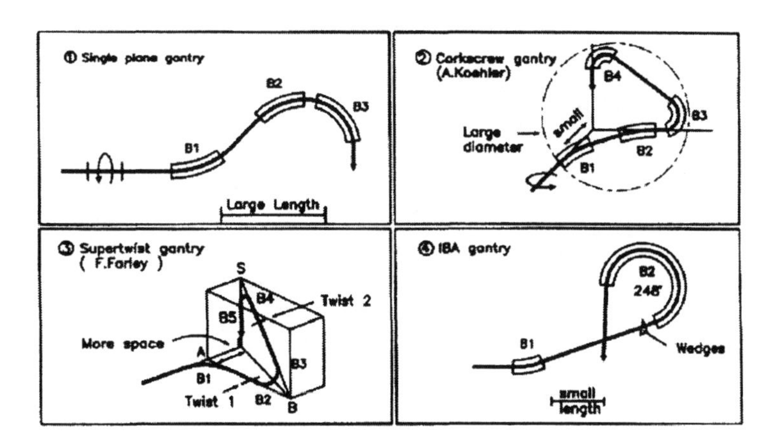

Figure 4.2. Various proton gantries; after F.J.M. Farley [4.7].

4.1.5 Beam Delivery Systems

A proton beam delivered by an accelerator is a thin beam with mono-energy and their Bragg peak localizes in a small region. On the contrary, tumors are different in size and shape. Moreover, the density of normal tissue which protons passe through to attain to a tumor is not uniform. Thus, the proton beams of an accelerator must be modified to fulfill the following requirements: to expand the beam in a transverse plane to irradiate the whole tumor, to adjust the proton energy according to the tumor depth, to increase the energy spread to irradiate thick tumor uniformly, to make corrections for tumor shape and non-uniformity of normal tissue which the proton beam passes through. The beam delivery system, located at the end of the beam line usually in the treatment room, modifies and monitors the beam to achieve the prescribed radiation dose distribution inside the target volume [1.23].

A typical beam delivery system is schematically shown in Fig. 4.3. Beam spreading devices, range modulation devices and field shaping devices are identified. A *scatterer* is used to expand the proton beams. The proton beams are scattered with a lead plate and spread laterally. A *collimator* cuts far-axis protons so that protons distribute almost uniformly given diameter circle at the patient's treatment bed.

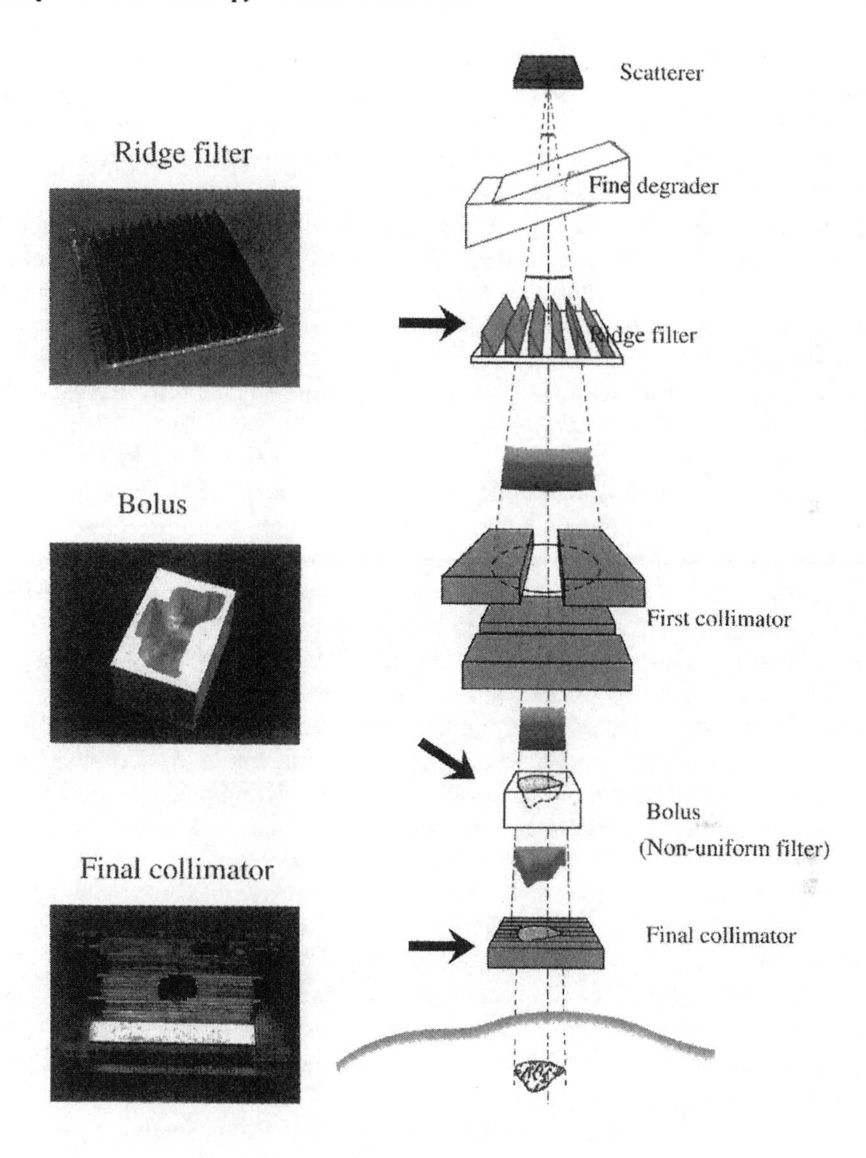

Figure 4.3. A typical beam delivery system; after [1.23].

A *fine degrader* adjusts the proton energy according to the tumor depth. It consists of two thick acryl edges which are facing each other. It can change the proton energy continuously by changing overlapping acryl thickness which proton, pass through. Since the protons lose their energy in the acryl edges, it is possible to produce a Bragg peak in a tumor of any depth. A *ridge filter* expands the energy spread of protons to the extent which corresponds to the tumor thickness. It is an array of metallic bars, which have step-wise thicknesses. The protons, which pass through metal of different thickness, produce Bragg peaks at different depths. Thus, it is possible to make an appropriate spread-out Bragg peak for each tumor. A sort of absorber, *bolus*, is machined from a wax-like material block. It is put on a final collimator and corrects the maximum proton energy distribution in the lateral plane following the tumor configuration [1.23]. It is a *passive method of Bragg peak modulation.*

The above mentioned shortcomings of the passive system can be overcome by employing magnetic control of the beam in the x-y plane in conjunction with an independent system of depth modulation control. The fundamentals of *scanning systems* have been described in Secs. 7.1.8 and 7.5.3.2 and the *systems used for deflecting proton and heavy ion beams* have been presented schematically in Fig. 4.4. In the *wobbler system* (Fig. 4.4b), the deflecting magnets are driven by a sinusoidal current 90° out of phase so that the beam travels in a circular path on the treated field. The radius of the circle can be controlled by changing the amplitude of the sinusoidal supply current. A *Lissajous scanner* (Fig. 4.4c), well known in the oscilloscope techniques, is another type of the wobbler. The deflecting magnets are driven by see-saw ways of independent frequencies. In a *raster scanner* (Fig. 4.4d) the beam is deflected by a fast single see-saw sweep of the current and a second magnet is also driven by much slower see-saw waves which ensure slow linear scanning movement. The system shown in Fig. 4.4e, called *pixel or spot scanning*, moves the beam from pixel to pixel, i.e. from one spot to the next. The beam can be maintained at a given pixel until the full predetermined dose is achieved, or it can be moved several times so that each pixel receives a partial dose during scan and then all the partial doses become added to give total dose [1.22].

An overview of the therapeutic options of the various scanning systems is given in Sec. 7.5.3.2. The common trend in the development of those systems is to adjust them to implementation of *3D conformal therapy.*

The required parameters of beam delivery systems are presented in Sec. 7.1.8, the introduction of the devices of the beam delivery system are presented in Sec. 7.5.3.

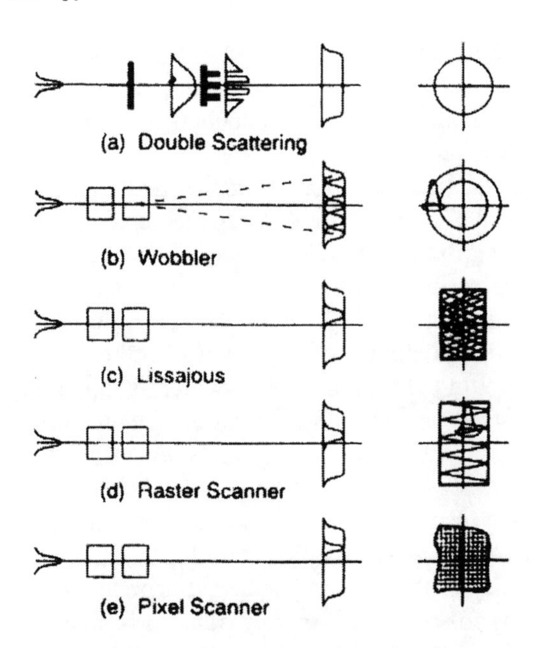

Figure 4.4. Proton beam modification (spreading) systems: a) double scattering foil, b) wobbler, c) Lissajous scanning, d) raster scanning, e) pixel scanning; after J.R. Alonso [4.5].

4.2 Historical Development of Hadrontherapy Centers

Studies of the properties of protons at the Berkeley laboratories led R. Wilson to pursue the idea of using protons for medical treatments [3.4]. This pursuit stimulated the clinical use of protons at pioneering facilities throughout the world (Table 3.1).

As a result of a half-century of research and more than 40 years of clinical use, Protontherapy is an accepted, proven form of radiation therapy, one that uses established radiobiological principles and adds significantly increased precision to this major cancer-treatment modality (Chapter 2 and 3). Proton radiation therapy is not investigational, its biologic effects on human cells, tissues and organs are comparable to those of X rays and electron beams and so are well known and understood. Its salient feature is superior controllability; this feature makes it possible for the physician to concentrate the radiation on the desired target more accurately than is possible with X rays or electrons. Its acceptance is shown in increased and accelerating use around the world [4.8].

Protontherapy centers are located in the United States, Russia, South Africa, Japan and in Europe. Almost all these centers utilize clinical beams with energy varying from 60 MeV to 250 MeV (Tables 3.1 and 4.1) and corresponding depths of penetration in tissue ranging from 3.1 cm to 38 cm, suited for the treatment of both eye and deep-seated tumors.

At present many accelerators used for hadrontherapy accelerated proton beams up to maximum extraction energy, which is much higher than the one necessary for clinical applications. Most of the centers have cyclotrons, which are fixed-extraction energy machines. As a consequence, the beam energy has to be reduced by *degraders* (passive beam energy degradation techniques) that are responsible for worsening the physical characteristics of the beam, for dwindling the useful particle flux and for increasing the induced radioactivity in the absorber material. The technological complexity of passive methods for beam energy degradation is smaller compared to the dynamic ones. Moreover the physicists have a larger experience with the passive methods. That explain why at Tsukuba (Japan) the 500 MeV proton beam is degraded to 250 MeV by means of a graphite absorber in spite of the availability of a synchrotron (i.e. a machine with a variable extraction energy). This is the reason why at present the extraction energy is not varied pulse by pulse in any of the centers having a synchrotron [1.4]. In 1989, the reconstruction of 680 MeV cyclotron was finished at the Joint Institute of Nuclear Research (JINR, Dubna), and patients have been treated using a degraded proton beam. At the Institute of Theoretical and Experimental Physics (ITEP, Moscow) for instance, the extraction energy can be chosen only among 5 fixed values of between 70 – 250 MeV. A more precise range adjustment is obtained by inserting variable-thickness degraders in the beam line. The same technique is operational at the Loma Linda University Medical Center (LLUMC, California, USA). At the Leningrad Institute for Nuclear Physics (LINPh, St. Petersburg) the 1 GeV proton beam is used without previous degradation in radiosurgical applications of the brain. In this case, the dose deposition in the target is performed on the plateau portion of the depth dose curve of a discrete number of narrow beams isocentrically focused converge to the tumor. The same technique is in use at the National Accelerator Centre at Faure (South Africa) with a 200 MeV proton beam [1.4].

The present proton radiotherapy centers are mainly equipped with horizontal fixed beams. As a result, patients must be treated in uncomfortable and hard to keep positions and this constitutes a constraint on the choice of the optimum treatment plan. PMRC at Tsukuba is the only center equipped with a fixed vertical beam. Nowadays, the rotating isocentric gantries are in operation at LLUMC, PSI and NPTC.

4.2.1 Hadrontherapy Centers Localized in the Research Facilities

First hadrontherapy facilities were localized in high-energy physics research laboratories.

LBL. The first clinical use of protons was initiated at Lawrence Berkeley Laboratory in Berkeley in 1954. During 1954 – 1957, about 30 patients were treated with proton beam from synchrocyclotron; after 1957, LBL devoted itself to light ion radiation therapy.

UPPSALA. The clinical application of proton beams for radiosurgery and for the treatment of malignant diseases using the former synchrocyclotron in Uppsala (Fig. 4.5) started in 1957. The institute lies amidst a conglomerate of scientific, less then 1 km from the University Hospital, which serves a population of 1.5 mln. Hence, the conditions were ideal for clinical tests during years 1957 - 1968. In order to provide a homogenous dose distribution in large target volumes, a number of physical innovations were made. The pencil-shaped proton beam was swept in a reticular pattern with a variable, crosses magnetic fields. This system allowed a transverse homogeneity of ± 5% over fields up to 20 × 20 cm fields. To rearrange the depth distribution of the Bragg curve in tumor therapy, two different systems were developed. First, a variable water absorber was used. Then, it was replaced by ridge filters, which introduces different thicknesses of absorbing material in different parts of the beam. Homogenous depth-dose distributions were obtained by transverse oscillation of the filter.

The horizontal beam-line for 185 MeV protons was used and any cross-section between 0.1 and 20 cm diameter could be selected. Depending on size of the field, the maximum dose rate varied between 1 and 100 Gy min^{-1}. Two irradiation sites were available in the beam-line, one for narrow fields and other for broad fields [4.9]. Up till 1976, the number of 73 patients with different cancer cases were treated at Gustaw Werner Institute (GWI).

The work continued until 1976, when the accelerator shut down for major modifications. The aim was the construction of a three-sector variable-energy synchrocyclotron. New buildings were necessary. In 1984, an area 650 m^2 for physics and biomedical research was completed. Most of the buildings below ground level and closely surrounded by a number of other university buildings.

In 1988, treatment has restart with narrow beams only. They are used for treatng intraocular melanomas, endocrinologically active pituitary tumors and other small inrtracranial targets. From July 1994, the availability of 180 MeV protons for therapy was increased from 6 weeks yearly to 10 weeks. The increased beam time increased the number of patients treated from below of 5 patients in 1991, to between 10 an 15 patients in 1993 and 1994, 29 patients yearly 1995, 48 patients in 1998 and 41 patients in 1999. Most patients have been treated with 4 fractions in 4 days [4.42]. During the last year, a new chair was installed allowing treatments in more than one plane and since fall 1997 also patients not only sitting upright [4.10].

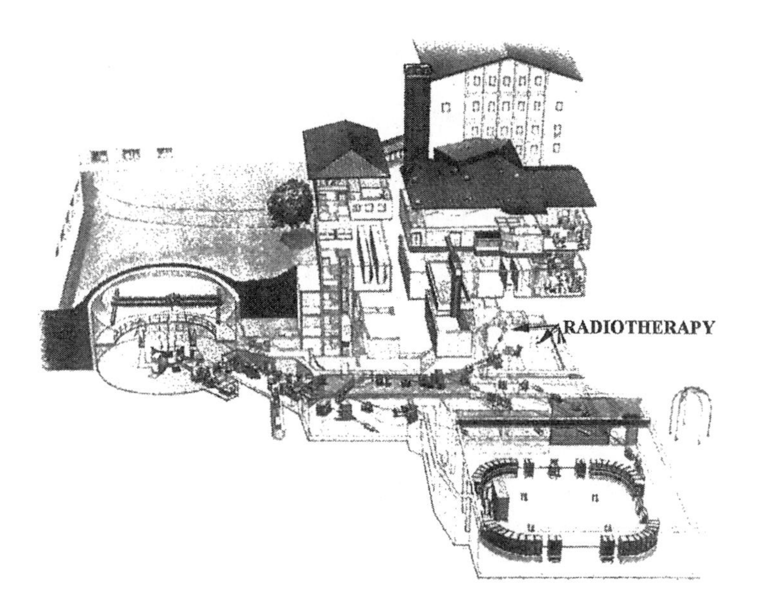

Figure 4.5. Layout of the TSL facility; after S. Graffman et al. [4.10]

Since 1989, the equipment has permitted treatment with a fixed, narrow, horizontal beam in one plane with the patient sitting fixated in a chair. During the two last years the technics for aligning the beam, and for daily dosimetry checks has been improved. Because of this more time is available for actual treatment, so that 8 patients can be treated during one day.

There were the plans to built the compact proton gantry allowing the full isocentricity. The development of a raster scanning system for protontherapy has taken place for some time [4.55, 4.58, 4.59].

The Figs. 4.5 and 4.6 show the layout of TSL facility and the beam lines for proton facility.

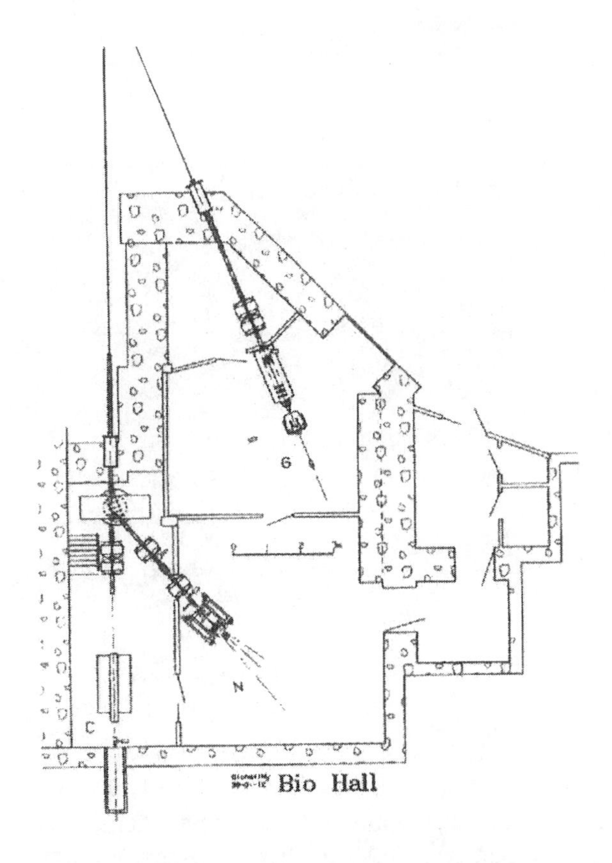

Figure 4.6. Layout of the beam line for proton facility at TSL: (G) the old narrow-beam facility, (N) the new broad-beam facility; after [4.8].

HCL. The largest and longest protontherapy experience in the world is that of Harvard Cyclotron Laboratory, which has treated ~ 7700 patients in conjunction with departments at Massachusetts General Hospital and Massachusetts Eye and Ear Infirmary. HCL uses a 160 MeV synchrocyclotron built in 1949 for physics research with horizontal beam lines available for treatment in two separate rooms. Fig. 4.7 shows HCL floor plan showing three treatment beam lines (A, B, C), a test (physics) beam (D) and two treatment rooms. The proton facility uses a passive scattering techniques to spread the beam out in the lateral dimension and specially designed rotating propellers or ridge filters to achieve an uniform dose distribution over the extended depth. At HCL uniform fields of 3 cm diameter and dose rates of 8 to 10 Gy × min^{-1} in one room, and up to 30 cm diameter with dose rates of 0.5 to 1.5 Gy × min^{-1} in the other room, can be achieved [3.7, 4.11].

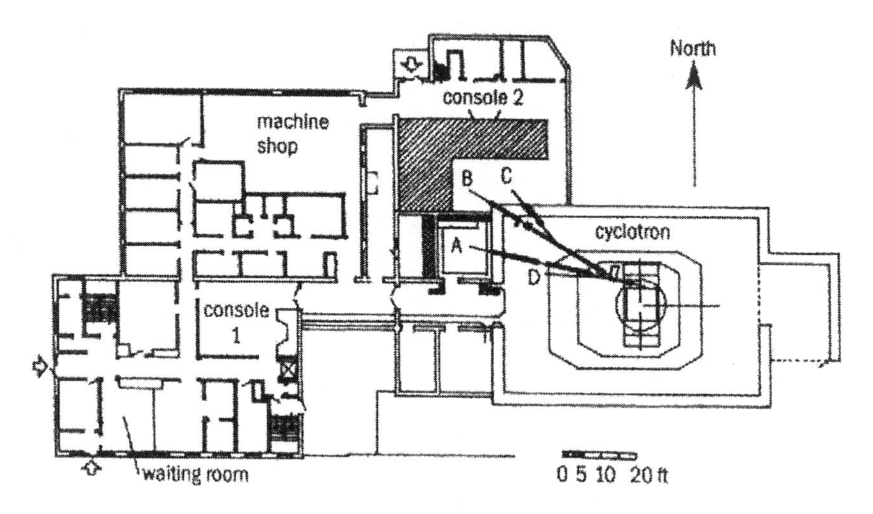

Figure 4.7. HCL floor plan showing three treatment beam lines (A, B, C), a test (physics) beam (D) and two treatment rooms; after B. Gottschalk et al. [4.38].

TRIUMF. The TRIUMF cyclotron has exceeded two decades of successful beam operation. Simultaneously, it delivers 500 MeV proton beams up to 200 nA for meson production, lower intensity variable energy beams (between 180 and 520 MeV) for experiments with protons, polarized neutrons and unstable nuclei, and beams from less than a 1 nA to up to

100 nA at energies between 65 and 120 MeV for isotope production and protontherapy.

In August of 1995 the first ocular melanoma patient was treated with protons with the recently commissioned Protontherapy Facility. This facility, which makes use of the existing 65 - 120 MeV proton beam. For eye treatments a beam energy of 70 MeV has been selected as optimum, with a spiral range degrader used to set the maximum treatment depth and a single lead scatterer and modulator to spread the Bragg peak over the treatment volume.

The patient chair was designed by a local company to have six degrees of motorized control. The treatment control system controls the beam on / off device, collect information from the beam monitors, and provide display pages for the operator, higher level controls such as beam scanning devices and to input patient and dose information. An independent hard-wired dose measurement system is provided as a backup. Special precautions are taken to ensure that reliable low intensity beams are delivered. The treatment beam intensity of 5 nA is achieved by adjusting the height of the carbon stripping wire [4.12].

Protontherapy Facilities in Russia. In Russia medical proton trials have been in progress in three institutes. Beginning in 1969, the Institute of Theoretical and Experimental Physics (ITEP) in Moscow has used proton beams from the synchrotron for patient treatments and since 1982 three treatment rooms have been available for clinical trials. The ITEP accelerator is a strong focusing l0 GeV proton synchrotron with internal beam intensity near 10^{12} protons per pulse and 22 acceleration cycles per minute. The external medical beam is extracted by a beam-kicker. Short pulse (15×10^{-9}s) of the kicker ejects only one of the 4 accelerated bunches of protons every fourth acceleration cycle. Since only one sixteenth of the accelerated protons are used for the medical beam, medical procedures in the external beam can be pursued independently and simultaneously with physical investigations at the accelerator. The energy of the external medical beam can varied from 70 to 200 MeV and depends upon the time of ejection. The proton beam brought out of the vacuum chamber is always monochromatic and has only a slight angular divergence, because there is no need to decelerate it by means or thick degraders. This creates appropriate conditions for forming clinically useful beams. The available intensity of the extracted beam provides a rate ranging from l to 100 Gy min^{-1}, depending on the need, on the size or the target volume and on the shape of the dose distribution. Patients have been treated at the proton accelerator since 1969. About 2000 oncological, neurosurgical, ophthalmological, and endocrinological patients have been treated as yet [4.14].

A general view of the medical facility in JINR is shown in Fig. 4.8.

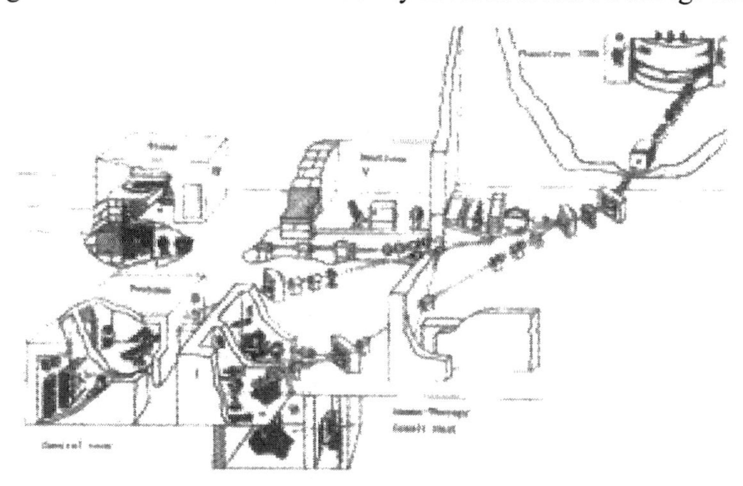

Figure 4.8. General view of the JINR medical facility; after V.M. Abazov [4.43].

In 1989, the reconstruction of a 680 MeV cyclotron was finished at the Joint Institute of Nuclear Research (JINR) in Dubna, and patients have been treated using a degraded proton beam. The clinico-physical facility consists of several medical channels: three therapeutic proton beams with energies from 100 to 660 MeV , a negative pion beam, a therapeutic neutron beam and a therapeutic γ-unit with the ^{60}Co source. The JINR accelerator is a proton phasotron with maximum energy 660 MeV. The intesity of external proton beam is about 3 μA ($1{,}9 * 10^{13}$ protons/sec). The protons beams with lower energy is obtained by deceleration of 660 MeV beam in carbon energy degrader. The method of Brag peak modification based on transformation of existing wide energy distribution of the decelerated proton beam by means of the magnetic analysis and a multicollimator system. After the magnetic analysis the proton beam is separated in space in accordance proton with proton energy. Each collimator from the multicollimator system, placed in a region with a certain energy of protons, may control the number of passed protons with this energy. After mixing the passed protons, an energy distribution allowed to form a modified shape of Bragg curve is obtained [4.1, 4.43, 4.44].

At the Central Research Roentgenoradiology Institute, they use the 1000 MeV proton beams from the synchrocyclotron of the Leningrad Institute for Nuclear Physics (LINPh) in Gatchina, near St. Petersburg. Intracranial diseases have been treated using plateau radiation. These Russian facilities have treated more than 2500 patients [4.1].

NIRS. Very active Protontherapy programs are in progress in Japan. Protontherapy at NIRS has been carried out using 70 MeV (to be upgraded to 90 MeV) proton beams generated by a cyclotron which, until 1993, had also been used for fast neutron therapy. For spreading out the beam, a spot scanning method has been used. So far, 96 patients were treated with protons, of whom ocular melanomas comprised the major part, and treatment results for them were satisfactory, almost identical as those obtained at world-wide facilities [4.14].

4.2.2 Second Generation of Research Hadrontherapy Facilities

The additional beam lines built in the existing physics research laboratories are the second generation of research Protontherapy facilities.

Clatterbridge Hospital. The protontherapy with new beam line started in the Clatterbridge Hospital. The Medical Research Council Cyclotron Unit at the Clatterbridge Hospital, (Merseyside, UK) uses 62.5 MeV protons from a fixed energy AVF cyclotron (Scanditronix MC 60) to treat ocular melanoma patients. Patients have been treated since its opening in 1989. Boosting the energy to 180 MeV using a drift tube LINAC (DTL) technology has been recently proposed. The energy of the proton beam may be increased to 200 or 250 MeV at a later date by adding more accelerating tanks.

NAC. The additional proton beam line was built in the existing accelerator complex in NAC Facility at Cape Town, South Africa; the clinical proton energy is 200 MeV. All the major facilities were locally designed and include a solid-pole light-ion injector cyclotron (SPC1), a variable-energy separated-sector cyclotron (SSC), capable of accelerating protons to a maximum energy of 200 MeV. The medical complex includes three radiotherapy treatment vaults, laboratories, offices, full medical physics and radiobiology facilities as well as a 30-bed on-site hospital.

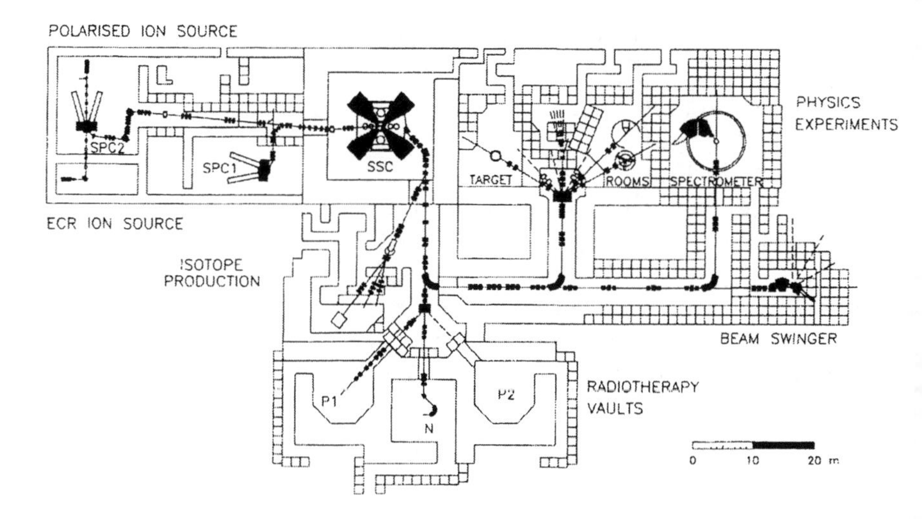

Figure 4.9. Layout of NAC radiation area. The horizontal Protontherapy facility is located in vault P1; after D. Jones [4.16].

Most of the components in the beam line are mounted on bearings on an accurate optical bench and their positions can be easily changed. The total distance between the vacuum window and the isocenter is 7 m. A double scatterer plus occluding ring system is used to flatten the proton beam. The beam delivery system is designed for a maximum field diameter of 10 cm. The first scatterer is located immediately downstream of the vacuum window and is a 1 mm thick lead plate. The occluding ring (50 mm thick brass) are mounted on the second scatterer, which is a 1 mm thick brass plate and 2.9 m downstream of the first scatterer. This beam-spreading system is currently used for all field sizes. The longitudal spread out of the Bragg peak is accomplished by using a propeller made up of different thicknesses of acrylic, which is rotated in the proton beam [4.16, 4.51]. The layout of NAC facility is shown in Fig. 4.9.

CPO. The 200 MeV sychrocyclotron localized in Orsay Nuclear Physics Laboratory was addapted to the unique needs of proton radiotherapy (Centre de Protontherapie d'Orsay, CPO). The first proton beam has been delivered in 1958 with energy 155 MeV and 2.5% ejection efficiency. Between 1975 and 1977 the machine has been upgraded to reach the energy 200 MeV with a 85% ejection efficiency. With a degraded proton beam of energies about 70 MeV the patients with occular tumors are treated. With the 200 MeV beam, deep lesions can be treated. First brain protons treatment starts at the end of 1993 [4.17].

CAL. The treatment of patients started in 1991 at the Centre Antoine-Lacassagne in Nice (France) which has reinstalled its Medicyc cyclotron to produce a 63 MeV protontherapy beam. Medicyc is a medical machine and this influenced initial design and further modifications as important as decision to switch from positive to negative charged ions (H^-) acceleration, in order to increase ion source duration of life and simplify extraction of proton beam. Eight major radiotherapy centers in southeastern France have formed a cooperative group to exploit this capability.

The beam is left freely from the last dipole magnet and is collimated before entering the treatment area. Once inside the treatment area, the proton beam continues to travel in air before being modulated by a rotating plexiglass wheel with variable thickness angular sectors. The diameter of the beam is kept at 35 mm by Al collimators mounted on the optical bench. At the end of the bench, it is finally collimated to the treated tumor shape. These arrangements follow close those of Clatterbridge. The protontherapy chair has been built in the workshops of the Medical Research Council in Clatterbridge [4.18].

PMRC. Since 1983 protontherapy using 250 MeV beams has been carried out at Proton Medical Research Center (PMRC), Tsukuba University, Tsukuba in conjunction with the booster synchrotron of the proton accelerator complex at the National Laboratory for High Energy Physics (KEK). Booster Synchrotron Utilization Facility (BSF) uses 250 MeV protons, which are produced by energy degradation from 500 MeV protons. It has two treatment rooms, one is equipped with a vertical beam, the other with a horizontal beam. Thus, it is possible to irradiate a patient through anterior and lateral portals. Unlike many other proton facilities, major efforts have been placed on the treatment of deep-seated tumors. So far, more than 500 patients were treated for a partial or full course with proton beam therapy. The distribution of the tumor sites or diseases treated are the liver, bladder, esophagus, lung, AVMs and others [4.19].

The new protontherapy facility is now constructed at PRMC (Fig. 4.10 a, b).

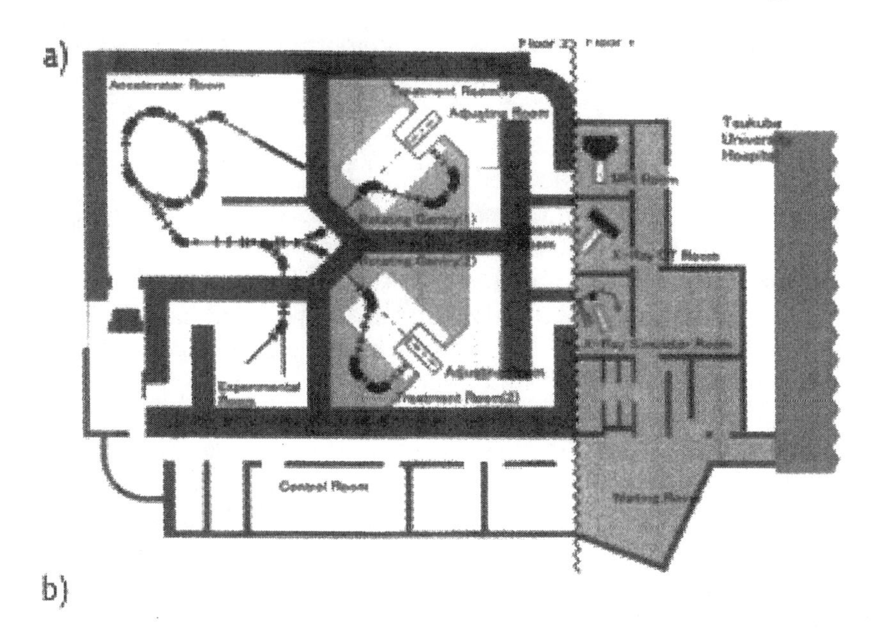

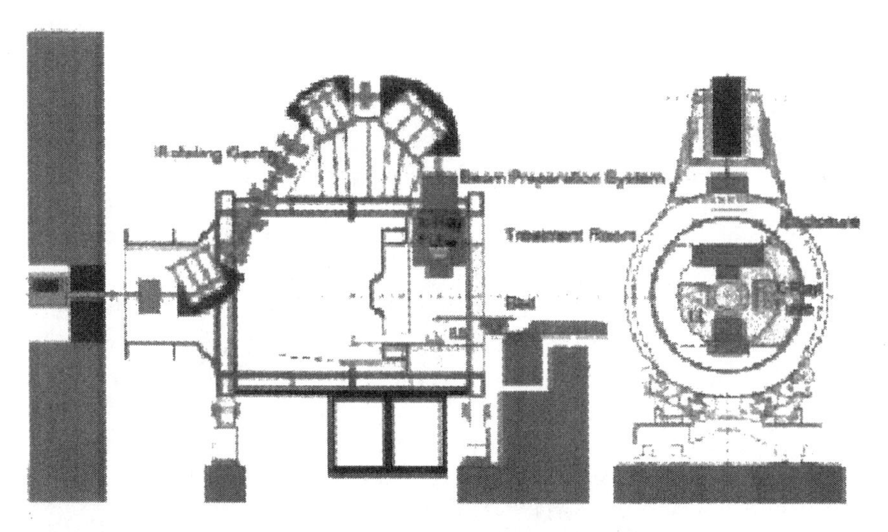

Figure 4.10. a) Layout of new Protontherapy facility at PRMC. b) PRMC rotating gantry design; after A. Maruhashi [4.69], private correspondence.

Movement of the target volume during exposure to radiation decrease control rate in radiotherapy. Especially, the respiratory of lung or liver cancer has been an inevitable problem in proton radiotherapy. In PRMC they developed a reliable control system, which synchronizes the irradiation to the respiration phase [4.18].

PSI. At the Paul Scherrer Institute (PSI, Switzerland) at the Opthalmological Therapy Installation (OPTIS) employing 72 MeV proton beam extracted from the cyclotron has been used to treat ocular melanomas. The patient is positioned in a stereotactic chair, his head being fixed in an individual mask with a bite block attached to the chair frame. He has to stare at a small light diode on a polar coordinate system, whose position is given by the therapy planning program. In order to limit the proton beam to the tumor outline one uses a copper collimator machined to the tumor Sec. with a 2 mm safety margin, which is put on the extremity of the beam tube during irradiation.

A new proton radiotherapy facility for the treatment of deep-seated tumors using a scanned proton beam has been assembled at PSI. Patient treatments started at the facility in December 1996. Due to its innovative approach, this project is the technologically most advanced in the field of Protontherapy nowadays. The major technical features, which characterize the project, are the following: the spot scanning technique, the compact isocentric gantry, patient handling with patient transporter, high quality assurance tools (computer optimized treatment planning for 3D conformal therapy, 3D-dosimetry for conformal therapy, proton radiography, dose error algorithms). The PSI proton gantry (Fig. 5.7, Sec. 8.2.3) with spot scanning is presently the only facility, which can deliver the dose using inverse planning methods with protons [4.20, 4.21].

Forty patients have been treated in three years using the gantry, 21 of these in the year 1999. All suffered from tumors near the base of the skull or the spinal cord and in the pelvis region. A treatment program for prostate ccancer was begun with a first patient. Three tumors near the spinal cord, two of which in children 7 and 8 years of age, were also treated. In one case, a so-called intensity modulated Protontherapy was carried out for the first time in the world. That means that defined dose distribution were superimposed from different directions [3.18].

Actually at PSI, they are working on the developing the concept of *Riesenrad gantry*. The basic idea is to deflect the ion beam with a single 90° doipole, which rotates around the incoming beam axis, and direct it towardsthe eccentrically positionned patient cabin. Inside the cabin similar conditions as existing in a classical iso-centric treatment room prevail [4.47].

One of the new important project at PSI concerns the overcoming the organ motion during the treatment (project TULOC). A novel magnetic tracking method is being developed. It shall enable the real time tumor tracking with high resolution in time and space during radiotherapy. In the first phase this method shall be used for tumor-position dependent beam gating. The goal of the second R&D phase is the fast adaptation of the radiation pattern to the tumor motion in the feed-back loop. The proper functioning of the magnetic tracking has been demonstrated recently. They simulated a moving tumor by putting an x-ray film in the phantom moving in the patient table by the help of a computer controlled device, and recorded the dose deposition in the film. The gating of the proton beam with their system re-established the sharp dose distribution seen before in a motionless film. This is world-wide first gating of a precision irradiation with a sensor which can be placed within or close to a tumor [4.48].

GSI. In 1994, a pilot project was started by the GSI Darmstadt, the DKFZ Heilderberg, the Radiologische Universitatsklinik Heilderberg and Forschungszentrum Rossendorf to install an experimental ion therapy unit at the accelerator complex of GSI. Fig. 4.11 shows a global view of the GSI accelerator complex with gray-coded path representing the accelerator chain involved in therapy.

The synchrotron of GSI is presently the only accelerator in Europe that can be employed for the treatment of deep-seated tumors with ion beams. GSI has decided to take full advantage of an active energy variation in the accelerator and to realise the range modulation by changing the energy of extracted beam in a *pulse-to –pulse mode*.

Another major achievement of the GSI is the development of the novel technique for lateral beam spreading – *the raster scanning*. In the raster scanning, a pencil-like beam (spot size diameter between 4 and 10 mm) is moved continuously in a preselected pattern over the target area, and a well defined number of particles corresponding to the desired dose is delivered to each line element. In order to modulate the dose, *the scanning speed is controlled by the intensity of the incoming beam.*

A treatment planning system for ion therapy has been developed using the universal VOXELPLAN system adding special software modules addressing the following aspects: interaction of the charged particle beams with matter including physical quantities and biological effects, generation and control data for the raster scanning system.

The first patients were treated with carbon ion beam at GSI in December 1997 [1.5].

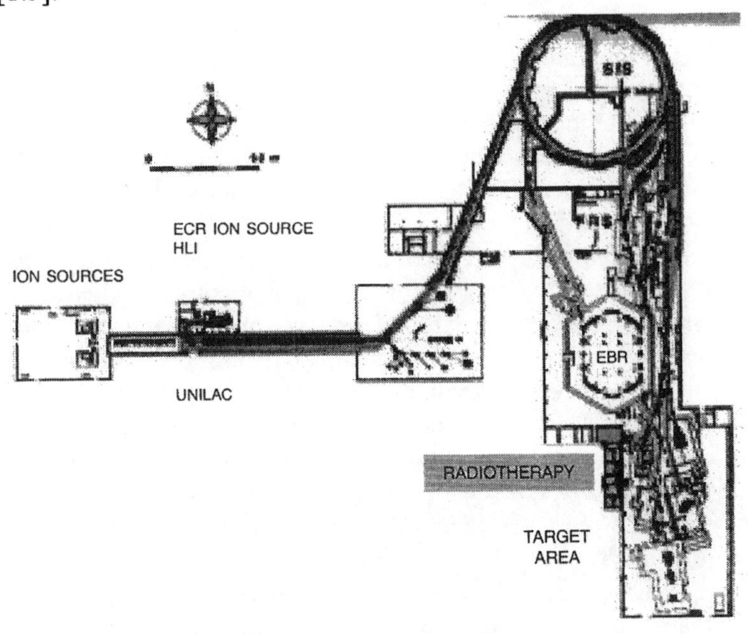

Figure 4.11 The GSI accelerator facility; after J. Debus [1.5].

4.2.3 Hospital-Based Dedicated Hadrontherapy Facilities

The most of the protontherapy centers are located in the environment of research laboratories originally built for physics research or even still running physics research programs in parallel. As a consequence, the accelerators used for therapy have not been designed in a custom-tailored way for medical purposes. It may cause some deviations from the optimum conditions for therapy. On the other hand, running the accelerator laboratory environment can profit from the presence of well-experienced staff and top-experts in many accelerator-related fields as well as from availability of specialized auxiliary equipment, which is not available in hospitals.

The research proton facilities of first and second generations do not ensure of providing needs for protontherapy because they are able to treat a limited number of patients. So far, ion beams have been delivered mostly by fixed beam lines. The exception was PSI with an eccentric compact proton gantry. Tumor targeting has been done exclusively by using passive methods. Therefore, the potential advantages of ion therapy beam have not been fully exploited so far [1.5]. These facts have influenced current trends in ion therapy, which will be widely discussed in this chapter and in the Sec. 4.3. Much attention is paid to the hospital-based dedicated installations and to the construction of modern advanced beam delivery systems.

LLUMC. Interest in developing a dedicated facility for treating patients began at Loma Linda University (LLU) in 1970. A study of the feasibility of such an undertaking indicated the deficiencies in developing a system that could fully exploit the potential for treating patients with heavy charged particles. One small but essential missing component was a technology that could accurately guide the external beam of radiation to the unobservable target within the patient. Beginning in 1975, Loma Linda investigators started working with investigators at the University of California, Berkeley, and then later with clinical researchers at Los Alamos Pi Meson Facility (LAMPF), to gain experience in using heavy charged particles. In 1984, some physicians from LLU visited Argonne National Laboratory to discuss options for developing proton accelerator. From a small number of physicians, physicists, and engineers interested in developing charged-particle treatment capabilities came the working group that would meet together at regular intervals for the purpose of defining the design requirements for the accelerator, beam transport system, beam delivery system, and the facility to house this hardware. The consortium became known as the *Protontherapy Cooperative Group* (PTCOG). The group divided his tasks into three major areas of interest and formed three subcommittees to investigate them: accelerator design, facility design and clinical studies group. As the time passed, the design of accelerator, beam transport system, the treatment room beam delivery system, and the facility layout itself began to take shape and to assume sufficient form that the feasibility for developing an engineering become evident. In 1986, therefore, the Loma Linda group propose Fermilab to develop engineering design of the accelerator and its beam transport system for LLU [4.22]

In 1990 the first dedicated 250 MeV proton synchrotron was installed at the Loma Linda University Medical Center in Loma Linda, California (Fig. 7.3). The accelerator was designed and built by the Fermi National Accelerator Laboratory. It is the smallest synchrotron in the world. It was designed from the start to be a patient-treatment machine. It was also the first dedicated proton accelerator facility built for a hospital.

The LLUMC facility, in a large hospital located on the Loma Linda University campus in Southern California, is divided into three areas: the accelerator, the beam transport lines, and the treatment rooms. The accelerator is a zero gradient synchrotron capable of accelerating protons to final energies between 70 and 250 MeV (Fig. 6.7). The proton beam is initially accelerated to 2 MeV by a radio frequency quadrupole (RFQ) and subsequently injected into the ring after a 180° bending magnet. Shortly after the ring is filled, the beam is accelerated to its final energy by an rf cavity, located diametrically opposite to the point of injection. After achieving final energy, the beam is resonantly extracted at the first ½ integer betatron frequency. This frequency, often called the tune of the accelerator, is regulated by quadrupole magnets in the accelerator. During resonant extraction, the portion of the beam crossing the electrostatic septum is deflected into a Lambertson extraction magnet which in turn deflects the beam out of the synchrotron into the beam transport lines, The pulsed beam is extracted with a typical repetition rate of 2 s with a 0.4 s spill duration.

The intensity of the accelerated beam of protons at the end of the acceleration process is about 1.5×10^{11} and the beam can be extracted uniformly in 1 s. The efficiency of beam extraction is higher than 95%. The diameter of the synchrotron magnetic system is less than 7.0 m, it means that the Fermilab accelerator is the smallest machine of all the existing proton synchrotrons.

The beam transport lines guide the beam into any one of four treatment rooms (Fig. 5.5). Three of the rooms contain isocentric gantries. In all gantries, there is about 3.4 m space between the last bending magnet and the gantry isocenter. This space contains the beam delivery system referred to here as the nozzle. It contains dosimetry instrumentation and beam spreading devices to monitor and control the dose to the patient. The beam entering the *nozzle* is approximately 1 cm FWHM with a divergence typically less than several milliradians. It is therefore necessary to spread the beam into a larger area with good field uniformity. Initially, field sizes up to 20 cm diameter will be required with field uniformity better than ~3%. The dose should also be as uniform as possible for target depths ranging from 1 to 15 cm.

The first patient was treated in the eye beam in October 1990, and the second beam line, the horizontal beam line with a 250 MeV beam was put into clinical use in March 1991. The patient treatments began in June 1991 using the beam delivered by one of the gantries. The proton treatment facility is designed to accommodate up to 100 patients per 10-hour day [4.22]. It can serve between 1000 and 2000 patients annually. In practice the statement that the facility is capable to treat 1000 patients per year, or, in the other words, to deliver about 16 000 fractions per year. In general, it corresponds to irradiations of 65 patients per day [4.23, 4.24].

At LLUMC, the average treatment time per patient is about two minutes and the time to switch treatment rooms less than one minute. Since the time to align the patient requires typically ten to fifteen minutes, the accelerator performance does not currenly limit the patient throughput. The Loma Linda facility has demonstrated patient throughput as high as 125 patients in a 16-hour day [3.16].

Patients with eye, head and neck tumors are treated with the stationary beam. Patients with tumors in other anatomic sites are treated with the movable beam [4.21].

Some of the characteristics of Loma Linda synchrotron are presented in Table 5.1.

NPTC. A proton beam therapy center is being built on the main campus of Massachusetts General Hospital (MGH). The facility will have state-of-the-art equipment and a full range of patient and research support services. It will be used to investigate new approaches to cancer therapy and to provide clinical services for treatments, which are considered to be of proven value (Chapter 3). The facility, called the *Northeast Protontherapy Center* (*NPTC*), will provide a unique regional resource to the greater New England area (250 miles radius with the center in Boston, 38 mln population) [4.3, 4.25].

The new facility, totaling 44,000 square feet, is located on the southwest corner of the MGH campus along Fruit Street, on the site formerly occupied by the Charles Street Jail. The facility is two stories, with a below-grade floor which will house all treatment activities, and a ground-level floor for offices, treatment planning, and other administrative and support areas (Table 5.2). The program space includes three treatment rooms (up to two with gantries and one with horizontal beams), a room for the cyclotron accelerator, beam transport lines, an integrated control system, immobilization, fabrication and storage areas, mechanical, electrical and vacuum shops, treatment planning facilities, offices and conference rooms, and storage. Although the treatment floor is underground, there win be skylights and a two-story lobby / waiting room area which will provide generous natural light to this space. The facility is being designed so as to be capable of future expansion to four treatment rooms, three of which can contain gantries [4.25].

The protontherapy equipment is comprised of a 235 MeV isochronous cyclotron, an energy select system, a beam transport and switching system, a large-throw, in plane, isocentric gantry, fixed horizontal beam lines, patient positioning systems, an integrated control system, and safety systems. The accelerator is the first Compact Isochronous Cyclotron in the world able to to accelerate protons to 235 MeV. Its design is directly derived from superconducting cyclotron technology. even its coils have been replaced by room temperature ones. The central magnetic field has been chosen equal to 1.76 T ensuring a complete magnetic saturation of the hills (3T) whose gaps have been made elliptic. This shape, adapted to the saturated state of the hills, reduces drastically the radial extension of the fringing field, so permits the "simplified ejection" which consists in an electrostatic deflector followed by a magnetostatic corrector spanning 140° in azimuth only (more than 300° are necessary in all the other constant hill gap machines). The accelerator is able to deliver beams of up to 1.5 mA, but hardware limited at 300 nA in order to limit the maximum possible dose rate to the patient. The isocentric gantries allows the proton beam to be directed at a patient from any angle, thereby ensuring the greatest possible ability to avoid critical organs in the course of irradiating the tumor [4.3, 4.25, 4.26].

The facility has the capacity to treat at least 1000 patients per year, or approximately 16 000 treatment fractions per year, at the rate of about 65 patients per day [4.25].

The missing element at NPTC is the *therapy control system* (*TCS*) – a sophisicated computer control system which has mandated by the hospital's specifications for automated and very efficient operation. This, in turn, was at least partly motivated by the need to transfer the existing and flourshing clinical program at the Harvard Cyclotron Laboratory as soon as the NPTC became operational. IBA's first aproach to the design and coding of the TCS ran into difficulties and the development had to be out-sourced to a company specializing in control system software development. This introduced a substantial delay into the project. For July 2000, the design, coding and testing of the software is substantially completed, as is the software integration phase (a complex matter of cumulatively assembling software modules and testing the software / hardware inteerfacces). Once validation tests are complete, the tests results will be submitted to the FDA for per-market clearance – the so-called 510(k) process. Shortly thereafter, acceptance tests will be performed, followed by a period of clinical commissioning, leading up to a first patient treatment on the first gantry. They anticipate that the machine should be accepted by the end of the year 2000, and the first patient treatment should take place by the end of the first quarter of 2001. The clinical program at HCL will be transferred to the NPTC over the next several months [4.61].

NCC. The proton treatment facility at the National Cancer Center Hospital East (NCC, Kashiwa) is the first dedicated proton facility in Japan. The building is connected with the hospital building through passage way. The prime contractor of the equipment is Sumitomo Heavy Industries Ltd.

A 235 MeV isochronous cyclotron installed at NCC is the same as that of NPTC was installed. Energy selection system reduces the 235 MeV beam extracted from the accelerator to 190, 150 and 110 MeV. There are two isocentrically rotating gantry treatment rooms (Fig. 4.12). Accuracy of isocenter is ± 1 mm, accuracy of stop angle is ± 0.5 degree, rotational speed is 1 rpm and they were confirmed [4.35].

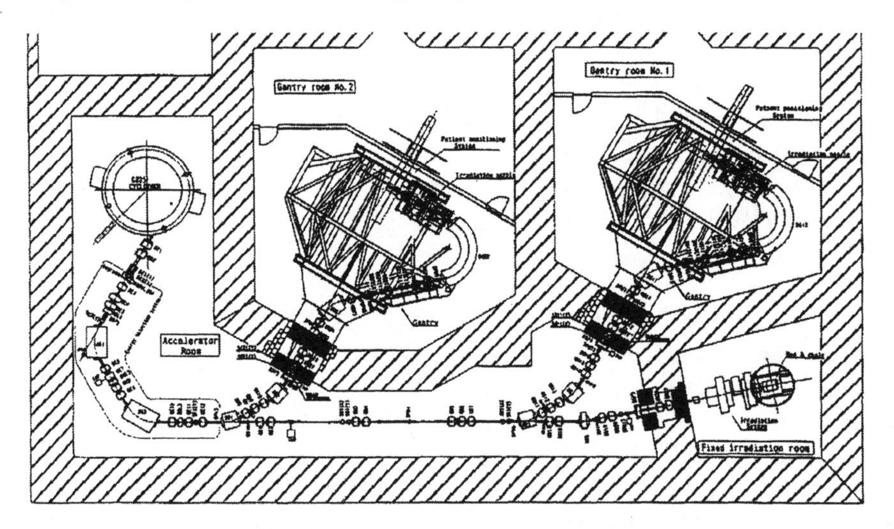

Figure 4.12. NCC Protontherapy system – general layout; after T. Tachikawa et. al. [4.36]

NIRS (HIMAC). A plan of the project was started in 1989 at National Institute of Radiological Sciences (NIRS) as part of the national ten-year plan to combat cancer, and in 1993 the construction and installation of all facilities were completed. In 1994, the clinical trial of heavy ion therapy was begun at the NIRS using carbon ions generated by a medically dedicated accelerator (HIMAC, Heavy Ion Medical Accelerator in Chiba, Japan). The HIMAC is the first heavy ion accelerator complex dedicated to medical use in a hospital environment. It consists of two types of ion sources, radiofrequency (RFQ) and Alvarez linacs, dual synchrotron rings, high energy beam transport lines, and beam delivery system for radiotherapy and experimental work (Fig. 4.13).

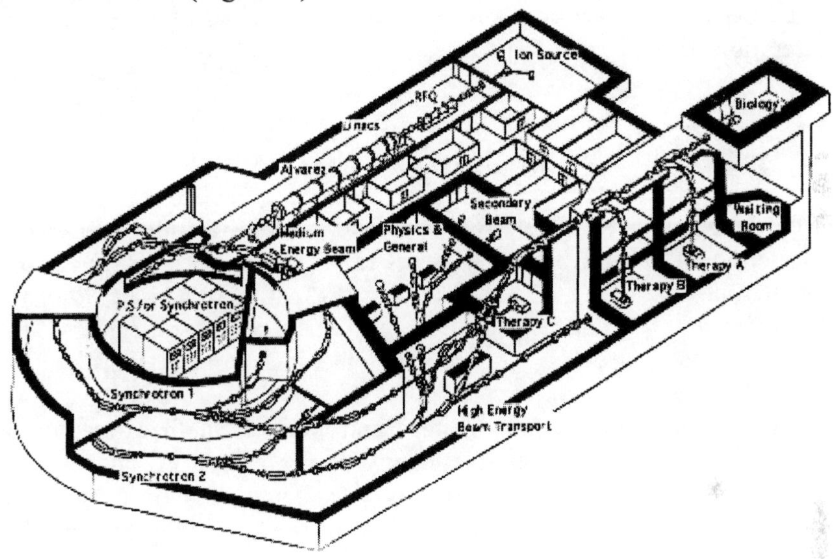

Figure 4.13. Bird's eye view of the HIMAC facility; after S. Yamada et al. [4.34]

There are three treatment rooms with fixed apertures for vertical, horizontal and vertical / horizontal beam lines as well as rooms for physics and biological research studies (Fig. 4.13). The beam modulation system for shaping the original Bragg peak to a target shape consists of wobbler magnets, beam scatterers, ridge filters, multileaf collimators, and beam monitoring devices.

The maximum energy of HIMAC is designed to be 800 MeV / u for light ions with $q/A = ½$ so that silicon ions reach 30 cm deep in a human body. Ion species of He, C, Ne, Si, etc. are reqired for the clinical treatment.

They have two types of ion sources. A PIG source is a hot cathode type and operated in a very short pulse with a relatively long time interval. Such an operation mode increases an arc impedance resulting in a high arc voltage. An ECR source is a single stage type and energized with a 10 GHz microwave source of 2 kW max. The source is equipped with a sextupole permanent magnet of 9 kG at a pole tip. The output beam intensities and emittance are satisfactory for treatments. Since ECR source is very stable and easy to operate, the ECR source is preferable in the daily operation for the clinical treatment. Both sources are installed independently on high voltage decks of 60 kV max and provide ions with 8 keV / u to the next acceleration stage [4.34, 4.68].

4.3 Planned Hadrontherapy Centers

Discussed below are some of the clinical proton facilities either under construction at various existing accelerator laboratories or planned for dedicated medical use. J. Sisterson anticipates that there may be 30 or more protontherapy and 3 ion facilities operating early in the next century [4.27]. Available data concerning the proton therapy projects worldwide are presented in Table 4.1.

NAC. At the National Accelerator Centre, Faure, South Africa, a proton facility with a 200 MeV variable-energy sector cyclotron, and a clinical beam line ended with the proton gantry is being developed (Fig. 4.14). Personnel from the Groote Schuur Hospital of the University of Cape Town plan to treat patients by the end of 2001 [4.16].

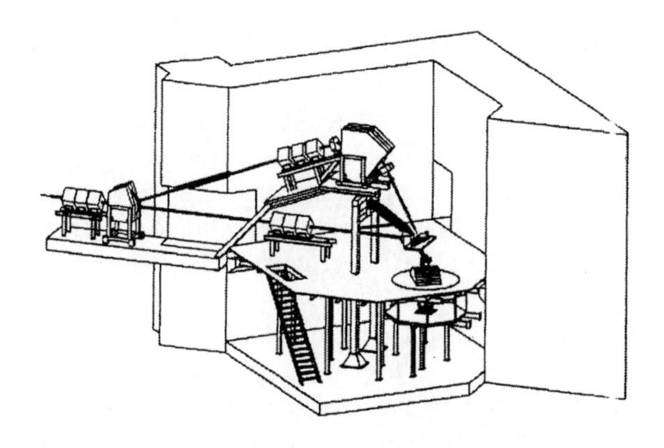

Figure 4.14. Schematic representation of the proposed 200 MeV NAC protontherapy facility; after [4.37].

Who	Accelerator	Energy [MeV]	Date of the first / last treatment	Beam delivery system
First generation of research protontherapy facilities				
Donner Lab., Berkeley, USA	Synchrocyclotron	340	1954-57	horizontal
Werner Inst. Uppsala, Sweden	Synchrocyclotron	85	1957-76	horizontal
Harvard Cycl. Lab. , USA	Synchrocyclotron	160	1961	horizontal
JINR, Dubna, Russia	Synchrocyclotron	90-200	1964-74	horizontal
ITEP, Moscow, Russia	Synchrotron	70-200	1969	horizontal
LIJF, St. Petersburg, Russia	Synchrocyclotron	70-1000	1975	horizontal
NIRS, Chiba, Japan	Synchrocyclotron	70	1979	horizontal
PMRC, Tskuba, Japan	Synchrotron	250	1983	vertical, horizontal
PSI, Villigen, Switzerland	Isochronous cyclotron	70-590	1984	1 horizontal, 1 gantry
JINR, Dubna, Russia	Synchrocyclotron		1987	horizontal
Werner Inst. Uppsala, Sweden	Synchrocyclotron		1989	horizontal
TRIUMF, Canada	Cyclotron	180-520	1995	horizontal
Second generation of proton research therapy facilities				
Clatterbridge, UK	Isochronous cyclotron	62	1989	horizontal
UCL, Louvain-la-Neuve, Belgium	Cyclotron	90	1991	horizontal
CAL, Nicea, France	Cyclotron	65	1991	horizontal
CPO Orsay, France	Synchrocyclotron	70-200	1991	horizontal
NAC, South Africa	Cyclotron	200	1993	horizontal
MPRI, IN, USA	Cyclotron	200	1993	horizontal
UCF-CNL, CA USA	Cyclotron	68.5	1994	
PSI, Villigen, Switzerland	Cyclotron	200	1996	1 eccentric gantry, spot scanning
Berlin, Germany			1998	
Dedicated protontherapy centers				
LLUMC, Loma Linda, USA	Synchrotron	70-250	1990	1 horizontal, 3 gantries
NCC, Kashiwa, Japan	Cyclotron	235	1998	2 gantries

Table 4.1. Existing protontherapy facilities world-wide.

Planned Protontherapy Facilities in the USA. *MPRI (IUCF).* The Indiana University Cyclotron Facility (IUCF) is planning a conversion of the space formerly used for nuclear physics experiments into treatments rooms for proton radiation therapy treatments under auspices of the Midwest Proton Radiation Institute (MPRI), an organization being developed by IUCF in conjunction with the Advanced Research Technology Institute (ARTI) at Indiana University. The 210 MeV cyclotron IUCF will be used to provide protons to the treatments rooms. In addition there will be an experimental room for biological and radiation effects studies. The new beam transport system has been designed to bring the 210 MeV proton beam from the Indiana University Cyclotron to four Protontherapy treatment rooms.

There will be two treatment rooms with a fixed horizontal beam line. The first will be an upgrade to the existing eye treatment facility and the second will be designed for head, neck, and brain treatments including stereotactic radiosurgery. In addition there will be two rooms with iso-centric gantries. It is planned to use commercially available single-plane gantries [4.41].

RTMC. A consortium organized by the Texas National Research Laboratory Commission (TNRLC) under a Department of Energy (DOE) grant proposes to built and operate a Regional Medical Technology Center (RTMC) to function as a combined medical radioisotope production complex and proton cancer therapy facility using the linear accelerator assets of the cancelled Superconducting Super Collider (SSC). The SSC design included a 35 kV ion source, a low energy beam transport system (to convert diverging beam emerging from the source into the converging beam required by the 428 MHz radio frequency quadrupole (RFQ), a 2.5 MeV RFQ, a matching section into the drift tube linac (DTL), a 70 MeV DTL, a matching section into a coupled-cavity linac (CCL), and the 600 MeV CCL. The DTL consisted of four tanks resonant at 428 MHz, each with its own 4 MW klystron power source. The CCL consisted of nine modules resonant at 1284 MHz, each with its own 20 MW klystron poower source. The technical requirements for the RMTC linac are given in Table 4.2, which also lists the equivalent reqirements of the SSC linac, for which the hardware and facilities were designed. It is possible tosupport two users with this linac because the requirements of the users are highly complementary, and because the basic design provides the number of convinient features. The very small currents required to support the protontherapy will not be missed when they are diverted from the main flow to the radioisotope production facility. The transition between DTL and CCL occurs at 70 MeV, which is also the highest energy requiered for isotope production and the lowest needed for protontherapy [4.60].

	SSC	Radio-isotopes	Protontherapy
Particle	H^-	H^-	H^-
Energy (MeV)	600	30, 50,70	70-250
Peak current (mA)	25	50	50
Pulse lengh (μs)	2-35	90	5
Pulse repetition frequency (Hz)	10	360	10
Average current (μA)	<10	>1000	1-50
Transverse emittance (n, rms) (π mm mrad)	≤0.3	≤2	≤0.5
Energy spread (keV)	≤100	≤1000	250
Scheduled availablity (%)	≥ 98.8	≥95	≥95

Table 4.2. RTMC linac requirements; after L.W. Funk [4.60].

PTCA. In the United States, the FDA approves medical devices for "marketing". There are three alternative routes to obtain FDA approval to market devices. The pre-market approval application (PMA) process involves a full-scale review of a device's safety and efficacy by the FDA. The second route to market is a rarely used one known as the "custom device exemption". It is intended for devices used by a single practictioner (such as a custom surgical devices) and is not applicable to proton beam therapy. The third route to market is reffered to as "substantial equivalency" or the 510(k) process – a reference to the relevant section of the Medical Device Amendments. In 1987, LLUMC acting as manufacturer of its proton beam therapy system submitted a 510(k) pre-market notification for its protontherapy system. MGH for developpment of its NPTC, as the manufacturer of their system, they submitted a 510(k) pre-market notification to the FDA in 1988. Lastly, Protontherapy Corporation of America (PTCA), a subsidiary of Tenet HealthSystem, has contracted with Ion Beam Associates (IBA) of Belgium to manufacture proton beam therapy systems for facilities PTCA is developing across the country. IBA also submitted its pre-market notification to the FDA in 1998 [4.63].

Planned Japaneese Protontherapy Facilities. The medical community in Japan is aggressively pursuing acquisition of new dedicated proton facilities. The Agency for Science and Technology of the Japanese Government has decided to provide assistance to some regional authorities to establish dedicated protontherapy centers. They are negotiating with the Ministry of Finance to get their plan funded. The list of ongoing projects of charged particle therapy facilities are presented in Table 4.3, the geographical plan of implantation of hadrontherapy facilities is presented in Fig. 4.15.

Place	Machine	Max. energy	1st patient	Major facility	Status
NCC	cyclotron	p: 235 MeV	1998	2 gantries, 1 horiz.	1st beam obtained in 1998
Hyogo	synchrotron	p: 230 MeV	2001	2 gantries, 1 horiz.	under construction
		C: 320 MeV/u		1 horiz., 1 45 deg.	
Tsukuba, PRMC	synchrotron	p: 270 MeV	2001	2 gantries, 1 horiz.	construction star-ted
Wakasa	synchrotron	p: 200 MeV	2001	1 horiz.	under construction
Shizouka	synchrotron	p: 230 MeV	2001	2 gantries, 1 horiz.	funded

Table 4.3. Ongoing projects of Japaneese charged particle therapy facilities; after S. Yamada et al. [4.34].

A new dedicated medical synchrotron with an energy variable in steps of 120, 180, and 230 MeV is planned for construction at Tsukuba (PRMC). The designed beam intensity is 20 nA, which corresponds to 1.25×10^{11} protons per second. Two modes of extraction, fast and slow, are planned. Two treatment rooms with horizontal and vertical (up and down) beams are planned. Tsukuba changed its main accelerator from a synchrotron to a compact cyclotron in 1992. There are two reasons for the change : the cyclotron is supplied by IBA and Sumimoto with a guaranteed beam parameters, cyclotron cw beams are more convenient for a scanning system of beam delivery than that of a slow-extracted synchrotron beam. As yet, the Tsukuba group has not been granted the necessary funds [4.46].

The construction of a 250 MeV AVF separate-sector cyclotron is nearing completion at the Osaka University in Japan. The machine is planned to be used for medical sciences as part of an interdisciplinary research program. However, anticipating the beam-time demand for this machine by the physics community, another new facility is being considered.

There is also a proposal to build a 250 MeV synchrotron for medical use at Kyoto University. At present, the National Cancer Center has obtained funds for its plan to build a dedicated Protontherapy facility at its Kashiwa campus, about 30 km from Tokyo. The amount of money allocated to this project is some 8.3 billion yen, i.e. 83 mln US dollars.

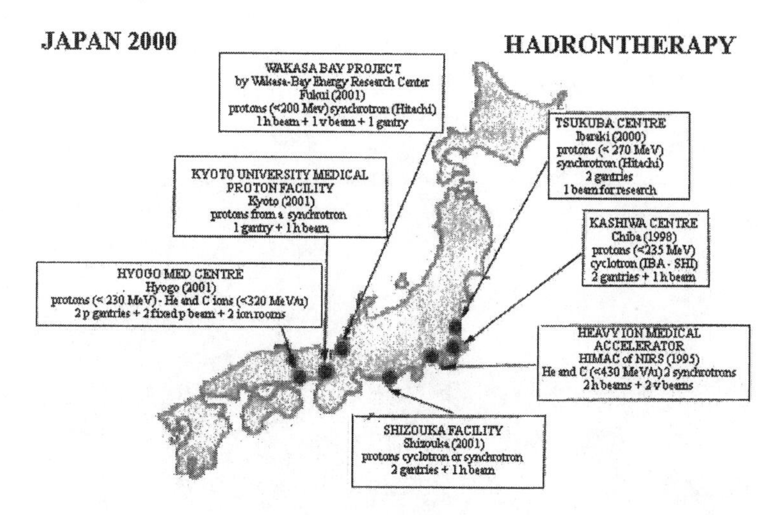

Figure 4.15. Planned Japanese hadrotherapy facilities; after U. Amaldi [4.65].

There is also a proposal to build a 230 MeV synchrotron for medical use at Hyogo, which will be equipped into 2 gantries (protons and carbon ions) and 3 fixed beam rooms. Kobe is the capital of the Hyogo prefecture. The local government is planning a particle beam facility there. This was the most advanced project implemented in Japan. Three kinds of particles: protons, helium and carbon ions were to be delivered, however, it is still possible that only protontherapy will be chosen. Its engineering design was to be started in early 1995, and the whole facility was supposed to be completed by the end of 1998. Although the officials of the Hyogo prefecture are enthusiastic, the damage caused by the earthquake may affect the timetable.

Another project concern the multipurposal accelerator in Wakasa Bay. For 2001 an opening of the medical center equipped with synchrotron (230 MeV), 2 proton gantries and 1 fixed beam is planned in Shizuoka Cancer Center. The layout of proton facility is presented in Fig. 4.16 [4.39].

Planned Russian Protontherapy Facilities. Several medical accelerators have been designed in Russia.

Specially notable are designs of a very compact synchrotron, accelerating protons to 200 MeV with a high magnetic field of 5-10 T, the orbit length of 47 m, and repetition rate of 10 Hz, developed at the Budker Institute of Nuclear Physics (BINS) in Novosibirsk.

Another interesting proposal is a negative hydrogen ion H^- synchrotron for Protontherapy facility, conceived at ITEP that permits multiple extractions to facilitate the beam sharing among six treatment rooms. ITEP has received funding for a dedicated medical proton synchrotron, and construction has scheduled to start in 1992. At the same time, ITEP is developing a smaller (less than 500 m^2) version of three-room Protontherapy facility [4.1].

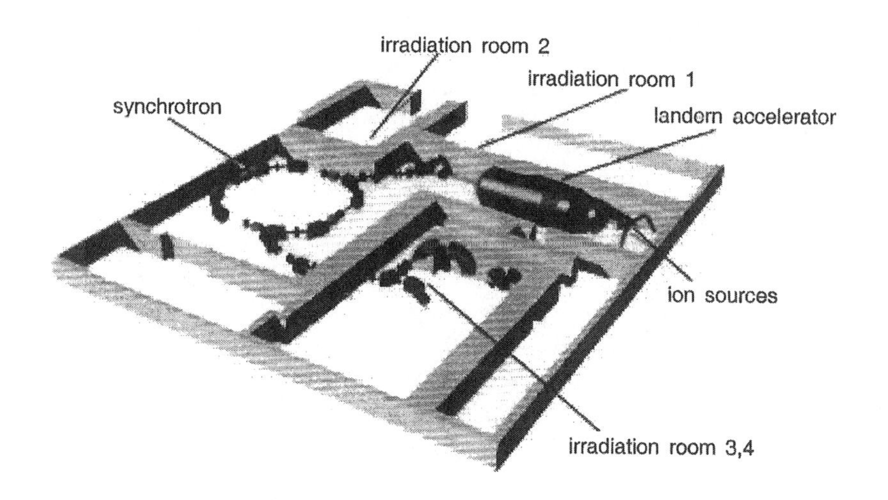

Figure 4.16. Layout of Wakasa Bay Protontherapy facility, after S. Fukuda [4.50].

Planned Italian Protontherapy Facilities. The ADROTERAPIA collaboration is a group of scientists and clinicians in Italy concerned with developing a hadrontherapy accelerator. The design work of the part of the Hadrontherapy Programme promoted by the TERA Foundation is organized in three projects:

1. The creation of an informatics and organisational network, called *RITA* (*Italian Network for Hadrontherapy Treatment*), which will connect the Associated Centres, distributed throughout Italy and abroad, and situated in the public oncological institutions and in private clinics, with the Centres where proton and ion beams will be made available.
2. The design and construction of small and possibly cheap novel hadrons accelerators, which could be built by Italian industries. A facility based on one of these accelerators should treat at least 300 patients a year with

a single beam line and double that number with an added treatment room. This is the *Programma Acceleratori Compatti per Oncologia* called *PACO.*

3. The planning and the construction in Milano of a National Centre for Oncological Hadrontherapy (*CNAO, Centro Nazionale di Adroterapia Oncologica*). It is a healthcare and research structure of excellence, which will be the focal point of all the hadrontherapy activities and, being equipped with proton and ion beams to be used in parallel in many treatment rooms, will treat with protons and carbon ions more than 1000 patients per year. This is the CNAO Project [4.28].

The CNAO will be equipped with a 22 m diameter H^- synchrotron, accelerating both protons (energies 60 – 250 MeV, average intensity 10 nA) and carbon ions.

The Fig. 4.17 represents the first CNAO lot. The facility will be provided with two treatment rooms equipped with an isocentric gantry capable of transporting protons up to 250 MeV; one treatment room equipped with one horizontal beam and one vertical beam pointing downward also for 250 MeV proton beams. One room will be equipped with two horizontal beam lines, one for irradiations of eye tumors and one mainly devoted to head and neck treatments. One room with one horizontal beam will be devoted to experimental activities with both protons and light ions (dosimetry. radiobiology, calibrations. etc.). One room will be devoted to future light ion treatments. Two smaller rooms served by the 11 MeV proton beam from the injector: one for the production of positron emitting radionuclides for PET uses, the other for thermal neutron production for boron neutron capture therapy (BNCT).

The complex will consist of two buildings: an underground, heavily shielded area (the "bunker") housing the accelerators and the treatment rooms, and a surface building above ground with conventional facilities and office space. The bunker will have a surface area of about 3500 m^2, the maximum height will be about 15 m to accommodate the gantry rooms. Besides the basement of the gantry rooms, the lowest floor houses the rooms for BNCT and for radionuclide production [4.27].

The idea of a compact accelerator for protontherapy applications bore within TERA Collaboration, having realized that a design should also have been developed for a small proton accelerator that could be housed in already existing hospitals, requiring then a limited space and limited shielding. It is because they want to fully exploit the potentiality of the new treatment at a National level, apart a design for an integrally new hospital facility hosting a relatively large accelerator having also light ions capability.

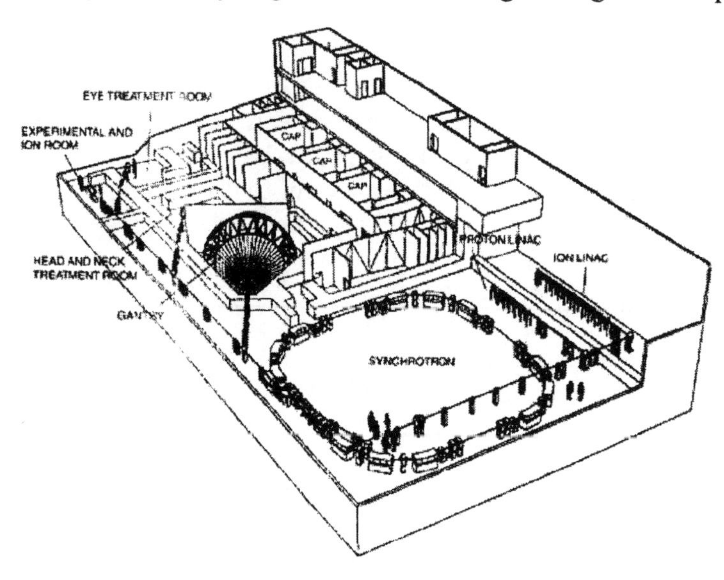

Figure 4.17. The open perspective view for the Center for Oncological Handon-therapy, after U. Amaldi and S. Rossi [1.15].

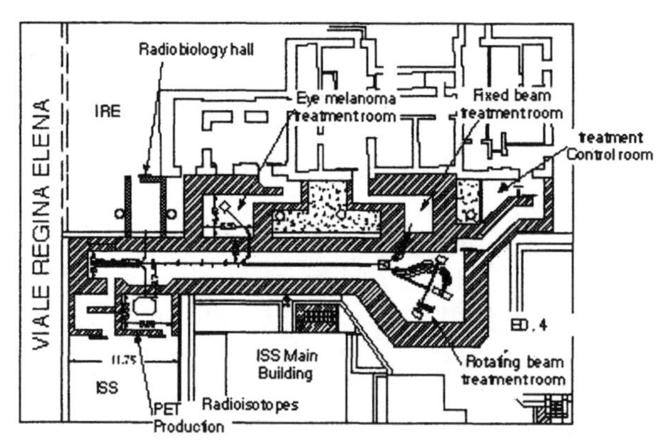

Figure 4.18. Layout of the ISS TOP LINAC; after S. Frullani [4.29].

This concept was adopted by *Italian National Institute of Health* (*ISS*) and they started a project *Oncological Therapy with Protons* (*TOP*). They has started in December 1995 the construction of a compact proton linac (total length 25 m, Fig. 4.18). The general concept is to use the accelerator for protontherapy (65 – 200 MeV), radioisotope production and radiobiology studies [4.29].

Planned German Hadrontherapy Center. Based on a successful pilot project at GSI, the Radiologische Universitatsklinik Heilderberg, the Deutsches Krebsforschungszentrum Heilderberg (DKFZ) and the Gesselschaft fur Schwerionenforschung Darmstadt (GSI) in cooperation with the Forschungszentrum Rossendorf jointly propose to built a hospital based ion beam facility for tumor therapy in Heilderberg. This facility will close a medical care gap in Germany for established indications and will also allow the evaluation of new indications in large-scale clinical studies. It can be constructed in five years by the participating institutions in cooperation with industry.

The planned therapy center will be housed in a building covering an area of 60 x 70 m^2. The heart of the facility is the therapy accelerator, comprising a compact linear accelerator as injector and a synchrotron for the acceleration to the desired final energy, which is variable from pulse-to-pulse. On the injector side, state-of-art. developments are being exploited in the fields of ion sources and accelerator structures (ECR ion sources and RFQ/IH Linac). After acceleration to the synchrotron injection energy of 7 MeV per nucleon, the ions pass through the stripper foil where the entire electron shell is stripped off. In the synchrotron, which have a diameter about 18 m and a magnetic rigidity of up to 6.6 Tm, the particles are accelerated to the desired energies, which are ranging from 50 to 430 MeV per nucleon. This coresspond to the penetration depth of the particle beams from 2 to 30 cm.

The synchrotron feeds into a beam transport system which guides the ion beam into three different treatment rooms. One room provides a horizontal beam, the other two are equipped with a gantry system allowing rotation of the therapeutic beam about the patient. The construction of the gantry systems relies on new technology. Special challenges are proposed by the integration of the raster scan method, which is indispensable for a precise dose application, within the gantry, and by the adaptation of the PET method of beam monitoring to a rotational beam transport system [1.5, 1.24].

Planned Polish Hadrontherapy Centers. ***Kraków.*** The aim of the project is to establish in Kraków a modern center of radiotherapy, which would exploit beams of protons and neutrons (i.e. hadrons) from the AIC-144 isochronous cyclotron of the Institute of Nuclear Physics in Kraków. That would exploit 60 MeV proton (one horizontal beam) and 15 and 30 MeV fast neutron beams (isocentric beam or two: horizontal and vertical). The Center of Hadron Radiotherapy created as a result of this Project would be the only center in Poland in which clinical work in this area would be carried out. This center would satisfy national needs for treating ocular melanoma and clinically indicated fast neutron radiotherapy, and produce selected medical radioisotopes. The Center would also offer research and training opportunities for about 50 physicians, physicists, biologists and engineers and allow them to actively develop modern techniques of treating cancer in close cooperation with laboratories abroad. They plan to treat about 300 patients a year and to produce small quantities of selected medical radioisotopes. Beam layout of the Hadron Radiotherapy Center at the Institute of Nuclear Physics in Kraków is presented in Fig. 4.19 [4.30].

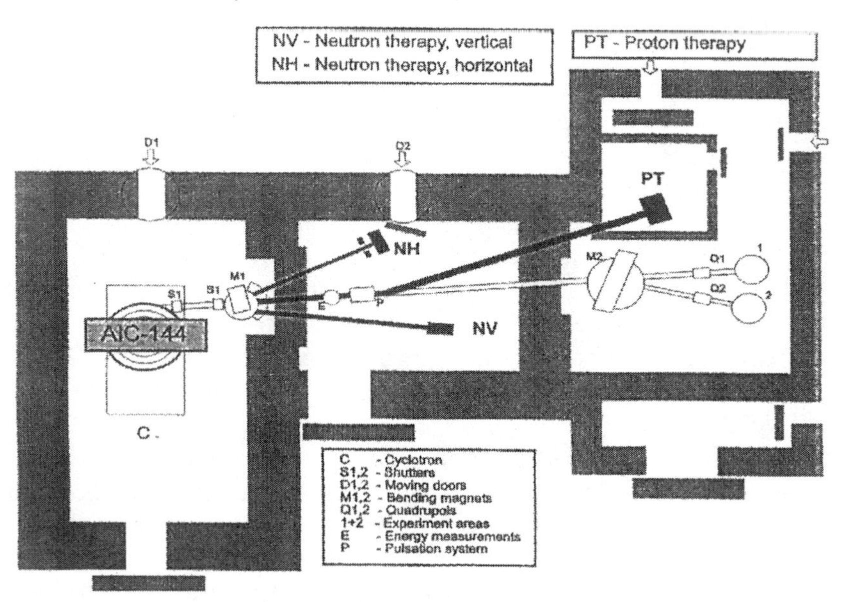

Figure 4.19. Beam layout of the Hadron Radiotherapy Center at the Institute of Nuclear Physics in Kraków; after M.P.R. Waligórski at al. [4.30].

Warsaw. There is a proposition of full-energy dedicated hospital Protontherapy center. Preliminary studies of such a center have been made by W. Wieszczycka in thesis Preliminary project of Polish dedicated Protontherapy facility [1.26].

4.4 International Collaborations for Design an Universal Protontherapy Center

EULIMA. A European initiative for implementation of a light ion accelerator into an existing hospital environment developed in the late '80s into *European Light Ion Medical Accelerator* (*EULIMA*). EULIMA was a prototype accelerator for cancer therapy to be built in Europe. Its installation should allow the treatment of a few thousand patients per year with fully stripped ions ranging from carbon to neon and energies from 140 to 400 MeV / u. This project was financially supported by the *European Community Medical and Health Research Programs* and the studies were based on radiation research centers and accelerator laboratories in France, Belgium, Germany and Switzerland. They were focused on the choice of a proper accelerator and beam delivery system. A feasibility study has been undertaken at CERN for a light ion medical synchrotron.

Two approaches had been explored to design a more compact medical accelerator: a *separated function machine* (*SFM*) and a *combined function machine* (*CFM*). The principle of the first one was maintaining the separated functions (pure dipole bending magnets, focussing and defocussing quadrupoles, discrete sextupoles). Weak focussing allowed to reduce the size of the synchrotron by 25%, the circumference was 59.08 m. The design of the SFM was close to the *LEAR* (*Low Energy Antiproton Ring*) at CERN, and therefore the accelerator could be flexible for multipurpose uses. A further reduction in the size was possible with a CFM. The fucussing and defocussing gradient was added to the dipole component of the bending magnets. Also the correcting elements for Q tuning and the sextupoles for the excitation of the extraction resonance were integrated in the bending magnets. A compact machine with a circumference of 48.6 m was obtained [4.31].

The EULIMA project was terminated at the end of 1991, because no European government wanted to provide the construction funds for the accelerator and associated equipment.

PIMMS. Some kind of ELIUMA program continuation is today *Proton-Ion Medical Machine Study* (*PIMMS*). PIMMS was set up following an agreement between the *TERA Foundation* (Italy) and the *MedAUSTRON* (Austria) to join their efforts in the design of a medical synchrotron that could later be adapted to national needs. CERN agreed to host this study inside its PS Division and to contribute one full-time member to the study team. The study group has worked in collaboration with GSI (Germany) and was more recently joined by *Onkologie 2000* (Czech Republic). Work started in January 1996.

The agreed aim of the study was to investigate and design a generic facility that would allow the direct clinical comparison of protons and ions for cancer treatment. Generic in this context meant a common design that could be modified with the minimum of effort to meet the particular requirements of each partner in the study. The machine was to be designed primarily for high-precision active beam scanning for both protons and carbon ions, but was also to be capable of delivering proton beams by passive scattering.

A synchrotron with slow extraction offers the flexibility needed for dual-species operation and the variable energy needed for active scanning. The principal design requirement is that of a smooth spill. This leads to an extraction technique that maintains all transverse optical parameters constant while using a betatron core to accelerate the beam into the resonance. In addition, the machine will be made less sensitive to power converter ripple by using an empty rf bucket to channel the particles more rapidly into the resonance. The choice of a synchrotron has the consequence that the extracted beam has unequal transverse emittances and that a more advanced technique is required for matching to the rotating gantry. This technique uses a device known as a *rotator*. The slow-extracted beam is not only asymmetric in terms of emittance, it is also asymmetric in terms of the shape of its footprint in phase space. In the vertical plane, the beam occupies the usual elliptical area, whereas, in the horizontal plane, it is a narrow bar. The positive aspect of this behavior is that it provides an independent handle on the control of the beam size in the horizontal plane. This is just one aspect of the flexibility needed for the treatment planning. In the vertical plane a different technique is used to control beam size. The extracted beam intensity can be varied either by controlling the initial intensity in the synchrotron or by varying the acceleration rate of the betatron core. The scanning speed of the beam spot can be varied over a wide range and a chopper is included to cleanly switch the spill on and off. One extra peculiarity of the slow-extracted beam is that the asymmetric phase-space footprints affect the way in which the beam scanning should be performed. Finally, the high magnetic rigidity of the carbon ions led to an investigation of an alternative gantry design. Thus, the extraction and beam delivery systems for a synchrotron with slow extraction differ significantly from the more conventional designs used with cyclotrons.

The performance parameters are defined by the clinical needs. Thus, the maximum energy of the machine is set to 400 MeV/u for carbon ions and 250 MeV for protons. The beam intensities and repetition rates have been adjusted so as to deliver a single treatment, or fraction, in about two minutes. It is understood that in the rare instance that a dose level, or volume, causes the capacity of the machine to be exceeded, the treatment time can be extended. For active scanning, it is assumed that a nominal fraction would be 2 Gray in 2 liters (or equivalent combination) delivered by 60 spills in 2 min. For passive scanning, it is assumed that a nominal fraction would be 2 Gray in 7.5 liters (or equivalent combination) delivered by 120 spills in 2.5 min [4.31, 4.32, 4.49, 4.53, 4.54].

LIBO. There are many cyclotrons of energy around 60 MeV in hospitals and physics laboratories, but their usefulness for medical purposes is limited by their energy; 60 MeV protons are suitable for treating eye melanomas, but cannot reach deep into the body. *LIBO* (*Linac BOster*) is a proton linac, intended to post-accelerate the proton beam of a cyclotron from 60 MeV to 200 MeV. This is the energy needed to treat deep-seated tumors (up to 27 cm deep), using proton intensities of about 3 nA. The design of such a linac was started by the TERA Foundation in 1993, and in 1998 a collaboration was established between CERN, TERA, and the INFN (in Naples and Milan) to realize a prototype of the first LIBO accelerating module (62-74 MeV). The intention is to test the module with rf power in the LIL machine at CERN. Once tested successfully, the technology will be transferred to industry for commercial production.

A review of proposed hadrontherapy facilities is presented in Table 4.4. Table does not describes the international projects proposed by CERN and accelerators companies.

Who	Place	Planned first irradiation	Comments
NPTC Harvard	USA	2000	protons, at MGH, cyclotron 230 MeV, 2 gantries + 2 horizontal
INFN-LNS, Catania	Italy	2000	70 MeV protons; 1 room, fixed horizontal beam
Hyogo	Japan	2001	protons, ions, synchrotron 70 – 230 MeV (protons, carbon ions), 2 gantries, 2 vertical, 1 horizontal, 1 45°; under construction
NAC, Faure	South Africa	2001	new proton treatment room with beam line 30° off vertical
Tsukuba	Japan	2001	270 MeV protons , 2 gantries, 1 research room, under construction
CGMH	Northern Taiwan	2001 ?	250 MeV proton synchrotron or 230 MeV cyclotron, 3 gantries, 1 fixed
Wakasa Bay	Japan	2002	multipurpose accelerator, building completed mid 1998
Bratislava	Slovakia	2003	protons, ions, cyclotron 72 MeV + BNCT + isotope production
IMP, Lanzhou	PR China	2003	C-ion from 100 MeV/u at HIRFL expand to 900 MeV/u at CSR, clinical treatment, biological research, no gantry, shifted patients
Shizuoka Cancer Center	Japan	2002 ?	synchrotron 230 MeV ? 2 gantries, 1 horizontal, funded
Erlangen	Germany	2002 ?	protons, 4 treatment rooms some with gantries
CNAO, Milan & Pavia	Italy	2004 ?	protons, ions, synchrotron , 2 gantries, 1 fixed beam room, 1 experimental room
Heiderberg	Germany	2005?	protons, ions
AUSTRON	Austria	?	2 proton gantries, 1 ion gantry, 1 fixed proton, 1 fixed ion, 1 experimental room
Beijing	China	?	synchrotron 250 MeV
Central Italy	Italy	?	cyclotron, 1 gantry, 1 fixed
Clatterbridge	UK	?	upgrade using booster linear accelerator to 200 MeV?
TOP project ISS, Rome	Italy	?	70 MeV proton linac, expand to 200 MeV?
3 projects in Moscow	Russia	?	protons including 320 MeV, probably no gantry
Krakow	Poland	?	60 MeV proton beam
Moscow	Russia	?	320 MeV compact proton accelerator, probably no gantry
Proton Development N.A. Inc.	IL, USA	?	300 MeV protons, therapy & lithography
PTCA, Tenet Health System	USA	?	Several systems throughout the USA
RMTC	USA	?	combined medical radioisotope production complex and Protontherapy facility using SSC linac

Table 4.4. Proposed new facilities for proton and ion beam therapy, July 2000; after J. Sisterson [3.6].

4.5 Firm Project of Protontherapy Centers

At present, some of specialist and commercial companies are proposing, or working on a system specially dedicated to Protontherapy. *Ion Beam Application S.A.* (*IBA*), Belgium, proposes commercial equipment specially designed for in-hospital operation [4.23]. The basic configuration of the IBA Protontherapy system comprises the following elements: (1) a 235 MeV isochronous cyclotron, able to deliver beams of up to 1.5 μA, but hardware-limited at 300 nA in order to limit the maximum possible dose rate to the patient; (2) an energy selection system transforming the fixed energy beam extracted from the cyclotron into a variable energy beam (235 to 70 MeV range) provided with energy spread and emittance limitations and verification; (3) a beam transport and switching system connecting the exit of the energy selection system to the entrance points of a number of gantries and fixed beam lines; (4) three gantries fitted with a nozzle, and a system consisting of two horizontal beam lines, the large field one being equipped with a nozzle. For beam spread-out on the gantries and on the large field line, both beam scattering and beam wobbling are available; (5) a global control system including, in addition to an accelerator control unit, three independent, but networked therapy control stations; (6) a global safety management system independent of the global control system; and (7) a robotic patient positioning system [1.3, 4.5, 4.66].

IBA protontherapy system was installed at NPTC and NCC facilities, detailed description of those facilities is presented in Sec. 4.2.3.

Chapter 5

REQUIREMENTS FOR HADRONTHERAPY CENTERS

The higher ballistic selectivity of protons as compared to photons and electrons used in conventional radiation therapy requires that stringent specifications on the dose delivery be in order to achieve a full 3D conformal therapy. The advantages of protontherapy, as well as basic clinical parameter of proton beam were already presented in Secs. 3.1.1 and 3.1.2. In this Chapter a complete list is given of the clinical performance specifications: beam range, Bragg peak modulation, range adjustment, dose rate, field size, field homogeneity and symmetry, lateral penumbra, distal dose fall-off, source-to-surface distance, displacement of the beam axis from the isocenter and gantry rotation. The Sections showing performance specifications on the accelerator and the beam transport and delivery systems that needed to meet the clinical requirements will follow this Chapter (Secs. 6.1, 7.2, 7.3 and 8.1).

5.1 Medical and Technical Performance Specifications for the Clinical Proton Beam

The proton facilities must have the following capabilities:

- to treat most of the tumors (adequate penetration depths of the proton beams and achievable field sizes),
- short treatment time (adequate beam intensity),
- desirable physical characteristics of the proton beams related to sharp delineation of the treated volume from the surrounding healthy tissues that have to be protected, or the uniformity of the dose within the treated volume,
- versatility of the treatment delivery (fixed and movable beams to provide entry of the beam into the patient from any arbitrary direction specified by therapy planning),

- optimized patient flow through the treatment facility,
- patient's safety (accurate dosimetry and reliable control system) [5.1].

5.1.1 Beam Range in Absorbing Medium

Range in patient is a function of the initial beam energy from the accelerator and the amount of material in the beam between the accelerator and the patient as vacuum windows, beam monitoring devices and any devices associated with beam spreading and range adjustment. A 250 MeV proton beam has a range of 38 g/cm^2 in water. The full range is not required for all field sizes; for larger field sizes ($\geq 22 \times 22$ cm^2) the minimum acceptable range is 22 g/cm^2. Also the smallest required range for extracted beam is about 3.5 g/cm^2 corresponding to an energy of about 70 MeV.

However, the effective range available for treatment will be less because of the energy suffered by the beam in any material encountered before reaching the patient. The main cause of energy loss are be represented by passive beam spreading devices. Beam monitor for diagnostics and dosimetry induces additional energy losses, but these represent a minor contribution as compared to beam spreading devices. The minimum energy of 70 MeV should guarantee the minimum range of 3.5 g/cm^2. Lesser ranges, required for the treatment of superficial tumors of the head and neck, can be provided through the use of energy absorbers. The maximum energy of 250 MeV should ensure the range in patients of 28 g/cm^2 required for deep-seated tumors. These range values correspond to energies at the entrance in the patients of about 50 MeV and 200 MeV, respectively [1.4, 5.1].

Currently there is one major application for low-energy protons: the treatment of ocular melanoma. A minimum proton energy of between 55 and 60 MeV is required for this application. Their conclusions were based on the proton ranges required for the first 94 patients treated using the Harvard Cyclotron (Boston, USA). That facility has a fixed horizontal beam line incorporating beam scattering foils, range modulators and dose monitors. The patients are treated seated on a motor driven chair and restrained by a bite block and head restraint or face mask.

The second group of applications involves the use of proton energies in excess of 160 MeV. Proton ranges greater than 17 cm in water are needed for treatment of more deep-seated lesions. There is currently many debates concerning the optimum specifications and choice of accelerator for use in high-energy protontherapy. The essential specifications arrived at for a

Protontherapy facility by the Protontherapy Co-operative Group (PTCOG). The maximum energy should be about 250 MeV (penetration in tissue – about 37 cm). The energy should be continuously variable or selectable in a few steps starting at 70 MeV and the average beam current should be at least 10 nA for reasonably short treatment times. A comparison of accelerator types for use in Protontherapy is given in Chapter 6. The three principal options for the accelerator are the cyclotron, the synchrotron and the linear accelerator.

5.1.2 Bragg Peak Modulation

The extension in depth of the SOBP should be variable in steps of 0.2 g/cm^2 for the ranges of the beam lower than 5 g/cm^2 and 0.5 g/cm^2 for the ranges greater than 5 g/cm^2. The greatest flexibility in dose delivery is achieved if this variation in range can be achieved by rapidly varying the energy of the beam extracted from the accelerator. If mechanical beam modulating devices such as metallic or plastic *range filters* or *propellers* are employed, the resulting step size for the range modulation may be practically limited to 1.0 g/cm^2 [1.4, 5.1].

The examples of SOBP possible to produce by NPTC facility is presented in Fig. 5.1.

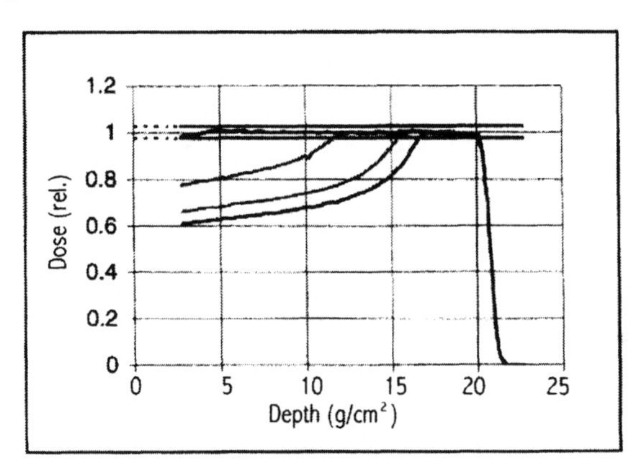

Figure 5.1. Spread out Bragg peaks with partial range modulation, after J.B. Flanz from NPTC [4.26].

5.1.3 Range Adjustment

The residual range of the beam in the patient should be adjustable in fine step sizes. Different steps of SOBP translation in depth are required according to the extension of the SOBP: steps of 0.1 – 0.3 g/cm^2 for extensions higher than 5 cm and extensions 0.05 – 0.2 cm or less for extensions lower than 5 cm. The adjustment can be made either by varying the energy of the beam from the accelerator or by degrading the energy of the beam with absorbers, or by both techniques in tandem. To minimize neutron production and preserve beam quality (minimize emittance growth), it is most desirable, that energy variation from the accelerator be used as much as possible and that absorbers be used only for fine-tuning the beam range [1.4, 4.1, 4.26, 5.2].

As an example, the minimum range adjustment possible to achieve in NPTC is shown in Fig. 5.2.

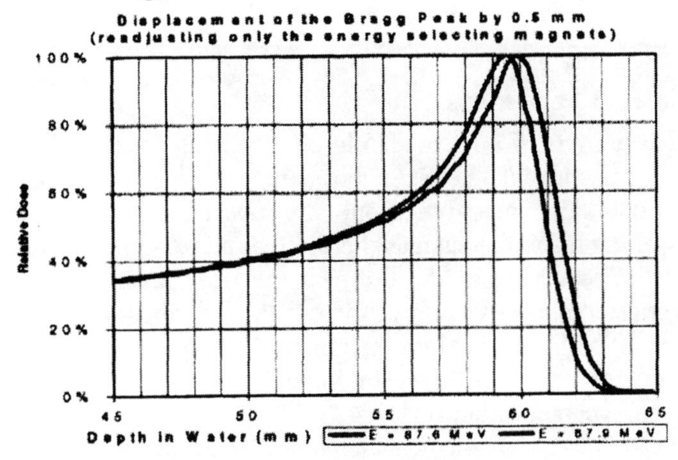

Figure 5.2. Depth dose measurements at NTPC showing achievement of minimum range adjustability (beam energies 87.6 MeV and 87.9 MeV); after J.B. Flanz [4.26].

5.1.4 Field Size

The protontherapy center should be equipped in the fixed beam rooms and gantry rooms. For the fixed horizontal beams, which would be exclusively used for ocular treatments, the required field dimensions and shape are determined, for every patient, by means of personalized collimators with a diameter ranging from 5 mm to 34 mm. For the horizontal beam used to treat head and neck pathologies, the required field size ranges from 2×2 cm^2 to 15×15 cm^2. The other horizontal and vertical beams should provide the same performance as the gantry beams. For these beams the required field size ranges from 2×2 cm^2 to 40×40 cm^2 at isocenter. The field dimensions, for all beams, must be adjustable with 1 mm steps with an accuracy of $\pm$ 0.5 mm [1.4, 4.26, 5.2].

5.1.5 Field Homogeneity and Symmetry

The homogeneity of a field on orthogonal (R_t) and longitudinal (R_l) sections to the propagation direction of the beam must be such that $R_t \leq 105\%$ and $R_l \leq 111\%$. The field symmetry is consequently less than 105%.

5.1.6 Lateral Penumbra

The lateral penumbra of a beam should be as low as possible, but not more than 2 mm for each side. The penumbra is measured at the entrance surface of the water-equivalent phantom as only the contribution given by the design parameters of the machine (such as the finite dimensions of the virtual beam source and the scattering of the beam in the beam modifying devices) is taken into consideration. At the treatment depth a further contribution to the penumbra is given by the intrinsic penumbra, which is due to the scattering of the beam interacting with the biological tissue. This contribution can neither be removed nor reduced [1.4, 4.22].

With a passive system, the beam penumbra at the surface of the patient is due to intrinsic source, contribution from the angular confusion due to scattering in each degrader upstream of the collimator or the aperture, producing an increase of the size of the effective source. The penumbra effect increases with increasing collimator to skin distance and reduces with increasing source to collimator distance. The penumbra effect is also enlarged by scattering contribution from downstream degraders (boluses) [1.4].

With an active scanning system the lateral penumbra is mainly determined by the divergence of the pencil beam (fixed by the optics of the beam delivery system) and by the scanning pattern. Assuming that the dose distribution delivered by the scanner is a superposition of Gaussian beam spots, the lateral dose fall-off can be improved using an optimized pattern with a variable distance between spots and a variable gaussian amplitude (weight) at the edge of the treatment field [1.4].

5.1.7 Distal Dose Fall-off

The range of a monoenergetic proton beam penetrating in tissue is subjected to a spread due to the statistical nature of the energy loss process (range straggling). *Proton range straggling*, defined as the standard deviation of the Gaussian of the stopping distribution, is approximately 1.1% of the proton range. Therefore, at 250 MeV it is about 4.5 mm, at 150 MeV - about 2 mm and at 100 MeV - about 1 mm.

The other factors contributing to the distal dose fall-off are the intrinsic *momentum spread* $(\Delta p/p)_i$ of the beam extracted from the accelerator and, with a passive beam spreading device, the additional momentum spread corresponding to the straggling in the materials. In the worst case, using a passive system for dose delivery and starting from the clinical requirement of 3 mm for the distal fall-off beyond the intrinsic spread, at 250 MeV the maximum acceptable contribution to the dose fall-off due to $(\Delta p/p)_i$ is 2.2 mm [1.4, 5.1].

5.1.8 Average Dose Rate

Average dose rate in the target volume depends on the beam intensity entering patient, the volume of the target, the shape of the target, the beam spreading technique, the beam delivery system design, the transmission efficiency from accelerator to treatment room and the technique for energy variation. Overall efficiency of particle use can approach 50% for scattering systems, but might be significantly less for heavily collimated beams. The beam intensity required is that which is needed to treat the largest commonly used field sizes (approximately 40×40 cm^2) at the maximum required depth (32 g/cm^2) to a dose of 2 - 10 Gy min^{-1}. For field sizes less than 5×5 cm^2, an average dose rate of at least 10 Gy min^{-1} at a depth of 3 g/cm^2 is required. In a special case of ocular tumors, which require very short irradiation times ($\leq 20 - 30$ s), very high dose rates in the SOBP are necessary (about 30 – 40 Gy min^{-1}), in order to guarantee a greater accuracy in the dose administered to the target volume [1.4, 5.1].

5.1.9 Dose Accuracy

The absolute accuracy of the delivered dose should be better than ± 2% of the specified dose [5.1].

The example of lateral distribution of the dose of wide field at 12.4 cm depth for NPTC beam is presented in Fig. 5.3.

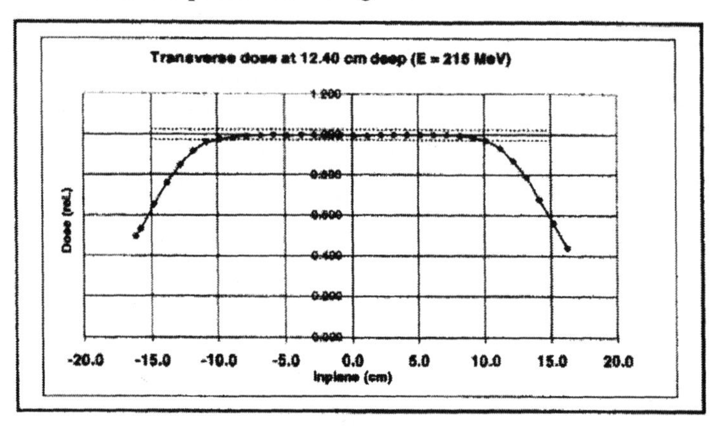

Figure 5.3. Lateral distribution of dose of wide field at 12.4 cm depth, for NPTC beam; after J.B. Flanz [4.26].

5.1.10 Beam Intensity Requirements

The extracted beam intensity is determined by the requirements in terms of dose rate and field size, using a passive beam spreading device - or maximum cross-section of the tumor, using a scanning system. The number of protons necessary to deliver a dose of 1 Gy over a volume of 1 liter, at a used energy of 200 MeV, varies in the range $(5 - 15) \times 10^{10}$, according to the shape of the volume. In fact, the number of particles that the accelerator has to provide will be larger. The beam current required is strongly dependent on the type of beam spreading device, scanning or scattering.

With a magnetic scanning system (Sec. 7.5.3.2) one should be able to achieve a full 3D conformal therapy by accurately distributing the particles within the target boundaries. In case of the synchrotron for example, the beam current, that has to provide to deliver a dose rate of 2 Gy/min in a volume of two liters, is about 2.6 nA, if one assume an efficiency of 70% for the scanning system (perhaps slightly conservative for raster scanning) and 90% for the beam transport system.

With a passive dose delivery system (Sec. 7.5.3.1), the beam intensity is not essentially determined by the target volume but by the field size (produced by the double scatterer) which circumscribes the effective area to be treated and by the maximum treatment depth. A compromise has to be made in choosing the number of scattering systems (scatterers and modulator). The highest intensity is required when treating the distal layer of the tumor volume, and rapidly decreases when "pulling back" the Bragg peak to treat the other layers. These have in fact already received part of the dose during the irradiation of the deepest layers (this also applies to a scanning system). Assuming an average efficiency of 40% for the double scatterer and 90% for the beam transport system, the beam currents that the synchrotron has to provide to deliver a dose rate of 2 Gy min^{-1} should range from 2 to 17 nA [1.4].

5.1.11 Source to Axis Distance (SAD); Source to Surface Distance (SSD)

The SAD influences the skin dose / target dose ratio. For a monoenergetic beam the geometrical contribution to this ratio is the ratio between the area irradiated at the Bragg peak depth and the area irradiated at the patient surface, i.e. $d^2 / (d - r)^2$, where d is the SAD and r is the proton range. In the case of a SOBP this dependence must be convoluted with the corresponding values of the ratio between the entrance dose and the dose at the Bragg peak for the different energies and with the values of the intensity delivered for each energy. One can also express the skin dose / target dose ratio as a function of SSD rather than SAD, as the two are strictly related. The skin dose / target dose ratio decreases with SSD (Fig. 5.4).

Space constraints in the beam delivery system in practice limit thc value of SAD. In a gantry, in particular, a SAD larger than 3 m is not feasible. A value of SAD between 2 m and 3 m is acceptable. With a passive beam spreading device this interval further reduces to 2.5 – 3 m [1.4, 5.1].

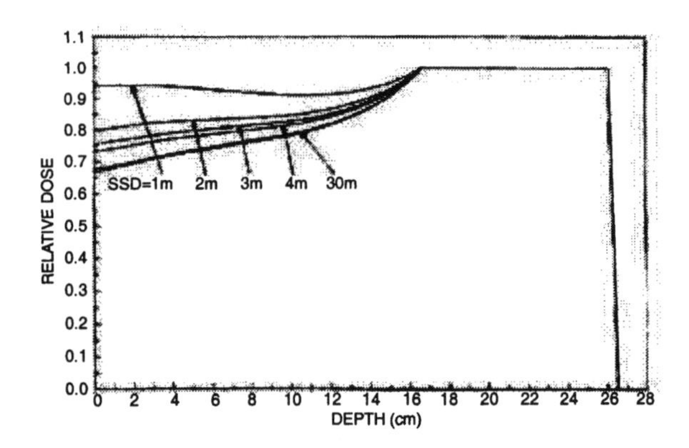

Figure 5.4. Depth dose curves for a 10 cm SOBP at a maximum energy of 200 MeV, for different values of SSD; after U. Amaldi et al. [1.4].

5.1.12 Displacement of the Beam Axis from the Isocenter

For a radiological facility in which the therapeutic beam comes from different angles (isocentric unit), the *isocenter* is defined as the center of the smallest sphere traversed by the beam axis. For an isocentric unit the maximum displacement between the beam axis and the isocenter must be kept within a 2 mm diameter sphere. This uncertainty on the beam direction can cause an error on the incidence angle of the beam of about $\pm 0.1°$ [1.4].

5.1.13 Gantry Rotation

The rotation speed of the gantry should not be greater than 6°/s, corresponding to one revolution per minute. The angular positioning accuracy should not be worse than $\pm 0.3°$. It should be possible to change the position of the gantry in 1° steps [1.4].

5.1.14 Time Structure of the Extracted Beam

The dose shaping with a passive beam delivery system is not influenced by the time structure of the beam, both in terms of duty cycle and of intensity fluctuations, as well as by the extraction time when using a synchrotron (an isochronous cyclotron provides a continuous beam). On the other hand, a magnetic scanning system is very sensitive to the above parameters. The specifications of the scanning system and those of the accelerator must be strictly interfaced. With the present design of the scanning system, beam intensity fluctuations of ± 50% within a window time of 50 μs can be handled.

The spill duration has an upper limit imposed by the need for keeping the treatment time within acceptable values and a lower limit determined by the maximum frequency of the scanning system and the time response or the dosimetric and control systems. The specifications of the acceptable intensity fluctuations within one spill, as well as their time structure, are determined by the dynamic range, the time response of the scanning system and the related control and dosimetry [1.4].

5.1.15 Beam Abort Time

In the case of a malfunction of a subsystem during treatment, the beam must be switched off in a time short enough to ensure that overdose is maintained within the clinical specifications. The most stringent requirement is imposed by the use of a scanning system, where a malfunction can cause a hot spot, whereas with a scattering system the beam is spread over a large area. In general, the beam abort time should not exceed 60 μs [1.4].

As an example, clinical specifications and preliminary measurements for a new NPTC in Boston, USA, is presented in Table 5.1.

5.2 Performance Specification for the General Facility

5.2.1 Treatment Rooms

A dedicated proton medical facility should have two or three rotating gantry treatment rooms and at least one horizontal fixed-beam treatment room. If the construction proceeds in phased mode, at least one gantry room and one fixed-beam room must be provided in the initial phase. The initial facility must be constructed in such a way that it will be compatible with future expansion in a cost-effective manner [5.1].

As an example, a view on a treatment level of the first operating Protontherapy facility is presented in Fig. 5.5.

Clinical specification	Preferred specification	Minimum specification	Preliminary measurement (date)
Beam dose shape			
Range in patient	maximum 32 g/cm^2	maximum 28 g/cm^2	32.8 g/cm^2 (3/97)
	minimum 3,5 g/cm^2	minimum 5,0 g/cm^2	3.5 g/cm^2 (3/97)
Range modulation	steps 0,5 g/cm^2 or less over full depth (0,2 g/cm^2 or less for ranges $\leq$ 5 g/cm^2)	steps 1.0 g/cm^2 or less over full depth (0,3 g/cm^2 or less for ranges $\leq$ 5 g/cm^2)	0.5 g/cm^2 (6/98)
Range adjustment	steps 0,1 g/cm^2 (0,05 g/cm^2 for ranges < 5 g/cm^2)	steps 0,1 g/cm^2	0.1 g/cm^2 (3/97)
Distal dose fall-off (80% - 20%)	not more than 0,1 g/cm^2 above physical limit from range straggling	not more than 0,6 g/cm^2	< 1 mm (set by ESS) (3/97)
Field size	fixed 40 × 40 cm^2 gantry 40 × 30 cm^2	fixed $\geq$ 28 × 28 cm^2 gantry $\geq$ 26 × 22 cm^2	20 cm^2 scattered; 30 × 40 cm^2 wobbeled (9/98)
Lateral penumbra (80% - 20%)	not more then 2 mm over the penumbra due to multiple scattering in the patient	not more then 4 mm over the penumbra due to multiple scattering in the patient	To be measured
Dose uniformity	± 2,5% over treatment field	± 4% over treatment field	To be measured
Beam dose			
Average dose rate	a beam intensity sufficient to treat a 25 × 25 cm^2 at depth of 32 g/cm^2 to dose of 2 Gy in 1 min. or less	a beam intensity sufficient to treat a 25 × 25 cm^2 at depth of 28 g/cm^2 to dose of 2 Gy in 4 min. or less	300 nA extracted cyclotron current (3/97)
Dose accuracy	Daily accuracy and reproducibility ± 1.5%		To be measured
Beam trajectory			
Beam position	Direct a beam to the target to better then ± 1 mm. (Results from several components including gantry mechanical isocenter accuracy, and PPS, patient positioner accuracy.)	Gantry measured mechanically Isocenter accuracy of ± 6 mm (3/98) PPS reproducibility < 0.3 mm (12/97)	
Other specifications			
Spill structure	"scanning ready"	"scanning upgradable"	
Effective SAD (for gantry beam)	$\geq$ 3 m	$\geq$ 2 m	

Table 5.1. Clinical specifications and preliminary indications of achievement to date for NPTC facility; after G.P. Gall et al. [5.2] and J. Flanz et al. [4.26].

The horizontal fixed-beam room must be designed so that small-field treatments (e.g. for eye treatment) and the large-field (up to 40 × 40 cm^2) treatments can be performed. The equipment switch time between the eye treatment and large-field, must be less then 10 min [5.1]. The example of the equipment installed in the fixed-beam room is presented in Fig. 5.6.

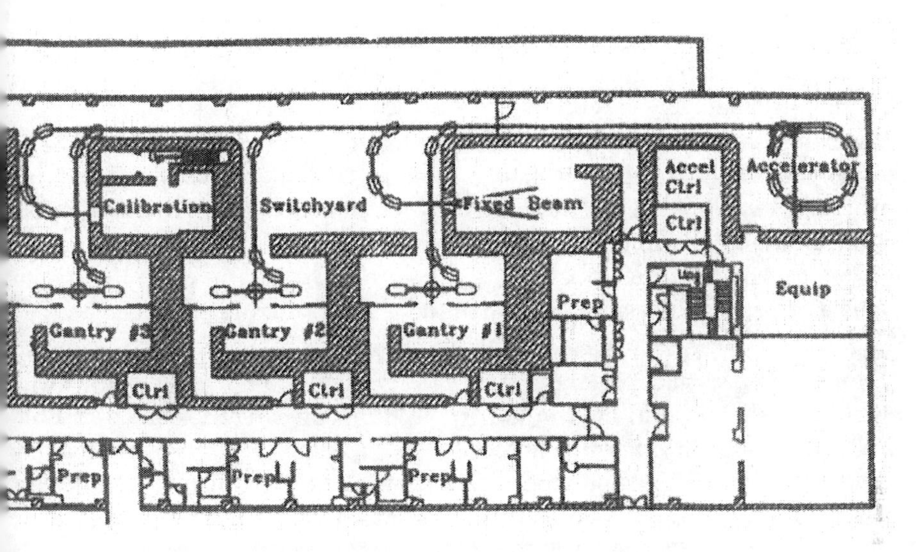

.5. Plan of the treatment level, Loma Linda University Medical Center; Coutracona [4.24].

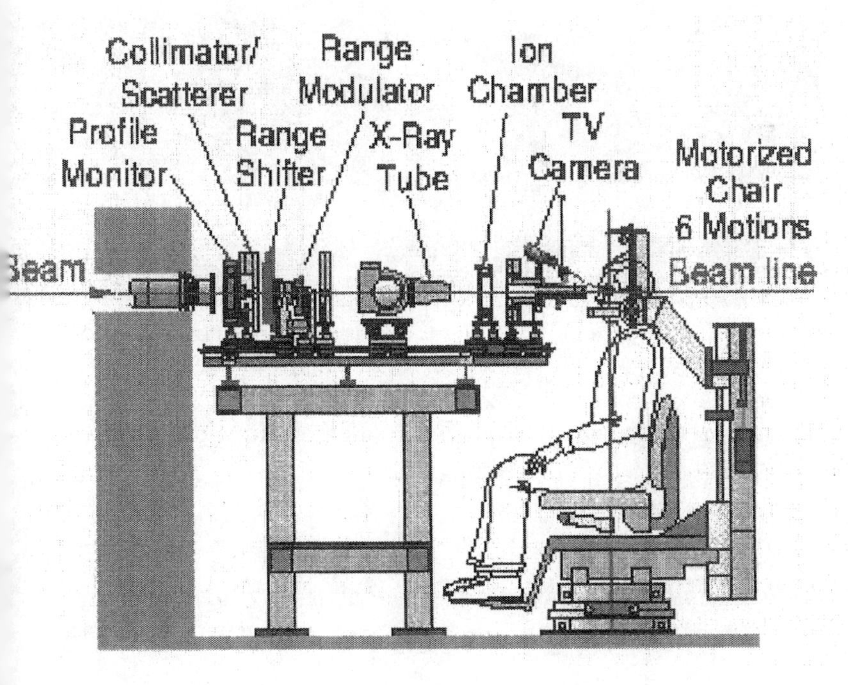

6. The common equipment of horizontal fixed-beam room on the example TRIUMF proton facility; after [5.3].

The rotating gantry room must be designed to provide at least 30×20 cm^2 fields. Together with the couch specification, the facility must be able to irradiate any part of patient from any direction [5.1].

Beam orientation should be fixed for horizontal beam and 4π steradian for gantry beam. A gantry rotating in a vertical plane, coupled with a movable patient couch (3 translations + rotation about vertical axis) can bring a beam into a supine patient from any angle, i.e. full 4π steradian beam delivery around the patient [5.1].

The example of possible rotation of the proton gantry is presented on the Fig. 5.7. It is the gantry system installed recently at PSI. Gantry rotation α by $\pm 185°$ is presented on the Fig. 5.7 (left). Stereotactic head treatments, which are usually performed in horizontal beam lines, are executed at PSI directly on the gantry. For this purpose a special support can be mounted on the gantry table. This support provides an additional computer controlled rotation of the patient couch in the horizontal plane. The rotation angle is selectable between 0° and 120° (Fig. 5.7 right) [5.4].

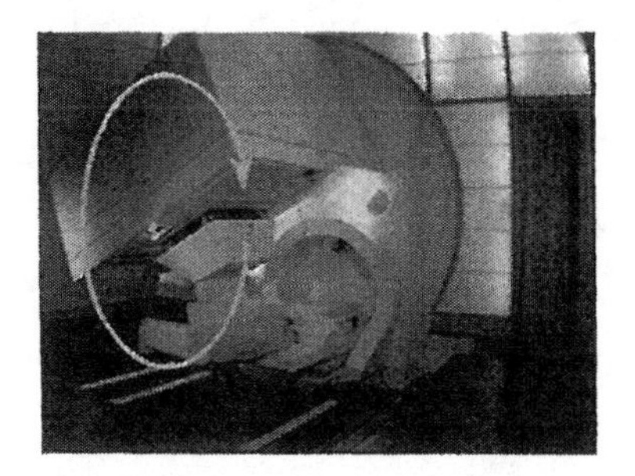
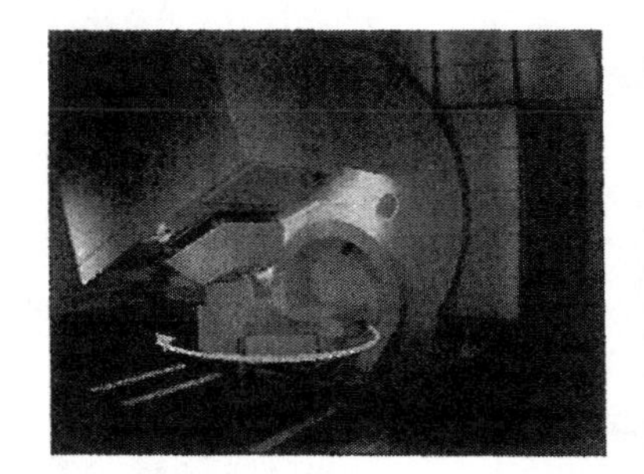

Figure 5.7. Degrees of freedom of the PSI gantry; after [5.4].

The treatment-room associated ancillary facilities. For each treatment room, there must be a patient setup room, a patient equipment storage area and a treatment console area. These facilities have been specified in the building specification chapter of the facility, which describes the size and locations of rooms including such rooms as patient waiting, dressing rooms, and other facilities [5.1].

5.2.2 Facility Availability

Reliability of individual components must be extremely high, maintenance procedures must stress early detection of weak components, and failures must be diagnosed and repaired quickly. Control systems should stress self-diagnosis of problems. The entire facility shall be available for treatment at least 95% of the time, that is, 95% of scheduled treatments should be given with a minor delay of no more than 5 minutes due to equipment malfunctions [5.1].

5.2.3 Treatment Beams

The choice of beam spreading system, both laterally and in depth, is an important element in the cost, complexity and flexibility of the facility. A scattering system, while substantially less expensive and less dependent on accelerator performance (with the exception of beam position stability), substantially limits flexibility in beam delivery and degrades lateral penumbra. Active magnetic beam deflection systems offer flexibility at the expense of increased dependence on spill structure and intensity control. It is envisioned that passive scattering can be used in the initial design, but the possibility of going to active scanning in the future must be part of the facility design [5.1].

5.2.4 Dosimetry Reproducibility

Daily reproducibility of the dosimetry system must be within ± 1.5% limits over the span of one day and ± 3.0% over a period of at least one week. Sufficient redundancy in dosimetry devices and measurements must be provided to ensure no loss of dosimetric information should any one device malfunction [5.1].

5.2.5 Control System

The proton medical facility control system must control and / or interface to the accelerator, safety systems, beam lines, treatment rooms, dosimetry systems, patient positioners, and beam modifying devices. The system should permit pulse-to-pulse and within-a-pulse control of the accelerator, pulse-to-pulse monitoring in dosimetry, rapid beam switching and efficient facility startup and shutdown.

The time and number of operations needed to operate all components of the facility should be minimized so that operators need only enter the minimum amount of information necessary to specify the desired state of the system. In general, the control system should permit the following operations to be performed in a timely manner:

- startup at the start of the day from standby status in < 60 min.
- startup from a cold start (with good vacuum) in < 120 min.
- shutdown at the end of the day to standby status in < 15 min.
- manual setup of all parameters needed for treatment in one room (excluding adjusting gantry angle and beam-modifying devices) in < 1 min.
- automatic setup of all parameters (from a pre-stored table) needed for treatment in one room (excluding adjusting gantry angle and beam-modifying devices) in < 0.5 min.
- time required to shut the beam off after a HALT is requested should be very short (~10 μs) and the time to restart treatment after a HALT should be < 2 s.
- time to terminate treatment ("soft" emergency off) must be the lesser of 1 s or 0.2 Gy delivered dose [5.1].

5.2.6 Radiation Safety of the Facility

Radiation Levels Inside and Outside Facility. Shielding designs must appropriately protect personnel working in the facility (radiation workers) as well as non-radiation workers and visitors. Patient exposure outside the designated treatment fields should be kept within regulatory limits. Principles of ALARA must be applied.

There are certain equipment design features that do impact the shielding requirements. Design considerations of concern in this regard include: techniques that call for large beam losses in the accelerator or transport system (such as degraders associated with fixed-energy accelerators or emittance filters), low-efficiency beam spreading techniques (large beam losses on collimators in the treatment room), techniques that call for large amounts of material in the beam inside the treatment room (inefficient scattering systems, range degraders near the patient). Where such beam losses are planned, local shielding should be designed and radiation levels outside the local shielding estimated. Neutron levels are of concern, as well as long-term activation of components and shielding [5.1].

5.2.7 Operation Costs

Operating costs for the facility must be kept at the lowest possible level (Chapter 13). The operating costs include the utility costs, maintenance costs (both personnel and parts), and operating personnel costs.

Radiation therapists are expected to control the beam to each room with a radiation physicist on-site to handle special problems. A full-time accelerator physicist should be on-call at all times but should not have to be routinely present during facility operations. A full-time engineer and technician, dedicated to operations and maintenance of the technical components, should be adequate for normal operations [5.1].

5.3 Protontherapy Equipment System

The core of a modular design is the division of the system into functionally distinct subsystems. Each subsystem performs well-defined functions, and they have straightforward and explicitly established relationships between them. In general, the following subsystems can be identified:

- accelerator,
- energy selection subsystem,
- beam transport subsystem,
- beam delivery subsystem (i.e. gantries, fixed horizontal beams)
- beam shaping subsystem (i.e. nozzles),
- patient positioners,
- control subsystem,
- safety subsystem.

Requirements, as well as existing and proposed solutions of above mentioned elements will be widely discussed in the Chapters 6 - 12.

5.4 Building Design

The proton facility building should house the protontherapy equipment and related program space such as clinical and administrative way [5.1]. But at the same time it is important to keep the clinical and technical areas separated. Although in all existing hadrontherapy facilities (with a partial exclusion of HIMAC) the technological (accelerator and beam transport) and medical parts are situated on the same level. A facility in which the clinical area is located at the different levels then the accelerator and services are probably the optimum. The advantages in terms of optimized patient flow and decoupled clinical and research / service activities should be balanced against the possible additional complexity of the building, shielding requirements, a more complex beam transport and overall costs [5.4].

As an example of the architectonic solution of Protontherapy center, we show the NTCP, which is located at Massachusetts General Hospital in Boston. The space of NTCP provide approximately 23 000 square feet of program space including three clinical treatment rooms, accelerator equipment, and offices. The treatment floor of the building is underground and pits to accommodate gantry rotation extend below that floor. The facility is designed so as to be capable of subsequent expansion to four treatment rooms, three of which can contain gantries, the remaining room being a fixed horizontal field room, which can be used for the treatment of head and neck sites and stereotactic radiosurgery.

The building layout is shown in Fig. 5.8. The lower level contains the accelerator, treatment rooms and some equipment support space. It also includes clinical space for patient examination and care. The ground level contains administrative, staff and miscellaneous support space as well as aspects of equipment support and clinical space.

Table 5.2 contains a summary of the gross square footage for the different areas of the building. This is compared to space requirements of other projects as reported in the TERA Project [1.4]. The percentage comparison of NPTC accelerator space does not include accelerator support labs. The NPTC space programming underwent a rigorous optimization procedure and the space remaining is probably as lean as possible. This is evident in the comparisons. However, the spread in the data is not large and it is interesting to note the similarly in relative program area among the projects [4.2].

The design of the *radiation shielding* walls around the equipment has a significant impact on both the cost and functionality of the facility. Reducing the mass of the shielding walls, while maintaining acceptable personnel and patient exposure levels, has the potential to substantially reduce the cost of the facility. The design of the treatment room and beam line entrances also effect ingress / egress times, costs and construction techniques.

Neutrons are the dominant secondary radiation field for high-energy proton facilities. The shielding walls must: (1) in some cases local stop the most energetic protons, (2) reduce to acceptable levels the secondary neutrons and gamma rays produced by the protons as they interact with the accelerator, degrader, beam analyzer, beam transport lines, gantries, nozzles, patients and shielding walls. Shielding is provided as part of a cost trade-off. Starling with the dose rate criteria established in the guidelines, and a model for the treatment parameters to be used at NPTC, the design of the shielding was an iterative process. Ultimately Monte Carlo based computer programs were used to refine the designs [4.3].

The examples of building designs are presented in Fig. 5.5 (Loma Linda facility) and 5.8 (NPTC).

	Area			
	NPTC	**HCL**	**LLUMC**	**TERA**
Treatment rooms	3	2	4 - 5	5
Experimental rooms	(1)	0	1	3
Beam lines	5	3	5	10
Footprint ft^2 (m^2)	22 000 (6 705)	15 400 (4 693)	51 700 (15 758)	38 000 (11 582)
Accelerator, BTS, gantry ft^2 (m^2)	4 580 (1 359)	4 305 (1 312)	7 750 (2 362))	13 455(4 101)
Treatment room ft^2 (m^2)	3 035 (925)	1 506 (459)	8 180 (2 493)	8 073 (2 461)
Total ft^2 (m^2)	44 000 (13 411)	23 250 (7 086)	82 882 (25 263)	88 264 (26 903)
Total / # beams ft^2 (m^2)	8 800 (2 682)	7 750 (2 362)	16 576 (5 052)	8 826 (2 690)
Accelerator / total (%)	10.4	18.5	9.4	15.2
Treatment room / total (%)	6.9	6.5	9.9	9.1

Table 5.2. Space comparisons of proton projects; after U. Amaldi [1.4] and J. Flanz et al. [4.3].

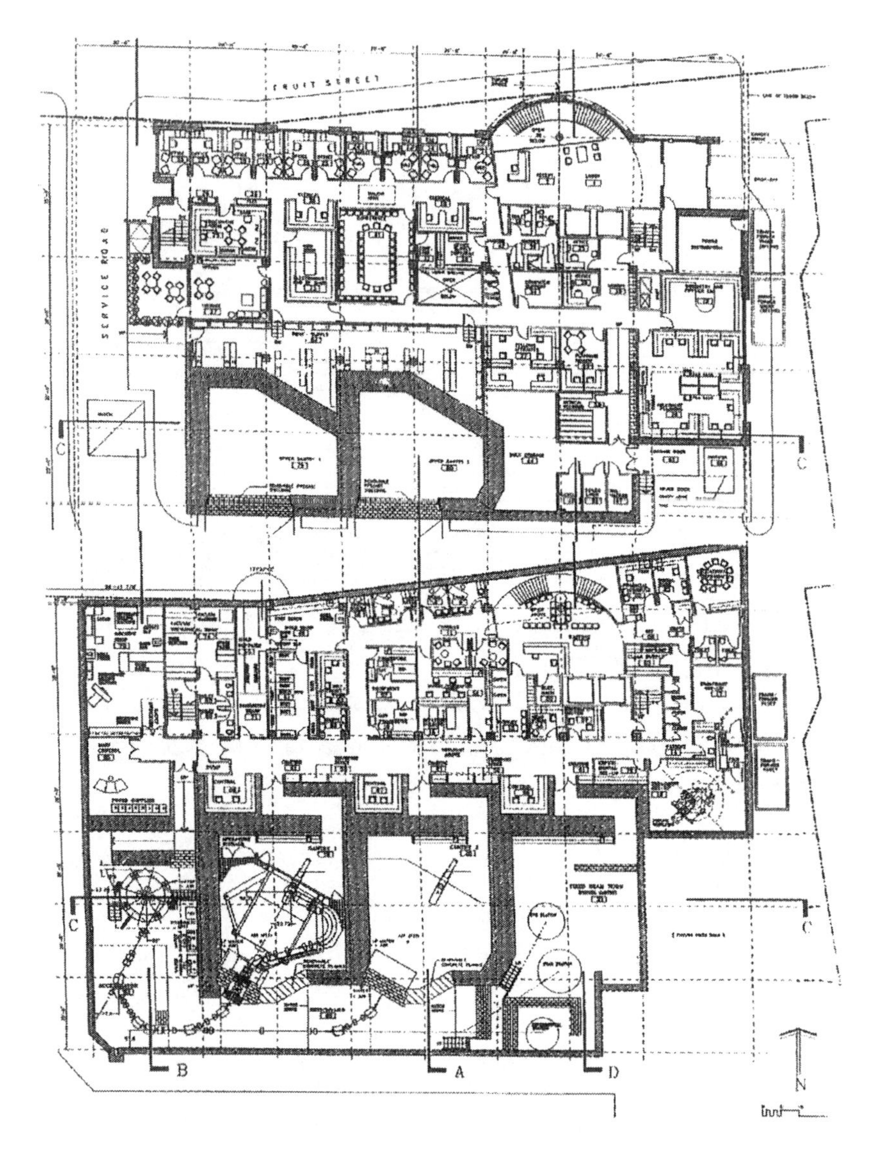

Figure 5.8. Layout of NPTC building including the lower level which includes the accelerator equipment and the ground level; after J. Flanz et al. [4.3]

Chapter 6

PROTONTHERAPY ACCELERATORS

First generation of accelerators used so far for proton radiotherapy was designed for high-energy physics studies. They have to be specially adapted to protontherapy, which often required quite complex technical procedures (Sec. 4.2.1). Accelerators of second generation employed for Protontherapy are those added to medical cyclotrons originally intended for the production of radionuclides and neutron therapy (Sec. 4.2.2). In addition to the existing accelerator facilities, a number of design studies have been undertaken for a *dedicated medical accelerator*. The three principal options for the accelerator designed specifically to operate in the hospital settings, are the cyclotron, the synchrotron and the linear accelerator. The recent increase of in the number of patients treated with this kind of radiotherapy reflects the increase in the number of operating dedicated facilities.

A hospital accelerator intended for protontherapy should deliver beam of accelerated protons with the maximum energy of at least 250 MeV. What is particularly important is that this energy should be easily controlled over a wide range of values because of the need to have the depth distribution quickly modified. For the dose rate of several Gy min^{-1} to be produced in the volume of 1000 cm^3 the beam intensity should be about 10^{10} p s^{-1}. Modern accelerator technology exists today to meet all of the clinical requirements (Chapter 5) within the reasonable budget for hospital-based hadrontherapy facilities. There is currently many debates concerning the optimum specifications and choice of accelerator for use in high-energy protontherapy. A hospital-based machine is rather different from a physics machine. The accelerator should run in a hospital environment without technical support services found in nuclear physics laboratories. Therefore, technological choices are particularly relevant.

The summary of main proton accelerator specifications will be presented in this Chapter. A comparison of parameters of accelerators used and proposed for protontherapy will be presented also.

6.1 Performance Specifications for Accelerators Systems

6.1.1 Energy

Energy Range. Based on the range in the patient and total amount of absorbing materials in the beam line (Sec. 5.1.1), the energy of extracted protons should be variable over the limits of 70 to 250 MeV at the exit of the gantry.

Time to establish new extraction energy. The treatment process may require that several stepped energies be available during a given treatment. The maximum treatment time is about two minutes and as many as 30 different energies may be required during a given treatment. For a synchrotron with a nominal cycle time of two seconds, taking even one pulse away from treatment for each energy to verify the new extraction conditions would add 60 s to the treatment, an unacceptable lowering of efficiency. Therefore, the newly selected extracted beam energy must be available on the first pulse at that energy. A method must be provided to verify that the correct energy has been provided, such as independent control of the beam transport system magnets with a design that has a reasonably small momentum aperture [5.1].

Energy Precision. The actual extraction energy of the machine must be accurate to within ± 0.4 MeV to satisfy the depth accuracy of the Bragg peak [5.1].

Energy Variability. The fineness of the energy variation affects the residual range of the beam in the patient. The range adjustment, obtained by varying the energy of the beam, must allow steps of 0.1 g/cm^2, or 0.05 g/cm^2 for ranges less than 5 g/cm^2 (Sec. 5.1.3). The beam energy may be determined by assuming the beam stops in water and applying the usual range formulas for protons. A resolution ± 0.4 MeV over the entire range should satisfy the range adjustment requirement [5.1].

Energy Spread. The energy spread should be ≤ ± 0.1 MeV FWHM at the exit of gantry at 100 MeV and up, measured with the beam monitors used during patient treatments. This requirement is driven mainly by momentum aperture of the beam transport system and gantry, although the diffuseness of the fall-off of the distal peak will also limit the acceptable energy spread [5.1].

Energy Variation of the Extraction. The energy dependence of the accelerator extraction system may cause a variation in the energy centroid during the extraction period if not carefully designed. The variation of the average energy of the beam during the spill or from spill to spill must not vary by more than ± 0.1% [5.1].

6.1.2 Beam Intensity

Beam Intensity. The average intensity should be a minimum of 10^{11} protons / s averaged over one cycle. Clinical specifications (Sec. 5.1.4) calls for a dose of 2 Gy in a 25 × 25 cm^2 field in 2 min. For a 2 Gy dose, about 5×10^{11} protons must be delivered to the most distal layer. If a scattering system is employed, about 1.8×10^{12} protons must be delivered within the specified field. Assuming a beam use efficiency of 20%, about 90 seconds are needed for the treatment at the specified maximum intensity. If the target volume is scanned (in all three dimensions) several times during the course of the treatment, fewer protons are delivered in one scan, and the required dose can be given to each layer in one pulse [5.1].

Time Structure. Dynamic beam spreading devices require a long duty factor. A CW regime or a duty factor close to 100% simplifies the controls of the accelerator and of the beam delivery system [6.1].

Macroscopic (Spill to Total Cycle Time) Duty Factor. Fraction of machine cycle in which beam is available should be ≥ 50%. Beam scanning and accurate beam intensity monitoring requires the highest duty factor possible to reduce the slew rate requirements of the scanning magnets and to minimize saturation effects in the beam monitors. A 10% duty factor is sufficient for a scattering system to avoid saturation effects in beam monitors and dosimeters [5.1].

Microscopic (rf) Duty Factor. Proton synchrotrons and cyclotrons produce a tightly bunched beam with a microstructure frequency of several MHz. Proton beams with rf structure are currently in use at LLUMC and at the Harvard Cyclotron, and the radiobiological response in tissue has been measured. The beam may be fully modulated by rf in the MHz range if it eases the design of the accelerator extraction system [5.1].

Undesired Beam Intensity. The raster scanning system may use variable scan velocity modulation to vary the dose over the treatment volume. The scanning system will sense instantaneous beam intensity and, along with required dose per voxel, will calculate the raster speed. The instantaneous beam intensity from the accelerator, in the most basic configuration, need not have an intensity modulation capability, but the undesired modulation of the beam must be held within certain limits described below. This will require fast feedback systems measuring the instantaneous beam intensity and correcting the extraction system parameters.

The modulation of the extracted beam is specified as a function of a windowing time. Raster variations are more permissible, as the finite size of the pencil beam integrates over the spatial variations of the dose. Slower variations are corrected by varying the sweep rate of the variable velocity sweep of the raster scanner. The limits of the sweep speed of the scanner, and the saturation effects in the beam monitors impose a maximum instantaneous extraction rate.

The limit on intensity excursions is given as a function of a time window in which the beam is integrated and compared to the average spill rate. For longer time windows specified in the Table 6.1, the integrated beam variation must be less than 20% in any time window, relaxing to larger variations permitted in shorter time windows. Ideally, one wants a dc beam with the instantaneous beam current within ± 2% of its average value during the spill. If a constant beam current were guaranteed, the modulation of the scan speed for shaping and optimizing the dose distribution could be calculated before the irradiation and performed without a feedback loop. For time intervals shorter than 25 μs a rapidly varying time structure is irrelevant since the beam spot will move less than 1 mm during that time at the highest scan speed. The scanning system specifications, which use a variable scan speed, should be controlled to ± 2% averaged over time intervals longer than 100 μs and to ± 20% when averaged over 25 μs to 100 μs. By measuring the current of the scanned beam's intensity fluctuations the dose may be controlled by varying the scan speed accordingly. The lower the undesired beam intensity fluctuations are, the easier it is to achieve the specified accuracy in the dose deposition.

The limits of the intensity excursion, or peak intensity, integrated within the specified window time for the instantaneous extracted beam rate are specified in the Table 6.1 [5.1]. No time interval of 25 μs may contain more than five times the average number of protons, or 5×10^6 protons, whichever is more, since neither the scanning system nor the beam abort can reliably react in less than 25 μs. The estimate for the tolerable number of protons is based on the assumption that if all the particles are deposited in one spot, it will not result in more than a 10% overdose in that spot for that particular scan. If a scattering system is used, the spill structure has no impact unless it prevents an accurate cutoff at the end of the treatment, or spikes in the spill, which lead to dosimeter saturation. No one millisecond interval may contain more than 10^9 protons or 0.1% of the total number of protons for the treatment, whichever is less.

Window time	Maximum excursion
Scanning method requirements	
> 20 μs	± 20%
200 μs – 100 μs	linearly rising to ± 100% at 100 μs
100 μs – 25 μs	linearly rising to excursions 5 × average spill rate, ior less than 5×10^6 particles, whichever is more at 25 μs
< 25 μs	no specification – will be contrlled by the r.f. structure
Scattering method requirements	
1 ms	no more than 10^9 protons, or 0.1% of the total number of protons in the treatment, whichever is less

Table 6.1. Scanning and scattering method requirements; after W.T. Chu et al. [5.1].

Beam Intensity Modulation within the Pulse. The minimum modulation requirement is that there be no modulation capability. The dose variations over the tumor volume will be provided entirely by velocity modulation of the scanned beam.

A more desirable capability, obtained by providing an upgrade path to the modulation capability, will be to modulate the beam intensity extracted from the accelerator over a range of 100:1 with a frequency range of dc to 5 kHz. A modulation capability within a beam pulse would enhance the capability of the scanning system to deliver specified dose distributions. The modulation capability would be nominally downward from maximum intensity, but with the ability to keep the feedback loops closed at the minimum intensity. The 100:1 range can include, if possible, some upward modulation in its range, such as a factor of 3 upward, and 30 downward over normal spill, for example. This modulation capability would extend over the range of 10^9 to 10^{11} protons / s peak, so the lowest controlled rate would be 10^7 protons / s at the lowest rate at the lowest peak intensity.

A modulation capability would not be useful with scattering treatment delivery systems [5.1].

Pulse-to-Pulse Selection of the Beam. The circulating beam intensity in the synchrotron must be variable over a 1000:1 dynamic range on a pulse-to-pulse basis with a $\pm 10\%$ accuracy at the 10^8 protons / s spill level, increasing to 2% accuracy at the 10^{11} protons / s average intensity level.

If a scattering system is employed, it will permit starting the treatment at a low intensity and to correct the beam centering at that intensity level before irradiating with full intensity.

Data will be provided to the accelerator control system of the required circulating beam intensity at extraction, no later than 0.1 s before the next linac injection pulse.

Various methods can be applied to reduce the circulating intensity, such as temporarily reducing the rf bucket size during acceleration, or by reducing the length of the injected beam pulse.

Beam Abort Time. In general, the beam issuing from the accelerator should cease within 10 μs after receipt of a trigger signal. This corresponds to approximately 100 turns of the circulating beam in the synchrotron ring.

This is an important safety parameter when using a scanning system. At full beam intensity, about 2.5×10^6 protons are delivered into an approximately 1 cm^2 spot area within 2 μs, corresponding to about 0.5 cGy. If the beam current during that time were 100× the average maximum value, a treatment could still be safely aborted. At most, a small spot would have received an unwanted dose of 5 cGy.

A beam abort time of 1 ms is sufficient for a scattering system since the beam is spread over a much larger area [5.1].

6.1.3 Quality of the Extracted Beam

Transverse Emittance. The transverse emittance of the beam determines the size of the magnets in the beam transport system, which is of particular interest in the gantry, where weight must be minimized, and in determining the size of the beam spot at the isocenter.

The beam emittance from the accelerator is determined by the nature of the extraction process in the extraction plane, and largely by the adiabatic damping of the beam from injection in the perpendicular plane. The unnormalized rms emittance of the beam extracted from the synchrotron would normally fall in the 0.2 - 0.5 π cm mrad range [5.1].

Position and Angle Stability of Extracted Beam. The variations in position and angle of the extracted beam must not cause position and size variations of the beam at the isocenter by more than 10% of the rms beam size at the isocenter either during a single pulse or from pulse to pulse at constant energy. To satisfy this, the position and angle of the transverse beam centroids exiting from the accelerator must not vary by more than ± 1 mm, or by ± 1 mrad during an entire spill cycle [5.1].

6.1.4 Accelerator Beam Monitoring

Monitoring of the beam circulating in synchrotron. The dynamic range and accuracy of the circulating beam monitor ensures safety and accuracy of the applied dose. The patient treatment system requires an accurate measurement of the available proton intensity for each spill to prepare the scanning for that spill. In addition, it is desirable to measure the beam centroid position in each straight Sec. of the accelerator for closed orbit measurement and control, for tune measurements, and for other diagnostic purposes. The usual single-turn monitors will be provided for initial tune up and injection studies [5.1].

Time to Recover from Various Shut-Down Conditions. Time to start up or shut-down conditions are specified in Table 6.2.

Item	Startup / shutdown time
Facility startup from total shutdown	1 day
daily operation startup to point where dosimetry can be done	1 hour
Control system startup so start and check computer	30 min.
Daily operation shutdown time to safe mode	15 min.
Facility shutdown and secure time	4 hours

Table 6.2. Time to recover from various shut-down conditions; after W.T. Chu [5.1].

6.1.5 Physical Characteristics of Accelerators

Size and Weight of the Accelerator. The size of accelerator fixes the size of the shielded room to house it, and the small footprint of the accelerator is certainly an advantage. The point might become relevant for the widespread development of protontherapy in the future where hospitals will look for accelerators, which could fit within the architectural plan of the radiotherapy department. Reducing the weight of the components of accelerator is certainly a challenge for the accelerator designers.

Acoustic Noise. The noise should be reduced in the hospital environment and power supplies and power transformers are a source of noise, which should be seriously considered.

6.1.6 Costs of the Dedicated Proton Accelerator

Costs of Dedicated Protontherapy Accelerator. It is estimated that the costs of the project of protontherapy facility should be 15% of the total costs of the center. The accelerator construction and installation costs should not exceed 15 – 30% of the total. In case of NPTC, roughly 26% was spent for the accelerator complex, the budget of the project was US$ 46.1 M [4.5]. The costs of construction the replica of LLUMC is estimated for US$ 55.4 M; 19% of this money is planned to be spent for accelerator design and construction [6.12].

6.2 Operating Principles of Proton Accelerators

6.2.1 Cyclotron

Conventional Cyclotron. The operating principle of a conventional cyclotron is shown in Fig. 6.1. An ion source located in the central part of the machine emits a beam of ions, which move inside two semicircular electrodes, called *dees*. The accelerated particles move in a nearly spiral trajectory which consists, strictly speaking, of two semicircles. The dees are installed inside a system of solid electromagnets. In a system of this kind, the condition of the equilibrium of forces is fulfilled between the Lorentz force and the centrifugal force according to Eq. (6.1).

$$vBq = \frac{mv^2}{r} \tag{6.1}$$

where r is orbit radius, B is the magnetic induction, v is ion velocity, q is the particle charge and m is the total ion mass.

The orbiting frequency of particles is then determined by Eq. (6.2).

$$f_R = \frac{v}{2\pi r} = \frac{(q/A)B}{2\pi m_n} \tag{6.2}$$

where A is the mass number and m is the mass of the nucleon.

Under the effect of an rf alternating voltage, an approximately uniform horizontal electric field or accelerating field is generated in the gap between the dees (Fig. 6.1). On the other hand, a constant magnetic field with induction of about 1.5 Wb m^{-2} acts throughout the whole chamber, bending the paths of the accelerated ions. The frequency of the accelerated voltage is constant, usually within the range of 10 – 30 MHz. The particle motion must be, naturally, fully synchronized with this frequency or its harmonic. Thus the single orbiting period T_0 must be also constant and independent of the particle energy:

$$T_0 = \frac{1}{f_s} = \frac{2\pi m_n}{(q/A)B} = const, \tag{6.3}$$

where f_s is the accelerating voltage frequency, synchronous with the particle motion, m_n is the mass of the nucleon, q is the ion charge, A is the mass number and B is the magnetic field induction.

Unfortunately, with an increase in ion energy a relativistic increase of mass occurs, which causes the process of acceleration to become desynchronized. The reason for this is that the accelerating voltage frequency in a conventional cyclotron is constant, or independent of time. As a result of the relativistic increase of mass, the particle becomes increasingly heavier and, consequently, cannot catch up with the changes in the accelerating voltage. The practical criterion of the desynchronization of the acceleration process is based on the kinetic energy gain E_k equivalent to an increase of about 1% of the rest mass. Taking into account the values of rest masses of various particles we can obtain the maximum energy to which protons may be accelerated in a conventional cyclotron $E_0 \approx 10$ MeV [1.22].

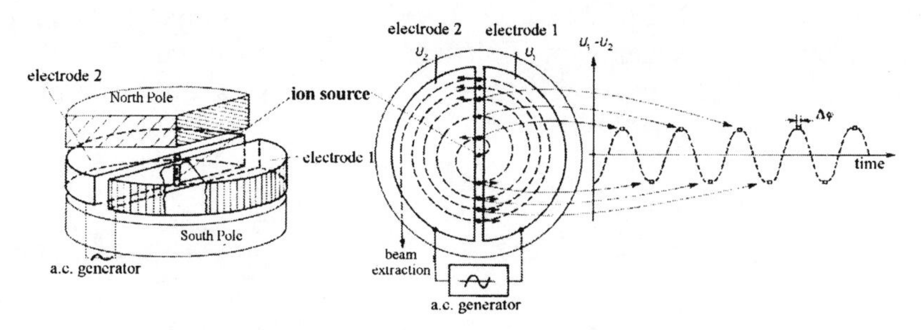

Figure 6.1 Operating principle of a conventional cyclotron; after W. Scharf [1.22].

The introduction of superconducting windings in the technology of the compact isochronous cyclotrons enlarged the capabilities for two reasons: 1) by overpassing the previous upper limit (1.4 T => 100 MeV p) to go in the range1.7 T to 5 T corresponding to 200 MeV to 450 MeV p, 2) by delivering cylindrical magnetic fields which can have the exact radial gradient needed to the proton mass increase, thus opening the door to very high proton energies machines and when you add a pair of coils in Helmholtz position you open the door to the Universal Compact Isochronous Cyclotron which AGOR is the only one representative.

Isochronous Cyclotron. In a conventional cyclotron, ions move in a uniform magnetic field. Therefore, at each point in an orbit of a radius r, exists the same induction B_r, independently of the value of the azimuth. In 1938, L.H. Thomas put forward an interesting modification of the cyclotron method, namely that of a totally different distribution of the magnetic field. In this new technique, the pole pieces are provided with segments forming *magnetic ridges* with gaps between sectors making up *magnetic valleys*. An ion, moving in an orbit of a given radius r, encounters sections whose magnetic induction is higher in the ridge than in the valley. Such a distribution of the magnetic field causes deformation of the particle trajectories compared with a spiral orbit in a conventional cyclotron. As a result, isochronousness is obtained over a much wider energy range. Even when a one percent increase of the relativistic mass is exceeded, condition is still valid and the ions remain synchronous with the accelerating voltage. In addition, the particles are subjected to a magnetic field with an alternating gradient and, consequently, strong focusing also occurs.

Since the particle orbit time is maintained constant, the *Thomas cyclotron* is often called an *isochronous* cyclotron. Cyclotrons of this type are denoted also by the acronym *FFAG* (*Fixed Field Alternating Gradient*), as the magnetic field is constant with time, although areas of the alternating field gradient occur along the azimuth.

The maximum unit energy or the energy, which can be imparted to a single nucleon is given by:

$$E_{kn} = k(q/A)^2 \tag{6.4}$$

$$k = \frac{R^2 B_{\max}^2}{2m_n} \tag{6.5}$$

where q is the ion charge, A is the ion mass number, R is the radius of the maximum orbit from which the beam is extracted, and B_{max} is the induction in the orbit *R*. The *k* coefficient, most often expressed in terms of MeV/A, has values usually ranging between 100 and 600. In the new designs of isochronous cyclotrons with superconducting windings, the *k*-coefficient can be as high as 1000 - 1500.

Pole pieces with magnetic segments or sectors, as in Fig. 6.2, are used in smaller isochronous cyclotrons with energies below 100 MeV. In larger cyclotrons, a segmental design is used in which spaces between the adjacent segments are void of the magnetic field, correspond to magnetic valleys. Instead of using an ion source located in the central part of the accelerator, in the segmental design, the beam, after being preaccelerated, is injected from the outside. The beam injection energy is normally of the order of several tens of MeV and is often obtained in another smaller cyclotron cooperating with the main cyclotron. A segmental isochronous cyclotron, i.e. one with separated sectors is denoted in the English language specialized literature by the acronym *SSC* (Separated Sector Cyclotron).

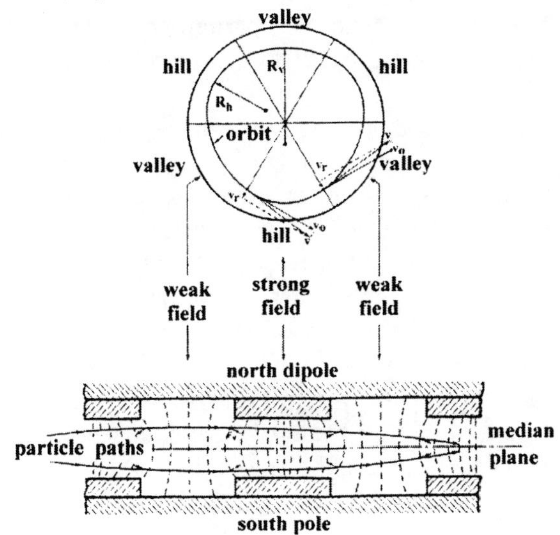

Figure 6.2. Operating principle of isochronous cyclotron; after T. Stammbach [6.2].

6.2.2 Proton Synchrotron

Fig. 6.3a shows a schematic of a *proton synchrotron*. Pre-accelerated particles are injected into an accelerating chamber equipped with a number of electromagnets (from several in small synchrotrons to several hundred in large machines) intended for bending the particle paths.

In synchrotrons, the acceleration proper is implemented with *resonators* (*gaps*) supplied with a rf voltage in a way similar to that in rf linear accelerators or cyclotrons. In proton synchrotrons the velocity of the protons or heavy ions which slowly gain speed varies widely during the acceleration cycle. It is for this reason that an alternating frequency of the accelerating voltage, i.e. frequency modulation has to be applied in addition to the magnetic field alternating with time. In contrast to synchrocyclotrons, in which the accelerating voltage frequency drops during acceleration, the proton synchrotrons feature an increase in frequency.

In a synchrotron (Fig. 6.3a), the particles are accelerated in an orbit with a constant radius ($r_s = const$). As the particle energy rises during the acceleration cycle, the magnetic field, which keeps them in a stable orbit, must also increase with time.

Fig. 6.3b is a graph of this field featuring a characteristic truncated triangle. This example refers to a small medical synchrotron with maximum proton energy of 250 MeV. The acceleration cycle is 2 s, and the acceleration proper lasts 0.5 s. During that time, the pre-accelerated protons, injected at t = 0 into in a stable orbit, gain a maximum energy of 250 MeV. At the time of injection the induction in the orbit is 0.118 T, and 0.5 s later, i.e. when the protons attain the energy of 250 MeV, it rises to 1.52 T. After this, the magnetic field remains stable for 1 s (the flat Sec. on the graph in Fig. 6.3b). This time, typical for synchrotrons, is referred to as the *flattop waiting period*, during which the beam of accelerated protons is gradually extracted (the bottom part of Fig. 6.3b). In the final stage of the cycle lasting again 0.5 s, the magnetic field falls from the maximum to its output value. Thus it is possible to obtain the repetition frequency of 30 cycles min^{-1}.

The conditions that have to be fulfilled for the synchrotron mode of accelerating particles can be presented in the following form:

$$B = f_1(t), \quad \omega = f_2(t), \quad v = f_3(t),$$

$$E = f_4(t) \sim \frac{B}{\omega_s}, \qquad r_s = f_5(t) = \frac{v}{\omega_s} \cong const. \tag{6.6}$$

where ω is angular velocity ($\omega = v/r$).

In large proton synchrotrons, the operation cycle ranges from several to several tens of seconds. Therefore, the frequency repetition of the subsequent acceleration cycles is rather small.

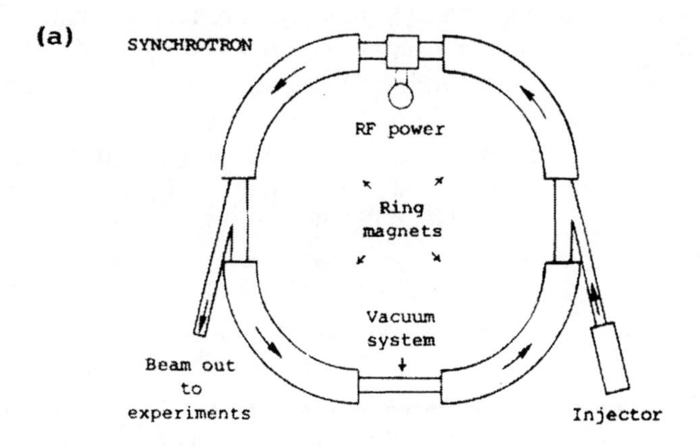

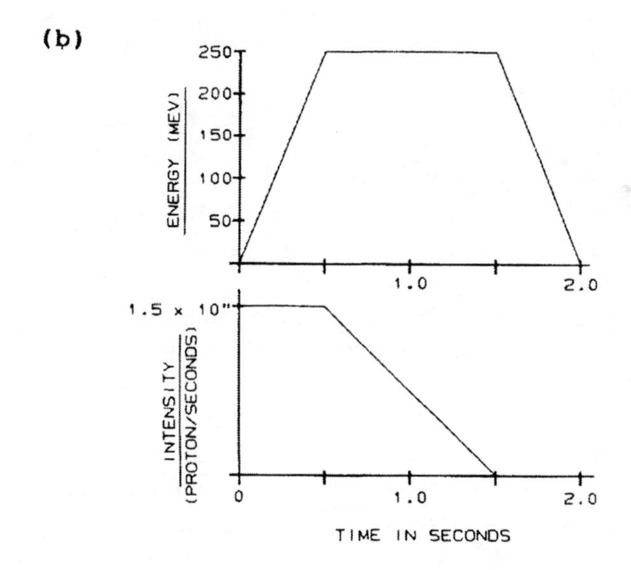

Figure 6.3. Proton synchrotron; a) magnetic system circuit, b) operation cycle of the 250 MeV medical synchrotron (Fermilab).

Although proton synchrotrons belong mostly to the domain of basic research, particularly that of high energy physics, they have also played an essential role in biomedical applications, and it seems that this role will become more important in the near future. Intensive work has been focused on the design and application of proton synchrotrons in radiotherapy. For protontherapy the maximum energy required is 250 MeV. In this range accelerated protons can be produced by both isochronous cyclotrons and proton synchrotrons. Heavy ion therapy will probably be conducted exclusively with synchrotrons since isochronous cyclotrons cannot ensure that high enough unit energies will be achieved. In proton radiotherapy, synchrotrons can be successfully competitive against isochronous cyclotrons, mostly because of the lower investment costs. Table 4.1 lists machines used up to now in experiments using heavy particles in therapy. All the facilities operated to date utilize the existing proton synchrotrons, initially intended for research in physics. They are, therefore, relatively complex in operation and designed for other than medical applications. Recently, this situation dramatically changed since, as a result of previous work, a small medical synchrotron has been built at the Fermilab for LLUMC (Sec. 4.2.3). Today, there are some projects of European and Japanese proton facilities (Table 4.3 and Fig. 4.14) based on proton synchrotrons.

6.2.3 Proton Linear Accelerator

A proton version of the conventional S-band standing-wave linear accelerator used for radiation therapy satisfies all the requirements for a dedicated protontherapy accelerator.

The proton beam produced in the source first arrives at a prebuncher. Prebuncher is intended for modulating the velocity of individual protons with the peak gap voltages so that the relatively delayed protons be accelerated more, and the protons with velocities higher than the average velocity be accelerated less. An rf resonator with symmetry corresponding to that of magnetic quadrupoles is used for this purpose. This system was referred to as an *rf quadrupole* and denoted by acronym RFQ. It combines the action of focusing and bunching the beam, in addition to acceleration proper. Focusing is ensured by a transverse electrical gradient, which is of particular importance since magnetic focusing is least effective for small velocities of ions. In an RFQ resonator equipped with four waveguides, longitudinal changes occur in the transverse gradient thanks to the wave-like modulation of the axial shape of the wave-guides. As a result, a longitudinal electrical component is formed, which ensures acceleration and bunching of particles. The acceleration is of continuous character and, therefore, it does not occur only in electrode gaps, or other structures. Equally strong is the effect of bunching, its efficiency being close to 100%. In practice, when the structure is 1 - 2 m long, protons can be accelerated from the initial energy several tens of keV to final energy of several MeV. RFQ resonators are now being increasingly often used in ion pre-acceleration. They constitute an injection system for subsequent stages where further acceleration is carried on.

The next component of the accelerator system is the proper accelerating structure. Linear proton accelerators are usually equipped with the standing-wave accelerating structures. A typical π/2-type structure equipped with circular diaphragms is originally employed in accelerators with a standing wave. A waveguide was constructed which had *side-coupled resonators* (*SCDTL*). These resonators act as elements coupling the accelerating resonators or cavities proper, hence the name - resonance coupling.

Up till now, any of the existing projects of the therapeutical proton linear accelerator was not practically realized. The Italian Institute of Health (ISS) decided on the construction of a proton linear accelerator for its TOP project; this machine will accelerate protons up to energies 7 – 65 MeV (Sec. 6.5.2.1).

6.2.4 Synchrocyclotron

Both in the conventional cyclotrons and in the isochronous cyclotrons, the accelerating voltage frequency is constant. The effect of the relativistic increase of mass can be compensated for, to some extent, by varying the frequency of the accelerating voltage synchronously with the increase in mass. The cyclotron of this type, in which the accelerating voltage frequency is modulated, is referred to as a *synchrocyclotron*, a *frequency-modulated cyclotron* or, in Russian terminology, a *phasotron*.

The fundamental equation, which describes the particle motion in a magnetic field, can also be presented in the following form:

$$\frac{v}{r} = qc^2 \frac{B}{E} = \omega_s \tag{6.7}$$

It follows from this equation that ions can be accelerated when:

$$B = \text{const}, \quad \omega_s = f(t), \quad E = f(t) \propto \frac{1}{\omega_s} \tag{6.8}$$

In the above ω_s denotes the particle angular velocity ($\omega = v/r$). Eq. (6.8) describes the operation of a synchrocyclotron.

Since the accelerating voltage frequency is time dependent, the particle moves in step with the accelerating voltage when its relativistic mass increases. In this way, the main energy limit imposed on a conventional cyclotron is eliminated. Although the maximum particle energies theoretically attainable in the synchrocyclotron may be as high as several billion electron volts, those attained in practice are mostly limited by economic reasons. This is because the maximum energy, E_{max}, in the synchrocyclotron is proportional to the square of the electromagnet radius and the large dimensions of the electromagnet involve increasing high costs. An additional disadvantage of synchrocyclotrons as compared with isochronous cyclotrons is the relatively low intensity of the accelerated beams. This is because the isochronous cyclotron operates in a continuous mode whereas the synchrocyclotron, due to the frequency modulation of the accelerating voltage, operates only in a cyclic fashion.

In spite of the fact that synchrocyclotrons do not seem to have much future in biomedical applications, they have played an important part in the past and are still quite popular. The Protontherapy centers based on synchrocyclotrons are the first generation facilities (Sec. 4.2.1, Table 4.1). The most important synchrocyclotron-based Protontherapy facility is HCL. Any of new projects of dedicated Protontherapy facilities is not based on synchrocyclotron.

6.2.5 Advantages and Disadvantages of Accelerators Used for Protontherapy

6.2.5.1 Advantages and Disadvantages of Protontherapy Cyclotrons

Since its early history in Berkeley, the cyclotron has been used for medical purposes. The choice of a fixed energy makes the design of cyclotrons relatively simple. The magnet can be optimized and trimming coils are not necessary, The accelerator is at a fixed radio frequency and all the settings of the beam lines are fixed. In order to reduce the dimensions of the cyclotron, a high magnetic field should be chosen. This choice has two important consequences: (1) it is impossible to accelerate negatively charged ions at high energies in a high magnetic field due to electromagnetic stripping, (2) superconducting magnets are attractive because they are lighter than room temperature magnets (about half of the weight) and the running costs are lower.

An isochronous cyclotron provides a continuous beam (cw) with ample intensity. This is a key factor for reliability. The beam intensity is easily controlled. The beam can be very stable and this is an important advantage for a dynamic beam spreading system. To fully benefit from these advantages, an external injection system makes the cyclotron very flexible.

In case of cyclotrons, no sophisticated controls are needed. In general, a simple programmable logic controller is sufficient. The operation requires less complexity and this reduces the manpower costs.

The cyclotron is a cw-machine in which acceleration of particles takes only 10 - 20 μs. Therefore the beam can be turned on and off rapidly by acting on the injection process. Furthermore, it is possible to setup dynamic control of the beam intensity with reference to a fast ionization chamber in the extracted beam, with the effect that the beam can be stabilized at any required level with response times lower than 100 ms. Therefore, the beam intensity could be strictly constant higher than 1% over short and long periods. This process could be fully automatic, a separate safety circuit ensuring that no malfunction of this servomechanism can overdose the patient. In the case of a detected failure, the beam can be turned off in less than 10 ms [6.1].

There are now an increasing number of projects for high-energy protontherapy aiming at treating any tumor types at any depth. The IBA company has recently installed two dedicated proton cyclotrons in NTCP, Boston, USA and in NCC, Kasiwa, Japan. We will present the examples of dedicated designs of cyclotrons in the Sec. 6.3.

6.2.5.2 Advantages and Disadvantages of Protontherapy Synchrotrons

The proven technology of synchrotrons ensures the reliability and a lot of expertise exists in several institutions throughout the world. A synchrotron produces pulsed beams at variable energies. The energy of the extracted beam can be varied from one cycle to the next in steps of a few MeV Hence, the necessary modulation of the Bragg peak to scan the target volume in depth can be achieved without absorbers. The weight of the different magnets (dipoles, quadrupoles) is relatively low. This facilitates the transport and the assembly. In general, no particular basements are needed in the building.

The required intensities of the proton beam can be achieved by a synchrotron fed by an adequate injector. Nevertheless, the traditional resonant extraction system is subject to small perturbations due to small variations in the excitation of all magnets that induce a time modulation of the extracted beam intensity. An alternative is non-resonant extraction as in H^- synchrotron, which uses a charge-exchange extraction process.

The operation of synchrotrons is complex because rapid magnetic field variations are requested. Moreover, undesired intensity variations require careful controls. Nevertheless, many sophisticated synchrotrons are running smoothly for high-energy physics reaching a high reliability level.

The synchrotron has variable energy. This eliminates the degrader with its associated scattering and momentum spread. Hence the target volume is discretized into slices of equal range. Each slice is scanned independently and then the beam energy is changed. With a synchrotron, this change could be made at the accelerator. The synchrotron beam comes in bursts and could be typically present for about half the time (slow ejection process). Therefore the scanning system should operate typically twice as fast as for the beam coming out of a cw accelerator (cyclotron). This requires more power in the magnets of the magnetic scanner.

6.2.5.3 Advantages and Disadvantages of Proton Linear Accelerators

For protontherapy in the 200 MeV range, the linear accelerator is the third candidate. Ambitious extrapolations of electron linear accelerator technology, hence using a high frequency structure, are proposed to shorten the length in view of reducing the costs for a hospital-based facility. The output energy from linear accelerator is fixed. Nevertheless it was proposed to adjust the energy by switching off the power of short-tank modules of the last section. of the linear accelerator structure. This concept requires a careful optimization of the structure, and detailed calculations of transverse emittances. As the levels high, the linear accelerator is pulsed and delivers short pulses. This time structure has important consequences for the beam delivery system, in particular for any dynamic beam spreading device. The spot scanning (discrete) is of course well-suited for a pulse length. In addition, the possibility of varying the output on and off the rf power in the short tank of the high-energy structure (the last section), opens the possibility of three-dimensional spot scanning, since the final output energy can be varied for each pulse.

Comparison of the main parameters of the accelerators proposed for protontherapy is presented in Table 6.3.

	Cyclotron	**Synchrotron**	**Linear accelerator**
Vendors	IBA (Belgium)	Fermilab / Otivus Technology (USA), Hitachi (Japan)	ENEA (Italy)
Intensity (protons s^{-1})	5×10^{13}	1×10^{11}	5×10^{14}
Possibility of intensity control	very good	average	good
Energy and energy variability	70–250 MeV fixed (energy degraders must be applied)	70–250 MeV continuous	70–250 MeV fixed (energy degraders must be applied)
Energy stability	not yet available	$\pm$ 0.1%	$\pm$ 0.1%
Impulse length (ms)	continuous	$(50\text{-}1)\times10^{-4}$	1000
Injection energy [MeV]	0,01-0,1	1-5	
Quality factor (>25%)	very good	very good	average
Beam emittance (unnormalized, 67% the beam)	large (10 π mm-mrad)	large (1-3 π mm-mrad)	small (0.1 π mm-mrad)
Average current[nA]	20-40	20-40	10-270
Power [kW]	~ 500	~ 150	~ 350
Duty factor (% beam on time)	100% or CW	20% at 0.5 Hz	0.1% at 300 Hz
Beam intensity uniformity (for scanned beams)	good	adequate [16]	good
Dimensions (m^2)	small (25) diameter ~6 m	medium (50) diameter ~10 m	big (150) length ~150 m
Extraction yield (%)	~ 40	~ 85 - 95	~ 100
Induced radioactivity	high	low	low
Control possibility	good	average	poor
Possibility of new options	big	big	poor
Safety	good	good	poor
Costs	average (~ \$US 10 mln)	average (~ \$US 10 mln)	large

Table 6.3. Comparison of protontherapy accelerators parameters; after [1.1], [1.4], [1.8], [3.16].

[16] Beam intensity uniformity for active beam generally requires intensity regulating feedback loops to achieve uniform intensity. First demonstrated use of the scanning beam system occurred at the Berekley synchrotron in 1992. Adequate spilll uniformity was achieved using intensity control feedback loops.

6.3 Examples of Cyclotrons for Protontherapy

6.3.1 A Room Temperature Design Cyclotron – Cyclone 235

An isochronous cyclotron is probably the simplest and most inexpensive way to produce 235 MeV protons. This technology is well-known and widespread as evidenced by the large number of similar and smaller machines installed.

The cyclotron operates cw (continuous wave). In contrast with a synchrotron, the machine parameters are constant in time and only five (possibly seven) of them are adjustable: main coils current, rf voltage, ion source current and voltage and electrostatic deflector voltage (possibly also two harmonic coil sets). None of these parameters requires operator tuning during machine operation or during the morning start-up: reference values are stored by the control system and fine-tuned using well-developed computer algorithms. Fast computer algorithms are not required, as there is no time-critical tunings but only the need to compensate for slow drifts.

The high intensity, continuous extracted beam can be intensity controlled from the ion source within 15 μs turn on / turn off time, and only 30 μs separate the time of the beam leaving the source and reaching the patient. This beam is, for these reasons, perfectly suited for beam scanning and wobbling. The high available intensity easily allows respiration gating without extending treatment times, even for very large fields.

The cyclotron extracted beam is characterized by a fixed energy, a low energy dispersion and a fixed, low emittance. The accelerator is unable to produce an extracted beam with incorrect energy, energy dispersion or emittance pattern.

The cyclotron splits at the median plane allowing easy maintenance: the upper yoke can be raised in minutes by hydraulic jacks, giving free access to all machine components. Thanks to the moderate vacuum requirements and the generous pumping system, full energy beam can be resumed less than 30 minutes after the start of pump-down.

The drawbacks of a fixed energy accelerator-degrader combination have been limited or eliminated as follows:

- **Energy straggling.** The inferior distal fall-off caused by energy straggling in the degrader has been eliminated by the addition of an energy selection section located just downstream of the energy degrader. Local movable shields are provided to reduce the radiation caused by the energy defining slits.

- **Emittance growth.** The beam emittance degradation at the energy degrader is controlled by divergence-limiting slits located just downstream. Those divergence limiting slits are also fitted with local, movable radiation shields. Large acceptance optics in the beam transport system and gantries contribute to minimize the beam losses.
- **Neutron production by lost beam.** Problems of secondary neutron fluxes and shielding, among others, were quantitatively addressed. They are strongly attenuated by a proper choice of equipment materials and concrete shielding walls thickness.

The design principle of Cyclone 235 is presented in Fig. 6.4. The summary of data for Cyclone 235 are presented in Table 6.4.

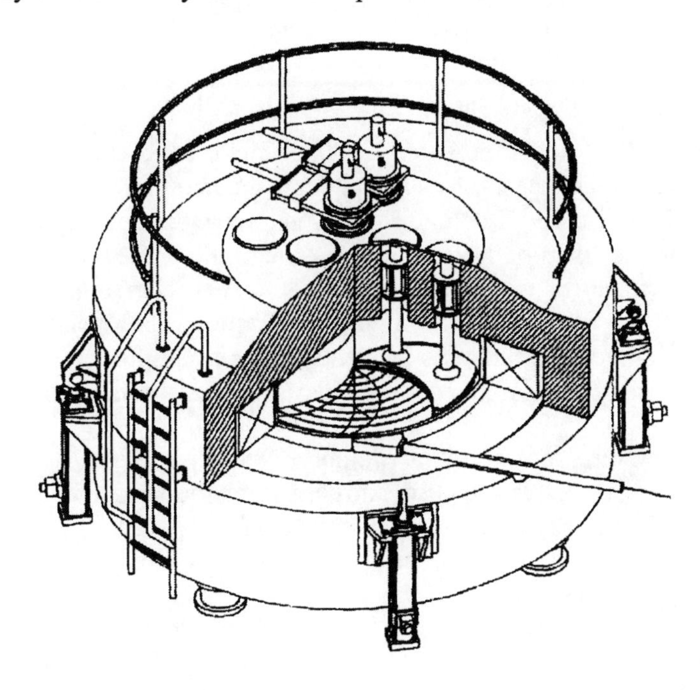

Figure 6.4. Ion Beam Applications 235 MeV room temperature cyclotron; after P. Mandrillon [6.1].

Energy (fixed)		235 MeV
Extracted beam current:	maximum minimum	300 nA 1 nA
Beam turn on/turn off time by ion source(20% to 80%)		15 μs
Transit time from ion source to patient		30 μs
Calculated extracted beam emittance unnormalised (less than)		3π mm mrad
Energy spread (one sigma) (Optimum machine tuning)		750 keV
External magnet diameter		434 cm
Total magnet height		210 cm
Total weight		220 tons
Electrical power consumption full beam extracted		420 kW
Vacuum on stand-by condition		8 kW
Control system allowing full automatic, unattended accelerator system operation, including automatic start-up, self-tune functions, and stop.		

Table 6.4. The Cyclone 235 summary data; after D. E. Bonnett [4.4].

In normal day-to-day operation of a fixed field, fixed energy cyclotron, only three parameters need to be adjusted to maintain optimum operation:

- The main coil current, defining the average magnetic field level in the cyclotron, needs to be fine-tuned to maintain the proper phase between the accelerated (or extracted) beam and the rf accelerating voltage. Although the power supply is stable and reproducible within a (1 to 2) $\times 10^{-5}$ range, fine tuning is needed to compensate for temperature induced mechanical variations, day-to-day hysteresis non reproducibility, etc.

  To lock the beam to rf phase, a non interceptive rf pick-up electrode is located on the extracted beam and reads both the beam current and the beam to rf phase. A dc voltage proportional to the phase effort is fed back as a correction signal to the main coils power supply.

  In order to automatically find the beam in the morning or after a long interruption, the P.L.C. automatically sweeps the main coils current around the expected value until the beam is found and the phase loop is locked.

- The rf dee voltage, defining the exact number of accelerating turns and the exact phase of the radial betatron oscillation (and therefore the exact radial turn profile at the entrance of the electrostatic deflector) needs to be fine tuned to optimize the extraction efficiency. A computer algorithm is used to keep the extraction efficiency peaked at all times.

- The proton beam current, and thus the dose rate, is quite proportional to the ion source arc current. Closing a feedback loop on the beam current is easy but presents risks in case of mistuning. While we keep this possibility open, we presently favor an approach in which the dose rate versus arc current table is frequently updated (for instance daily) and is used to accurately set the dose rate.

The continuous nature of the cyclotron operation and the small number of adjustable elements effectively simplify the design of the cyclotron control system.

To control the cyclotron, the beam transport and switching system and, possibly, the gantry magnets and their power supplies, a high level, multiprocessor, centralized control system is used, very similar to the system used on all IBA cyclotrons. The system handles both logical and analog parameters. Its main functions are the handling of all machine and safety interlocks, storing, retrieval and resetting of machine parameters, performance of closed-loop automatic beam tuning algorithms, machine logging and support of troubleshooting (including remote, modem-based troubleshooting). The system is designed to run completelys unattended and has demonstrated its ability to do so on existing cyclotrons.

The system can be operated either locally from a control room located next to the power supply room (used mainly for initial machine development and for maintenance), or from any terminal connected to the PLC through the network with appropriate access control. The interface essentially consists of graphical color screens, representing individual subsystems graphically, or more global system synthesis views. System control is essentially performed by means of a mouse although "virtual knobs" (knobs connected to optical encoders) are provided and can be software connected to various machine control parameters for those preferring the feel of manual tuning. The quality of the graphical presentations is a significant aspect contributing to the machine uptime: any abnormal condition is highlighted to immediately catch the operator's eye and cause a prompt and appropriate response.

In normal operation of the Protontherapy facility, the accelerator and beam transport system (BTS) will run autonomously. The cyclotron and BTS control system will then appear as a peripheral equipment, called by the irradiation control stations, delivering beam on a "first come, first served" basis with an appropriate request queuing routine [4.4, 6.7].

6.3.2 Compact Superconducting Cyclotron Design – CAL / Siemens

Fig. 6.5 shows a compact superconducting sector-focused cyclotron capable of supplying a continuous beam of 238 MeV protons. It has three spiral sectors with a supplementary groove along the center of each sector, making it more like a six-sector machine. Three rf cavities, galvanically connected in the central region, provide a large energy gain per turn in order to complete the acceleration in about 400 turns. Table 6.5 gives the main characteristics of this design, which was made by Siemens and The Cyclotron Laboratory of CAL in Nice. Protons are produced by an internal ion source [6.1].

A compact Pantechnik isochronous cyclotron PK210 (Fig. 6.5b) project is based on the high-temperature superconducting wires, produced by American Superconductor Corp., with a working temperature of 29 K, just surrounded by a multi-layer super insulation. Fig. 6.6 shows the basic principles of this system compared to a standard helium liquefier. The use of this technique is possible because Pantechnik has designed the cyclotron with relatively low magnetic field (2,01 T in the central region) in order to simplify the ejection problems. With such a field we are below critical current of the HTS (High Temperature Superconductors) for the temperatures in the range of 25 to 35 K. The cryostat around the coils, and all the peripheric equipment, as quench security valves, helium gas tank, are simplified [6.14, 6.15, 6.16].

In the classical cyclotrons a hot filament P.I.G. source introduced axially through the magnet yoke, is generally used. Pantechnik has patented the concept of an internal ECR (Electron Cyclotron Resonance) source for the beam production using main magnetic field of cyclotron. This source has a infinite life time and is very easy to operate with a very low power, high frequency, solid state generator connected to the source through a wave guide ended by a vacuum window. The main parameters of the PK 210 cyclotron are summarized in the Table 6.3.

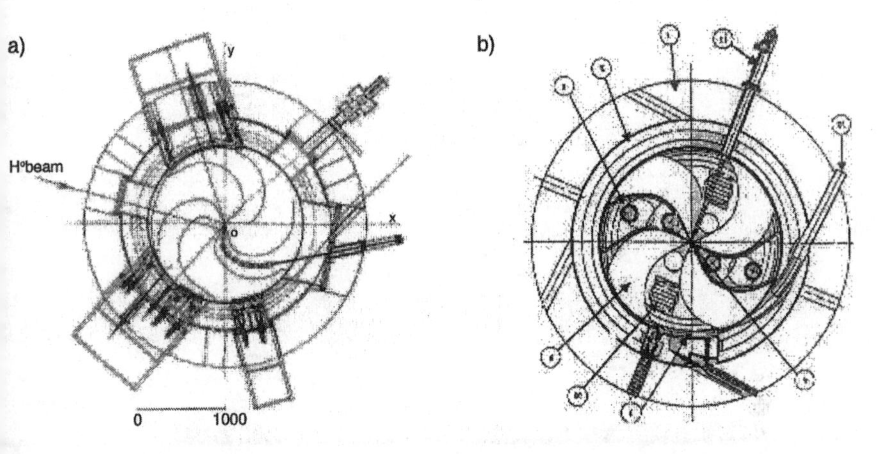

Figure 6.5. a) The CAL / Siemens – 238 MeV superconducting cyclotron; after P. Mandrillon [6.1]. b) View of Pantechnik PK 210 cyclotron in the median plan; after C. Beith, A. Laisne [6.14].

Parameter	CAL design	PK 210
Number of sectors	3, i.e. "pseudo six sectors"	
Yoke external diameters	3.1 m	3.26 m
Outer radius	1.6 m	
Height	2 m	1.72 m
Current density in superconducting coil	6800 A / cm^2	
Total weight	80 tons	88 tons
Power including refrigerator	35 kW	
Frequency / Harmonic mode	110 MHz / h = 3	123.4 MHz / h = 4
Dee voltage injection / extraction	80 kV / 130 kV	60 kV / 140 kV
RF power	120 kW	80 kW
Injection	two options: axial or radial (neutron beam)	ECR source

Table 6.5. The characteristics comparison of the Siemens and The Cyclotron Laboratory design of CAL cyclotron in Nice and PK 210 Pantechnik cyclotron; after P. Mandrillon [6.1, 6.13] and W.T. Chu [6.8] and C. Beith, A. Laisne [6.14].

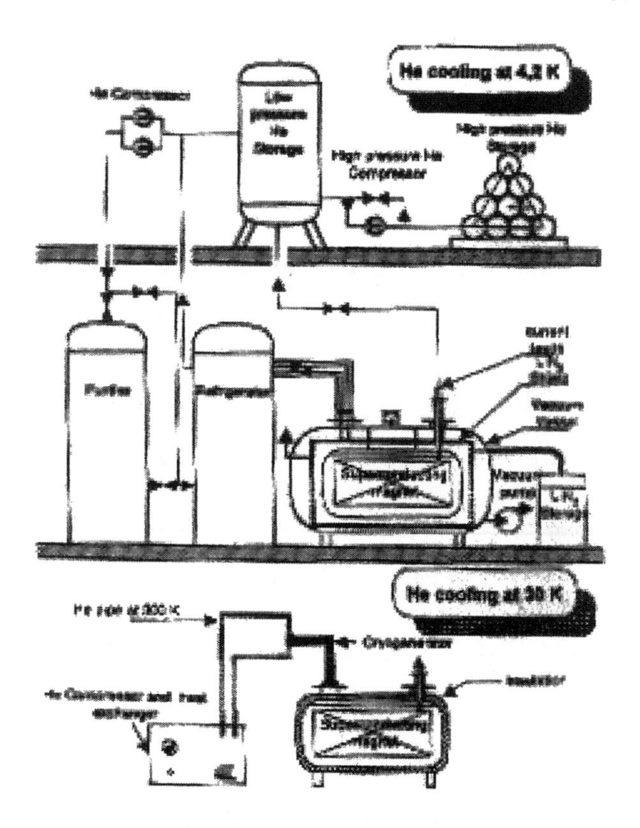

Figure 6.6. Principle of Pantechnik cyclotron cooling circuit compared to the standard helium liquefier; after C. Beith, A. Laisne [6.14].

6.4 Examples of Synchrotrons for Protontherapy

Weak Focusing LLUMC Synchrotron. The Loma Linda University Medical Center houses a proton synchrotron (Fig. 6.7) giving a variable energy from 70 up to 250 MeV (Sec. 4.2.3).

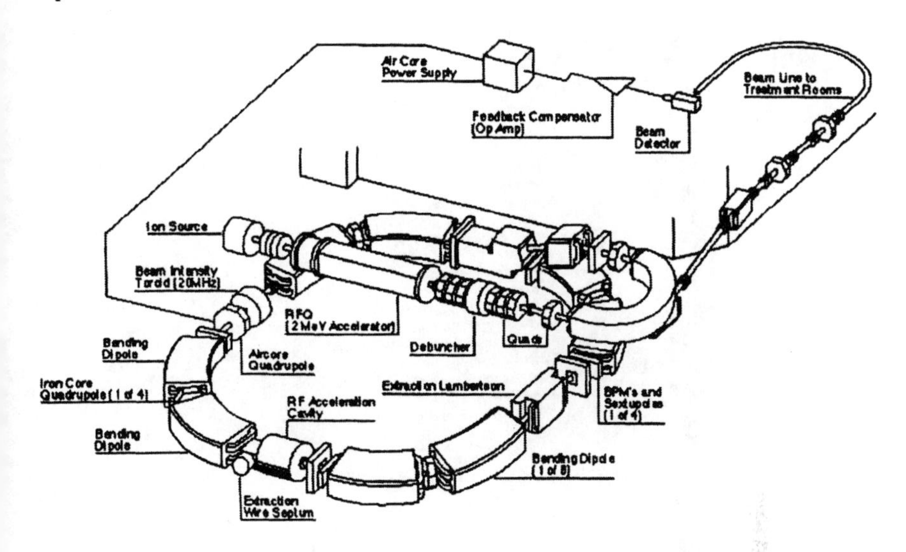

Figure 6.7. Loma Linda synchrotron with intensity feedback control; after G. Coutrakon et al. [6.3].

This zero gradient synchrotron has eight 45° dipole magnets arranged to have four straight sections providing space for injection, acceleration and extraction systems. The 30 keV proton beam from the duoplasmatron source is injected into a 2 MeV - RFQ operating at 425 MHz. Acceleration to 250 MeV is obtained via simple ferrite-loaded rf cavity, operating on the first harmonic. The beam is extracted from the synchrotron by using half-integral resonance. The cycle time is 2, 4 or 8 s, and the extraction time is variable from 0.4 to 10 s. The design intensity is 10^{11} protons / s. This design is a result of a collaboration between the Loma Linda University Medical Center, the Fermi National Laboratory and the Science Applications International Corporation (SAIC) Company [6.3].

H^- Synchrotron. A very compact synchrotron operating with pulsed high magnetic fields has been designed at the Budker Institute of Nuclear Physics in Novosihirsk (Russia). This design is one of the three possible candidates for a compact accelerator studied by the TERA collaboration (Fig. 6.8). The TERA collaboration in Italy also proposes a multitask synchrotron for its Hadrontherapy Center, which would be able to provide: (1) 250 MeV proton beams by accelerating H^-, hence simplifying the extraction but requiring a low magnetic field and a very good vacuum in order to avoid electromagnetic stripping of the H^- ions.

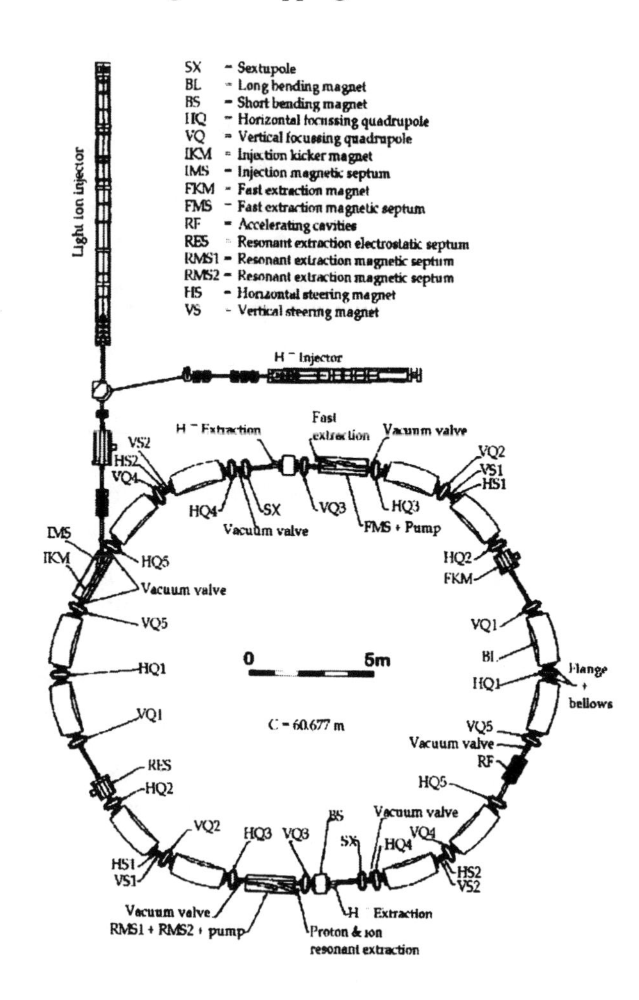

Figure 6.8. Layout of the TERA synchrotron; after G. Arduini et al. [6.10].

Table 6.6 summarizes main parameters of the existing and proposed proton synchrotrons for radiotherapy.

		Existing synchrotrons			Synchrotrons proposed								
Facility		LLUMC [1]	PMRC [2]	ITEP [3]	TERA [4]	KUMPF [5]	BPTC [6]		PMRC [2]*	Argonne [7]	PIMMS [8]	IUCF [9]	PRAMES [10]
Synchrotron type		weak focusing	strong focusing	10 GeV	type H^-	with multitask magnets	strong focusing	weak focusing, combined functions magnets	slow cycle, separated functions magnets	type H^-	slow extraction ion synchrotron	sector synchrotron with combined function	
Circumference (Diameter) [m]		20	24,95		60,677 (19.31)	22,9	34,4	20,776	34,939	43,039 (13.7)	74.04 (23.57)		41 (13.05)
Energy [MeV]	injection	2	5		11	10 keV	3,5	3,5	2		20	7	
	extraction	70 – 250	10 - 230	70-200	60-250/300	10-230	70-200	70-200	230	250	60 – 250	200	60 – 220
Injector type		RFQ	Linac	Linac	Linac	Linac	RFQ	RFQ	Pelletron	Pelletron		?	
Beam intensity [ions s^{-1}]		10^{11}	$6 \sim 12 \times 10^{10}$		$1,3 \times 10^{10}$	$1,3 \times 10^{11}$	$8,7 \times 10^{10}$	$12,3 \times 10^{10}$	$3,1 \times 10^{10}$	10^{10}-10^{11}	10^{11}		6.25×10^{10}
Repetition rate [Hz]		2	0.5		2	0,5	0,2-2	0,2-2	0,5-1	1		1	1
Average current [nA]		10	10		10	10	5-10	5-10	20			?	10
Extraction type		slow	slow		charge exchange resonant	slow	slow		slow			fast	slow

Table 6.6. Parameters of existing and proposed proton synchrotrons for radiotherapy, after [1.1, 6.9, 6.10, 6.11].

[1] Loma Linda University Medical Center, Loma Linda, California, USA
[2] Proton Medical Research Center, University of Tsukuba, Tsukuba-shi, Ibaraki-ken, 305, Japan
[3] Institute of Theoretical and Experimental Physics, Cheremushkinskaya, 25, 117259 Moscow, Russia
[4] TERA Foundation, Via Puccini 11, 28100 Novara, Italy
[5] Kyoto University Medical Proton Facility, Japan
[6] Bejing Protontherapy Center, China
[7] ACCTEK Associates, 901 S. Kensngton Avenue, La Graange, Illinois, USA
[8] PIMMS synchrotron parameters when protons are accelerated
[9] Indiana University Cyclotron Facility
[10] PRAgue MEdical Synchrotron
* planned

6.5 Examples of Linear Accelerators for Protontherapy

6.5.1 Standing Wave Design

6.5.1.1 AccSys Technology Project

The use of a conventional S-band standing-wave electron linear accelerator structure to accelerate low-current proton beams up to 250 MeV for protontherapy was first suggested in 1991 by the AccSys Technology Company [6.4] in the USA (Fig. 6.9). Specifications for this compact proton linear accelerator, the AccSys Model PL-250, are listed in Table 6.7.

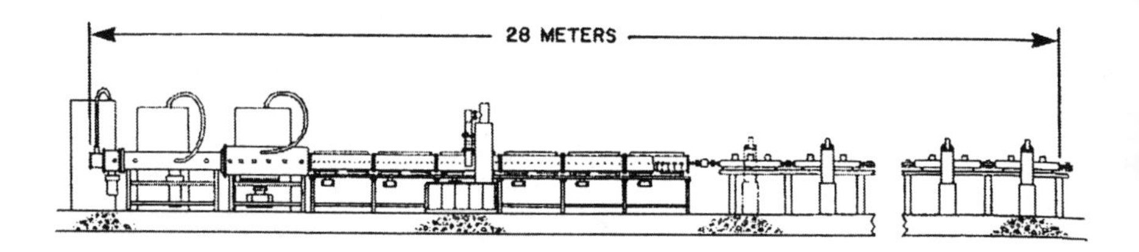

Figure 6.9. Schematic layout of Model PL-250 Protontherapy linear accelerator; after R.W. Hamm et al. [6.4].

Accelerated particle	H^+
Clinical beam energy	70 - 250 MeV
Peak beam current	100 – 300 μA
Beam pulse width	1 – 3 μs
Repetition rate	100 – 300 Hz
Average intensity	10 – 270 nA
Beam energy spread	± 0.4%
Maximum input power	350 kW
Stand-by power	25 kW
Accelerator length	28 m

Table 6.7. Preliminary specifications for a dedicated Protontherapy linear accelerator; after R.W. Hamm et al. [6.4].

6.5.1.2. TOP – ISS Project

The *Italian National Institute of Health* (*ISS*) has decided (Dec. 1995) on the construction of a proton linear accelerator for its *TOP* (*Terapia Oncologica con Protoni*) project. It is close to the compact high frequency linear accelerator, envisaged by the TERA Foundation for protontherapy. The TOP linear accelerator will be composed of a 428.3 MHz 7 MeV injector, comprising an RFQ and a drift tube linac (DTL) Sec. followed by a 7 – 70 MeV Sec. of the innovative 3 GHz side coupled DTL structure (SCDTL), a 180° bending magnetic system, and a 70 – 200 MeV variable energy side coupled linac (SCL) 3 GHz structure. This machine will be the first linear accelerator dedicated to protontherapy and the first 3 GHz proton linear accelerator ever built (Fig. 6.10, Table 6.8). One of the advantages of this high frequency linear accelerator is the small beam emittance (< 0.25 π mm mrad), which allows the use of a lightweight gantry and small transport elements. It should weight about 16 tons (compared to 90 tons gantry used in Loma Linda).

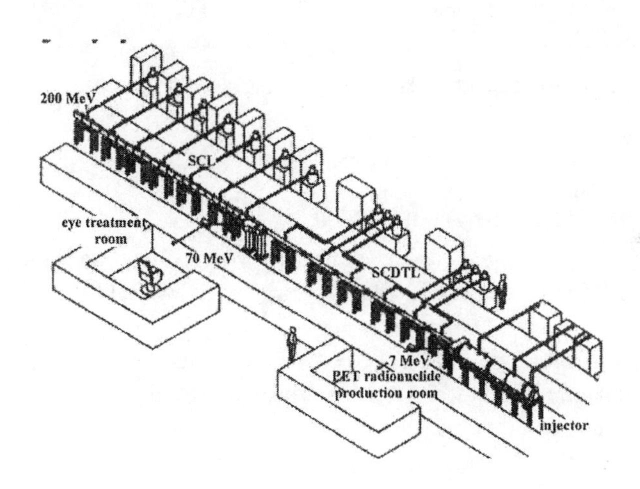

Figure 6.10. The TOP – ISS 3 GHz linear accelerator; after [5.5].

Main injector parameters			
	Protontherapy	PET	Radiobiology
Energy [MeV]	7	7	≤ 7
Maximum average current [μA]	0.1	100	0.001
Repetition frequency [Hz]	400	100	10 – 400
Pulse duration [μs]	5	100	5
RF frequency [MHz]	428.3 / 750	428.3	428.3

Main parameters of accelerating structures		
	SCDTL parameters	SCL parameters
Frequency [MHz]	2998	2998
Input / output energy [MeV]	7 – 65	65 – 200
Lengh [m]	9.98	11.4
Number of modules	4	7
Average field E_0 [MV / m]	12	15.5
Effective accelerating field E_0T [MV / m]	8.5	13.5
Average energy gain [MeV / m]	5.8	11.8
Peak power (25% margin included) [MW]	6.8	26

Table 6.8. Main parameters of TOP – ISS linear accelerator, after L. Picardi et al. [6.6].

6.5.2 Booster Linear Accelerator

Clatterbridge. A classical linear accelerator structure has recently been proposed as a booster for the 62 MeV cyclotron SCANDITRONIX MC60 of the *Douglas Centre* (Department of Oncology) in Clatterbridge, UK, permitting to implement high-energy protontherapy program at 200 MeV. The planned length of the designed structure is 19.9 m.

LIBO. Because several hospitals and laboratories possess proton cyclotrons with output energies of 60 to 70 MeV, a highfrequency (3 GHz) booster linac (LIBO) is proposed to upgrade the cyclotron beam to an energy of 200 MeV, sufficient to treat deep seated tumors. LIBO is a side-coupled linac (Fig. 6.11), which can produce beams with a variable output energy. LIBO studies are based on the Clatterbridge cyclotron, but the operating frequency has been pushed into the S-band in order to reduce the accelerator size and increase its breakdown limit. LIBO is a side-coupled linac (SCL) operating at 2998 MHz. It is composed of 36 tanks, separated by permanent magnetic quadrupoles (PMQs). Four tanks are grouped into a module, an RF unit, fed by its own RF chain. The average output beam required from LIBO is of the order of 10 nA, and it must be possible to vary its energy from about 130 to 200 MeV. This constrains both the number of tanks in a module and their length. The 3 GHz klystrons used to power the tanks operate in the pulsed mode with pulse lengths of up to 5 μs, but it is should be narrow enough to limit the distal fall off of the dose given to the patients to ≤ 2 mm. A beam pulse repetition rate of 400 Hz has been chosen to suit an active beam scanning method (e.g. pixel scanning). Preliminary tests have been performed to pulse the Clatterbridge cyclotron beam to study its use as a LIBO injector [4.64, 6.17].

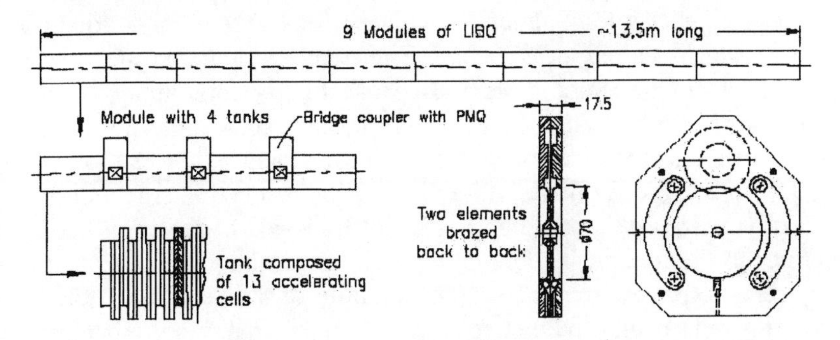

Figure 6.11. Schematic layout of LIBO; after U. Amaldi et al. [6.17].

Chapter 7

BEAM TRANSPORT AND DELIVERY SYSTEMS

The extracted beam is transported from the accelerator to the treatment room by the *beam transport system*, a series of dipole and quadrupole magnets. The arrangement of magnets, vacuum chambers, and diagnostic instrumentation is called a *beam line*. The rotating gantry is included as a part of the beam transport system. Because of his complexity it will be presented in Chapter 8. The beam delivery system, located at the end of the beam line usually in the treatment room, modifies and monitors the beam to achieve the prescribed radiation dose distribution inside the target volume.

Beam transport. A stable and efficient transport of the beam from the accelerator to the treatment room is required for reproducible dosimetry and thereby reliable patient treatments. The stability of the centroid of the beam position must typically be better than 1.0 mm. This requirement places constraints on the stability of the bending and focusing magnets needed to control the beam position and profile. The ease of adjusting the beam through the beam line, in what is called *tuning a beam line*, and the reproducibility of the beam-line tune are critical for efficient and reliable clinical operation. Beam tuning means adjusting the beam optics to transport the given beam to the desired location with the desired parameters at that location [4.1].

Beam delivery systems. The beam delivery system monitors the beam and shapes it in three dimensions to irradiate the target volume. Beam delivery can be divided into two categories: (1) beam widening techniques and (2) conformal techniques. Beam widening techniques give a homogeneous dose distribution in a field of the appropriate size and shape, combined with range modulators and shaped collimators to achieve a flat, regularly shaped high dose volume enclosing the target volume. For an individual beam it is unavoidable that also some adjacent normal tissue is included in the high dose volume. Conformal techniques give three dimensionally shaped high dose distributions conformal to the target volume (Fig. 7.1) [7.1].

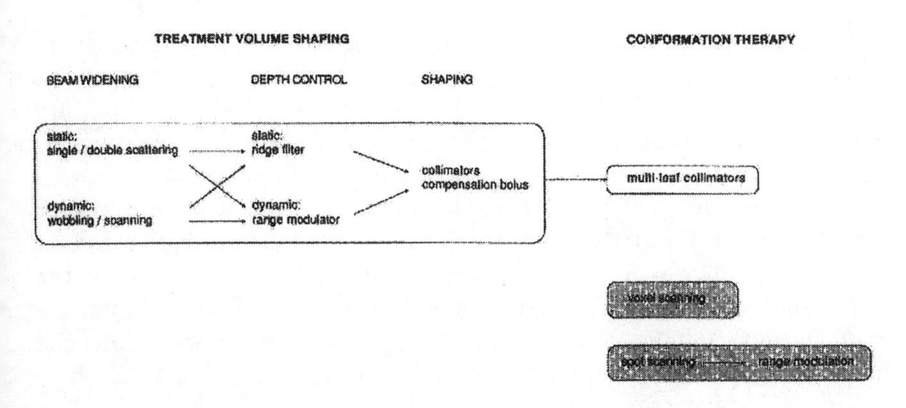

Figure 7.1. Beam delivery systems, after H. Blattmann [7.1].

The treatment volume can be fitted in two dimensions in the beams eye view to the shape or the target volume by various combinations of beam widening and depth control techniques. As the depth modulation is uniform over the whole field, some or the volume elements outside the target volume are exposed to the full target volume dose. The relevance of this unwanted exposure is reduced by multi-port irradiation. For conformation therapy the dose distribution is shaped in three dimensions for each port. For multi-port irradiation this results in an optimal reduction of the integral dose. With help of a multileaf collimator and a bolus to fit the distal shape or the dose distribution, the two dimension shaping or the beam cross Sec. can be developed into a three dimensional shaping. The same result can be achieved directly, without patient specific hardware, by voxel or spot scanning [7.1].

In a typical treatment room the beam, after passing through a vacuum window at the end of the beam transport system, travels through several kinds of devices and several drift spaces (path lengths the beam travels unmodified) before entering the patient. These devices change the beam range (Sec. 7.5.1), modulate the range (Sec. 7.5.2), spread the beam laterally and shape its lateral profile (Sec. 7.5.3). Dose measuring devices and beam monitoring devices (Sec. 9.2) are required for control of the radiation.

A general beam line layout places a set of dose detectors and beam monitoring devices near the patient and at a position before the beam spreading device, as shown in Fig. 5.6. The beam spreading devices are usually located as far from the patient as is practical to minimize the effective source size and the angular divergence of the beam. The drift space interspersed between the beam modifying and monitoring devices in a clinical beam line is typically 3 m or longer. One meter of a drift space in air is equivalent in stopping power to ~ 1.2 mm of water. Fragmentation of the heavier ions in this air path is negligible; however, multiple scattering can be significant for lighter ions, such as protons and helium ions.

The parameters of beam delivery systems in fact lead to competing requirements on the design and placement of the beam line elements. An optimization process involving several parameters with clinical tradeoffs is often required. In general, the maximum field size, the field uniformity, the treatment time, the beam divergence, the lateral dose penumbra of the field, the background radiation, and the beam fragmentation are important factors to consider. Furthermore, different clinical requirements necessitate different designs, which generally fall into two categories: namely, beam lines for small and large radiation fields. Large radiation fields require more complex systems for laterally spreading the beam. The small beam spot extracted from the accelerator must be modified to cover areas as great as 40×40 cm^2 with a dose uniformity within ± 2%, and the dose rate on the order of 1-2 Gy min^{-1}. The beam lines for small fields are simpler because they can often directly utilize the beam from the accelerator with minor modification, probably using a simple scatterer [4.1].

7.1 Effects of Material in the Beam Path

Any material in the beam path can potentially modify the beam because the material scatters the beam (multiple scattering), smears its energy (range straggling), and fragments some portion of it (nuclear fragmentation).

7.1.1 Multiple Scattering

Multiple scattering of the beam can be described in terms of small angle deflections of the beam particles due to interactions with the nuclei of the traversed material. These numerous small angle deflections lead to a divergence of the beam and to a lateral spreading of the particles away from the central trajectory. Elastic Coulomb scattering dominates this process with a small strong interaction scattering correction. The angular distribution of the scattered particles is roughly Gaussian for small deflection angles.

For protons, the widths of lateral falloff increase essentially linearly with the residual range of the beam (Secs. 2.3.3 and 3.1.1). The sharpest dose falloffs are obtained when the final collimator is at the surface of the patient and when the shortest range is used. For a small beam, it is rarely necessary for the final collimator to be further than 10 cm from the patient surface. For treating small targets, where the sharpness of the lateral dose falloff is essential, the choice of the heavy charged particle beam becomes important [5.1].

7.1.2 Range Straggling

Range straggling is the dispersion of the path length of a particle beam due to statistical fluctuations in the energy-loss process. The end result is to produce a smearing of the range of the stopping particle beam. For example if protons with a range of 30 cm pass through 25 cm of water, the variance in the range becomes 2.6 mm, or the distal dose falloff distance from 90% to 10% dose level is 4.4 mm, which may be unacceptable in certain clinical applications [5.1].

7.1.3 Beam Fragmentation

As a particle beam penetrates through matter the primary particles suffer fragmentation collisions, which decrease the number of primaries with the corresponding increase of lighter fragments. *Beam fragmentation* refers to the process whereby the beam particle, called the *projectile nucleus*, after suffering a nuclear collision with a target nucleus is broken apart into several daughter particles. The remnants of the projectile nucleus emerge from the absorbing material with similar momenta as that of the original projectile nucleus. The target nuclei may also fragment, but these fragments have relatively lower energy and momentum and do not travel with the beam. The probability that a particle will undergo such a nuclear interaction has an exponential dependence on the length it travels. A *nuclear interaction length* λ_I is defined as the length in the material in which an interaction will occur with a probability of 1/e.

For protons colliding with a water-like target material (e.g., tissue) the dominant interaction products are neutrons knocked out of the target nuclei. These neutrons are not easily stopped and contribute to the dose delivered beyond the stopping region of the primary projectile [5.1].

7.2 Performance Specifications for Beam Transport System

7.2.1 Beam Parameters

Beam energies and transmission. The transport system must be able handle beam energies from 60 MeV to 250 MeV. The transport system guides the beam to the tap-off points, where the beam is sent to the gantry or fixed beam room. The transfer matrix between these tap-off points is unity, so the beam parameters are identical at each tap-off point. This will allow identical optics in each beam delivery system. To ease the power supply regulation requirements and help insure a stable beam when changing energy, all bends should be fully achromatic. At the gantry entrance a suitable rotating vacuum seal allows a continuous vacuum with no material in the beam up to the exit of the last bending magnet in the gantry [5.1].

Continuous vacuum from accelerator to last window. All material in the beam line must be minimized to preserve the low beam emittance.

Beam parameters at the isocenter in treatment room. The beam characteristics must be within the specifications listed in the Table 7.1.

Description	Specifications
Spot size	$\sigma_x, \sigma_y < 3$ mm
Spot deviation from circle	$\lvert\sigma_x, \sigma_y\rvert < 0.3\ \sigma_x$
Divergence	$\sigma_{x'}, \sigma_{y'} < ½$ radian
Spot position accuracy	$\Delta x, \Delta y < 1$ mm
Angular accuracy (maximum allowable deviation of actual beam central axis from nominal axis	$< 0.2°$

Table 7.1. Beam specifications at the isocenter, after W.T. Chu [5.1].

Beam parameters at the gantry entrance. At the rotation point of the gantry entrance, the transport system must provide a round beam with identical emittance and betatron functions in both planes. This is to prevent emittance dilution and eliminate retuning when the gantry angle is changed.

7.2.2 Beam Switching and Tuning

Time to tune beam between treatment rooms. Beam at the correct energy must be available for treatment in a new room no more than one minute after treatment is completed in another room. In addition to requiring that the new parameters for the accelerator and beam transport system be rapidly called up and established, the accuracy of beam position in the new room must be suitable to ensure meeting the field uniformity specification [5.1].

Automatic beam tuning. After the initial daily setup and calibration no manual tuning or beam centering at the isocenter should be necessary when switching treatment rooms or changing the beam energy or gantry angle. It should be possible to check the tune of the beam transport system up to the gantry entrance beam stop. Automatic beam steering must ensure adequate centering to minimize beam loss along the transport system and compliance of the beam centroid and angle specifications at the gantry exit. The beam monitoring and beam abort conditions apply for automatic beam tuning. As no manual tuning is possible with the patient in place, any feedback system that centers the beam must operate very quickly and at low intensity. The entire beam transport system must track the accelerator pulse-to-pulse energy variation without retuning or test beams. This implies that the magnets may require field sensors. For any gantry design the reproducibility and stability of the beam tune require achromatic beam optics [5.1].

7.2.3 Beam Diagnostics, Monitoring and Safety

Beam diagnostics and monitors in beam transport system. Diagnostic elements for facilitating a fast and straightforward beam tuning procedure must be provided throughout the beam transport system. Parameters to be measured at one or more points are the beam intensity, the beam centroid, the beam profile, and the emittance. The details of the required monitoring are listed in the Table 7.2. Also indicated in the table is the need for floating jaws. Monitors must also be provided to verify the proper workings of systems, such as vacuum, radiation, temperature, and other systems [5.1].

The energy aperture of the beam line is likely to be fairly small due to the large dispersion at some points, and the energy measurement may be monitoring the beam centroid at the high dispersion points. The beam emittance growth must be minimized. Non destructive monitors that are active all the time are preferred. They should be enable auto-tuning as well as beam monitoring during treatments. Otherwise, beam monitors are required which introduce a minimum amount of material into the beam during tuning and can be withdrawn from the beam line during treatment. As the monitors will be frequently used, the mechanical insertion system must be designed for long lifetime (welded stainless vacuum bellows, for example) [5.1].

Location	Intensity	Centroid	Profile	Emittance	Jaws
Accelerator exit	+	+	+	+	
Exit of initial matcher		+			
nπ points along spine		+			
Collimator along spine	+	+	+		+
Max. dispersion points		+			+
Gantry rotation points		+		+	
Gantry exit	+	+	+		
Beam stop exit	+				
Beam dump	+	+	+		

Table 7.2. Locations of beam monitors and floating jaws, after W.T. Chu [5.1].

Materials in the beam line. The material in the beam line due to monitors must be minimized. The intensity, emittance and energy spread depend on the extracted energy. Monitors that place material in the beam increase the bare machine emittance and affect the beam spot size. The effect of every monitor in the beam line must be evaluated in terms of their accumulated effect on the total phase space at the isocenter and the ability to achieve the required beam-spot size [5.1].

Beam halt / abort. Several layers of patient protection must be provided by beam stops and treatment aborts. The most important safety feature is the ability to switch the beam off in less than 10 μs after an error condition has been detected. Independent operation of several treatment rooms requires mechanical beam plugs or beam stops which must completely stop the beam and render the treatment rooms safe for access. They should move into place in less than seconds, or preferably between pulses for a synchrotron-based facility [5.1].

Monitoring of beam in transport system. Monitors for a fast ($t < 10$ μs) detection of beam misalignment due to failures in the accelerator or beam transport system must be provided. Active collimators made of scintillator material at proper locations may be considered. All beam monitoring devices which are inserted in the beam transport system during treatments must be designed so that they do not increase the beam parameters beyond the maximum values compatible with the specifications in the Sec. 5.1.

In the treatment room, i.e., in the nozzle, the proton intensity is monitored in three locations independently, the centroid position of the beam is monitored in two locations, and the beam profile is monitored at the exit of the gantry and close to isocenter [5.1].

Monitoring of beam transport system components. The beam transport system must be monitored both by software based systems and by hardware at all times while a treatment or a calibration procedure is in progress [5.1].

7.3 Performance Specifications for Treatment Beam Line (Nozzle)

The nozzle comprises the devices, which laterally spread the beam and modulate the range of the particles and the measuring instruments, which serve as beam monitors, dosimeters and safety devices. The spreading of the particles, in depth and laterally, can be done in a variety of ways ranging from purely passive methods using scattering foils and range shifting devices to actively steering a narrow pencil beam by magnetically deflecting (scanning) it and changing the range of the particles many times during a treatment by changing the accelerated beam energy. If the treatment volume is scanned horizontally, vertically and in depth with the Bragg peak of a very narrow pencil beam and the scan is performed with a high enough precision, the target volume can be filled and the desired dose distribution delivered without using collimators or compensating boluses.

The scattering and scanning methods are fundamentally different. They require different beam monitoring, dosimetry and safety systems. Specifications for both methods are given in this chapter. Specifications are for a treatment room with a gantry unless otherwise specified.

7.3.1 Specifications for Nozzle Using Scattering

Scattering system. The specifications for the scattering system itself follow directly from the clinical specifications for treatment depth and width (Secs. 5.1.1-5.1.3), field size (Sec. 5.1.4), SAD (Sec. 5.1.11), dose compliance (Sec. 5.1.9), and penumbra (Sec. 5.1.6). The basic system consists of sets of two scattering foils. The first one is a "regular" high Z material scattering foil whereas the second one is a combination of a low and a high Z material shaped in such a way that the average scattering angle is largest at the center but the energy loss is independent of the distance to the center. Each set generates a uniform field of a given diameter for specific beam energy. The number of needed sets of scattering foils as well as the detailed design parameters depend not only on the clinical specifications but also on the number of operational beam energies, available beam currents, and the number of desired field sizes with less than the maximum diameter. For conforming the Bragg peak dose in 3D to the target volume and performing variable-modulation treatments a multileaf collimator has to be added to the scattering system. The conformation is achieved by irradiating the target volume in layers, which are shaped by the multileaf collimator and stacked in depth. The range shifting should be achieved preferably by varying the accelerator energy, however, in the initial phase of operation of the facility it may be accomplished using a mechanical device [5.1].

Beam monitoring. Beam positions, profiles, and other attributes must be measured and monitored at specified position on the beam line during the beam tuning, calibration processes and actual treatments as specified in Table 7.3.

Description	During tuning	During calibration and treatments
Beam position (centroid) at first scatterer	± 0.5 mm	± 0.5 mm
Beam profile at first scatterer (resolution)	1 mm	1 mm
Beam position at second scatterer (resolution)	± 0.5 mm	± 0.5 mm
Beam position (centroid of unspread beam) at a distance of less then 1 m from isocenter	± 0.5 mm	-
Compliance of the radiation field with the expected distribution measured at a distance of less than 1 m from isocenter	-	± 2%

Table 7.3. Beam attributes measured and monitored at specified position on the beam line when scattering system is used, after W.T. Chu [5.1].

Dosimetry. Three independent dosimeters that will provide continuous monitoring of the dose delivered to the patient are required. One dosimeter has to be located at a distance of less than 1 m from isocenter. One dosimeter will not saturate at the highest possible beam current focused into a $3 \times 3\ \mathrm{mm}^2$ area. At least two dosimeters must measure the total beam current, whereas the third one may measure a well-defined fraction of the total field. It should be the center portion and not be smaller than 5 cm in diameter [5.1].

Beam modifying devices. For variable modulation treatments the range of the protons must be changed during the treatment. The step down in energy from layer to layer is preferably done by changing the accelerator energy, but if that is not feasible in the first phase of operation, a range shifter is to be employed. The range shifter must cover 16 g/cm^2 in 0.5 g/cm^2 steps, and must be designed to satisfy the distal dose falloff requirement (Sec. 5.1.7). In order to achieve a good penumbra (Sec. 5.1.6) it is important that the nozzle design will allow for the closest possible positioning of the beam modifying devices like patient collimator, compensating bolus, and multileaf collimator to the patient. The range shifter must either be located upstream where the beam diameter is small in order to achieve a small apparent source size or it must be located immediately upstream of the patient collimator.

For fixed Bragg-peak modulation treatments mechanical devices such as a propeller (modulator wheel) have to be provided. Depending on accelerator and beam properties it may be possible to integrate the modulating devices with the first scatterer. Otherwise, the effect of the position of the modulator wheel on the penumbra must be carefully evaluated. The manufacture of propellers, in particular their design, size, modulation resolution and construction are strongly influenced by the design of the nozzle. Typically the construction material is water like. The size is larger than the beam size at the location of the device and may be as large as the entire radiation field. Example designs of filters for proton facilities are propeller-like with widths, which cover from 1 to 16 cm usually in 0.5 cm steps [5.1].

Safety. Whenever a beam monitor or dosimeter reading is out of tolerance action must be taken; in most cases the beam must be shut off immediately. The signal generation and therefore the readout of the beam monitoring devices and dosimeters must be fast enough to ensure that, under all circumstances, the beam can be cut off before more than 0.5% of the total treatment dose has been delivered anywhere in the treatment volume after the error condition has been identified [5.1].

Monitoring of positions of beam-modifying devices. All devices which can be physically moved, like scattering foils, beam monitors, etc., and can possibly interfere with the beam and alter the delivered dose distributions must be interlocked or monitored for correct positioning in software and hardware [5.1].

7.3.2 Specifications for Nozzle Using Scanning

The scanning system must be designed to modify the beam and deliver the treatment fields specified in the clinical specifications for penetration depth (Secs. 5.1.1-5.1.3), field size (Sec. 5.1.4), SAD (Sec. 5.1.11), dose uniformity (Sec. 5.1.9), and penumbra (Sec. 5.1.6). By employing a 3D scanning pattern the scanning system must be able to deliver a dose distribution, which comply with the prescribed (in general non-uniform in a given depth of treatment) dose distribution as closely as possible.

Dose compliance. Dose compliance is a measure of the difference between the prescribed dose distribution and that, which would be delivered in the absence of imperfections in the scanning system, the beam-current control, and patient motion. Dose compliance evaluation must take into account patient immobilization, the time needed for each scan, the number of redundant scans, the control of the beam position and of the beam intensity at the target, etc. The dose compliance must be such that for each point inside and outside the target volume, the dose is within $\pm 2.5\%$ of the intended value [5.1].

Permissible beam spot size. It is desirable to have the steepest possible dose gradient at the lateral edges of the field. The lateral dose gradient for a scanned beam depends mainly on two factors, the width of the scanned pencil beam entering the patient (σ_b) and the multiple scattering in material including the patient. Assume the beam has a Gaussian profile , and σ_b adds in quadrature to the σ_{scatt} due to multiple scattering:

$$\text{lateral penumbra (80\% - 20\%)} \qquad \sim [\sigma_b^2 + \sigma_{scatt}^2]^{1/2} \qquad (7.1)$$

The broadening of the beam due to multiple scattering is sensitive to the amount of material in the beam: the beam monitors, vacuum windows, and air, and also depends on the beam energy.

The maximum permissible beam diameters at isocenter are: $\sigma_{bmax} = 6$ mm at a beam energy of 250 MeV, and $\sigma_{bmax} = 10$ mm at 120 MeV where σ_b is the width (80% - 20%) of the beam in air at isocenter [5.1].

Length of scanning magnets. The shorter are the scanning magnets, the easier they can fit into the gantry nozzle and the longer becomes the SAD for a given drift space in the gantry. The clinical specifications for the field size must be satisfied by designing magnets and power supplies accordingly. In general, the total length of a scanning magnet must be less then 0.5 m [5.1].

Voxel scanning system. It will be a challenge for any voxel scanning system to satisfy the clinical requirements of irradiating the largest treatment volumes in the specified treatment time while providing a small enough beam spot for satisfying the penumbra and dose compliance specifications. The largest treatment volumes can contain up to 80,000 voxels of $5 \times 5 \times 5$ mm^3. This implies that the time available for moving the beam to the next location and depositing the dose is of the order 1 ms. The uncertainty in the beam spot position due to the scanning system must be less than 0.5 mm.. It must be possible to position the Bragg peak in the target volume in 5 mm or smaller steps in all directions, i.e., the available voxel size must be less than $5 \times 5 \times 5$ mm^3.

In general, the scanning system consists of two dipole magnets sweeping the beam in orthogonal directions, performing a lateral 3D scan. In one direction the beam is scanned rapidly and in the orthogonal direction slowly. The instantaneous scan speed in both directions should be variable (*scan velocity modulation*) in order to control the dose deposition as a function of the position of the beam centroid. The beam line vacuum has to be extended as far towards isocenter as possible in order to minimize the multiple scattering of the proton beam in air, which tends to increase the beam spot size. For the same reason, beam monitors and dosimeters must be as thin and placed as far downstream as possible. In this design example the vacuum extends beyond the scanning magnets. The vacuum pipe is followed by a helium bag and two ionization chambers. The dosimeters and other necessary devices can be telescoped back when space is needed for positioning the patient[5.1].. An example of scanning system devices is presented in Fig. 7.2.

Beam positions, profiles, and other attributes must be measured and monitored at specified position on the beam lime during the beam tuning, calibration processes, and actual treatments as specified in Table 7.4.

Scanning dosimetry. Three independent dosimeters that should provide continuous monitoring of the dose delivered to the patient are required. One dosimeter has to be located at a distance of less than 1 m from isocenter. At least one dosimeter must not saturate at the highest possible beam current focused into a 3 mm^2 area. A secondary electron emission monitor (SEM) fulfills this requirement. A dose detector with a two-dimensional position resolution of 5 mm or better must be provided. The detector must be able to generate a beam cut-off signal when any one spot has reached the specified dose limit. It can thus serve as one of the three independent dosimeters. All dosimeters scatter the beam and increase its lateral size. In order to satisfy the specifications for the maximum beam diameter at isocenter the detectors have to be made as thin as possible [5.1].

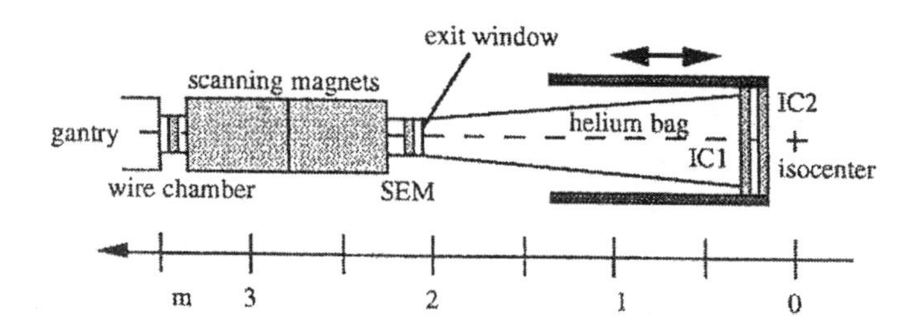

Figure 7.2. Components of beam scanning system at the end of the gantry: wire chamber, scanning magnet horizontal, scanning magnet vertical, total beam monitor (SEM), vacuum exit window, helium bag, ionization chamber serving as total beam monitor and used for beam centering, high resolution ionization chamber; after W.T. Chu [5.1].

Description	During tuning	During calibration and treatments
Beam position (centroid) upstream of scanning magnets	± 0.5 mm	± 0.5 mm
Beam profile upstream of scanning magnets (resolution)	1 mm	1 mm
Beam position at a distance of less than 1 m from isocenter	± 0.5 mm	-
Beam profile at a distance of less than 1 m from isocenter (resolution)	1 mm	-
Beam current of scanned beam	-	< 25 μs

Table 7.4. Beam attributes measured and monitored at specified position on the beam line wen scanning system is used; after W.T. Chu [5.1].

Safety. Whenever an "out of tolerance" condition is detected, the beam must be cut off before additional 5% of the treatment dose is delivered to any one spot in the treatment volume. Depending on spot size and scanning system design parameters, this specification requires the beam to be shut off in less than 25 μs [5.1].

7.4 Beam Transport Systems

The beam transport lines connect the different components of the accelerator complex (Chapter 6) and deliver the hadron beams from the extraction ports of the accelerators to the various treatment and experimental rooms. The beam is steered by magnets. There should be many safety interlocks in this Sec. . If any one of them detects that something is wrong, the beam should be stopped instantaneously. As an example of existing proton beam transport systems we will present the systems installed at LLUMC and NPTC.

7.4.1 LLUMC Beam Transport System

The extracted beam is created through half integer resonant extraction from the synchrotron. The extracted beam continues to make another orbit before it is sufficiently displaced horizontally from the circulating beam so that it may be vertically deflected by - 10° out of the ring by a Lambertson septum magnet. The beam is subsequently deflected by + 20° and - 10° by two small dog leg magnets so that it may be transported horizontally al a nominal height of 36 inches from the floor.

The beam is aligned and focused on a set of multi-wire ion chamber's (MWIC) specifically developed for the Loma Linda beam switchyard. The MWICs consist of two sets of horizontal and vertical wires with one millimeter spacing to simultaneously measure the beam profiles in both transverse planes.

Three quadrupole doublets with MWICs located between them are used to transport the beam to thc 90° bend towards gantry. When the two 45° dipoles which comprise this 90° bend are energized the beam is directed towards gantry. Subsequent focusing in two additional doublets transport the beam to gantry. MWICs are located between these two doublets and at the entrance to the gantry [8.3]. The layout of the LLUMC beam transport system is presented in Fig. 7.3.

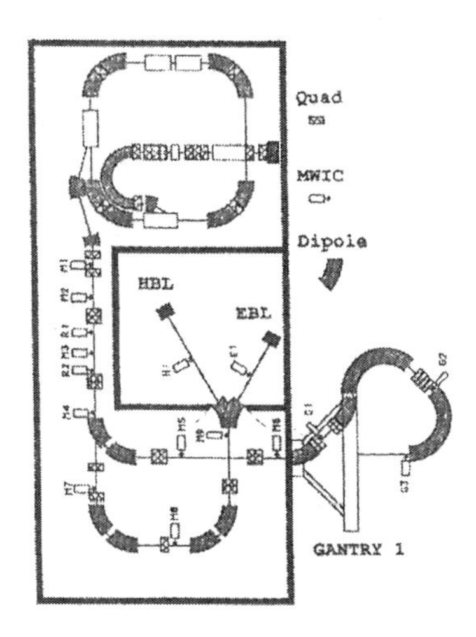

Figure 7.3. LLUMC accelerator and beam transport; after M.E. Schulze [8.3].

7.4.2 NPTC Beam Transport System

Energy selection system. The function of the Energy Selection System (ESS) is to transform the 235 MeV fixed energy beam extracted from the cyclotron into a beam having an energy variable between 235 MeV and 70 MeV, with a verified and controlled absolute energy, energy spread and emittance. The 235 MeV proton beam from the cyclotron is focused by a fast group of four quadrupoles into a small spot in the energy degrader (Fig. 4.1).

In front of the energy degrader, a horizontal-vertical non-interceptive beam profile monitor (NIBPM) allows measurement of the size and position of the beam spot and, through a control system algorithm, the automatic tuning of the upstream beam optics. These beam profile monitors are wire chambers operating in air at atmospheric pressure in the ionization chamber regime. The energy degrader is a rapidly adjustable, servo-controlled wedge of graphite. Just after the energy degrader, slits are provided to limit the beam divergence. The energy degrader is made of a high-density graphite (or synthetic aggregate) wedges. To guarantee the high accuracy required, the wedge is guided by high accuracy linear bearings and driven by a re-circulating ball screw.

Beam transport and switching system. After the degrader and collimator, the beam passes through an achromatic 90° bend. After the 90° bend that limits the momentum spread, the beam is focused to a small waist, fitted with a NIBPM used for the automatic tuning of the upstream analyzing section.

After the momentum defining Sec. , the beam is focused to a 5 mm waist in both x and y before going into either the 90° bend leading - to the first gantry, or into a straight Sec. that transports it to the 90° bend that diverts the beam - to the second gantry. The beam goes then into a long straight transport line that ultimately will lead to a fixed beam room. Switching magnets that can divert the beam into a gantry are located at three points along this line. The line is designed so that there is a waist with a 5 mm radius just before each of the switching magnets. Thus, the same tuning can be used in all the gantries for the same beam properties at the patient. The switching magnets deflect the beam through 45° from the straight line to direct it to a gantry. The deflected beam passes through three quadrupoles and another 45° bending magnet to make a 90° achromatic bend. The beam out of the achromatic bend is focused to a waist with a 5 mm half width in both transverse planes at the entrance to the gantry. Thus, the optics in the gantry do not change as the gantry is rotated. In addition to deflecting the beam 45° to a gantry, the last switching magnet in the transport line can deflect the beam by 300° or 0° to two sites in the fixed beam room. With this arrangement, a third gantry could be added after the fixed beam room is in use without disturbing equipment in the fixed beam room.

Transport system, the beam current and the horizontal and vertical beam position and beam profile are monitored with NIBMP. Through a control system algorithm the signal from these monitors can be used for automatic tuning of the up beam optics and for centering the beam by adjusting the bending magnets or trim steering magnets located at strategic locations in the transport system and the gantries.

The multi-wire chambers, operating in air at atmospheric pressure, are located at rather small waists so that the scattering of the beam by the monitors is small compared to the divergence of the beam at any particular location. One ionization chamber is located at the degrader, one at the waist just before each switching magnet, and one at the waist at the entrance to each gantry. Comparison of the beam size at three or more profile monitors with the transport calculations will provide a check on the emittance of the beam.

A beam stop is located at the exit flange of the cyclotron and at the input to each treatment room. When a treatment is being performed in one treatment room, the beam stops in the beam lines to all the other treatment rooms will be closed, so there is no possibility of inadvertently steering a beam into one of the treatment rooms that is not being used. The beam stop at the exit flange of the cyclotron serves as a beam dump when the cyclotron is being tuned up.

The beam stops supplied will stop a 250 MeV beam, will close in 1 s, and will close in the event of a power failure. In addition, they are insulated and are designed to give an accurate beam current reading when used as a Faraday cup. Fast turnoff of the beam is done by turning off the ion source in the cyclotron. Additional safety is provided in the case of an emergency by turning off the rf voltage to the dees in the cyclotron and by closing all the beam stops. There is two beam stops between the cyclotron and all the treatment rooms [4.4, 7.2].

7.5 Beam Preparation for Clinical Use

To cover an extended target volume with the Bragg-peak dose, the range of the beam must be modulated to spread out the Bragg peak, and the beam profile must be transversely broadened to cover the cross-sections whole area of the target volume.

7.5.1 Variable Range Shifters

Because the target volumes of clinical interests are usually thicker than the width of the Bragg peaks, the energy of the incident beam has to be modulated so that the Bragg-peak dose is deposited throughout the target volume from its proximal edge to the distal edge. This "stacking" of the Bragg peaks at different depth may be accomplished by changing the energy of the extracted particle beams from an accelerator. The entire beam transport system must also be changed in such a way that the tune and the positioning accuracy of the beam in the target volume remain constant.

Variable water column. Variable-thickness energy degraders may be made using various materials, such as water, oil, plastic plates, or metal plates. A variable water column is used to place a uniform, but adjustable thickness of water in the beam path. When the range of the incident beam is known, adjusting the range of the beam to an individual patient's prescription can be done simply and efficiently with such a device. This adjusted range is often called the *residual range of the beam* to distinguish it from the full range of the beam extracted from the accelerator. Water is often chosen as the medium for degrading the beam energy because of the similarity of its physical characteristics to soft tissue characteristics of the multiple scattering, range straggling, and for heavier ions, the fragmentation of the beam particles. However, other considerations may dictate the use of a different absorbing medium.

The water occupies the volume between the closed end of a cylinder and a movable piston. The motor-driven piston is sealed against the cylinder wall with O-rings, and, when operated, moves water in and out of a reservoir. This seal limits the rate at which the piston can be driven and hence the speed at which the water thickness can be changed. A rate of approximately 1 cm/s serves most radiotherapy needs. The mechanical linkage driving the piston is connected to an encoder for water-thickness control and monitoring. A relative accuracy of 0.1 mm in water thickness is achievable with magnetic resolvers or optical encoders. Since the two cylinder ends remain in the beam path, they introduce an additional range shift. Thin windows are therefore preferable, but must be sufficiently stiff, e.g., 5 mm thick Lucite, that their surfaces remain parallel to one another under the pressure of the water. The variable thickness is typically 30 – 40 cm which satisfies most clinical needs. Clinical requirements are that the absolute accuracy of the thickness of the column of water is ~ 0.5 mm over the entire range [4.1].

Binary filter. A device called a *binary filter* adjusts the range of the beam by means of a set of metal or plastic plates. The thicknesses of the series of plates double with each successive plate. The name binary filter derives from this power-of-two relationship among the plate thicknesses. Based on a set of ten plates, with the thinnest plate of 0.5 mm in water-equivalent thickness, any thickness from 0 to 511.5 mm can be achieved in steps of 0.5 mm. Because each plate is independently moved in or out of the beam, the time to setup the desired range change is independent of the incremental change in the thickness, in contrast to the cases with the water columns in which the time to change the thickness is roughly proportional to the incremental change. This feature permits range modulation with minimum loss of time. For protons and helium ions low-Z materials for the plates, such as Lucite or Plexiglas reduce multiple scattering for a given stopping power.

Double-wedge variable absorber. Alternatively, the thickness of absorbing material can be varied by using an absorber formed in a wedge shape. For a finite beam-spot size, not all particles would traverse the same thickness of the wedge, and therefore the range shifting would not be uniform. This problem can be corrected by using two wedges placed in opposing directions and moved in such a way that the particles in the finite beam spot traverse a constant thickness as shown in Fig. 7.4a. The overlapping area of the two wedges, which presents a constant thickness surface to the entire beam, must be larger than the beam spot. Consequently there is a minimum absorber thickness, greater than zero, attainable a system. Drawbacks are that for a small wedge the wedge becomes too large, and for a large wedge the minimum thickness attainable for a finite beam spot size becomes large.

As a variation, the wedge may be circularly shaped; each radius has a constant thickness, and its magnitude varies linearly with the angular displacement. In the example depicted in Fig. 7.4b, the beam goes through a circular wedge at an off-axis spot, and the range change is made uniform by a small compensating wedge placed in the beam path. The thickness of the absorbing material of the absorbing medium is varied by rotating the circular wedge on its axis, and its is monitored by measuring its angular displacement. Again the minimum attainable thickness is finite as in case of the linear wedge. system [4.1].

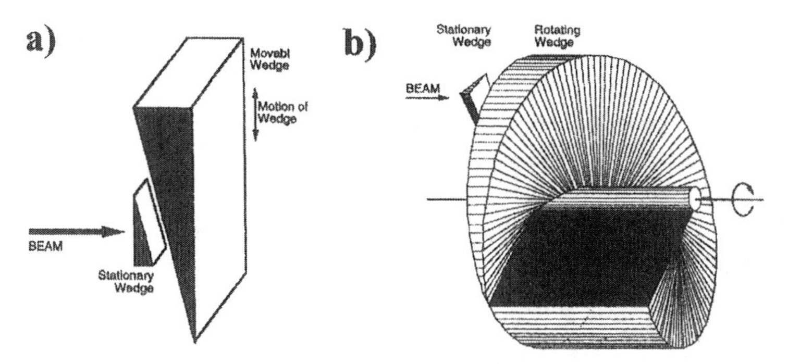

Figure 7.4. Schematics of a) double wedge system; b) circular wedge; after W.T. Chu et al. [4.1].

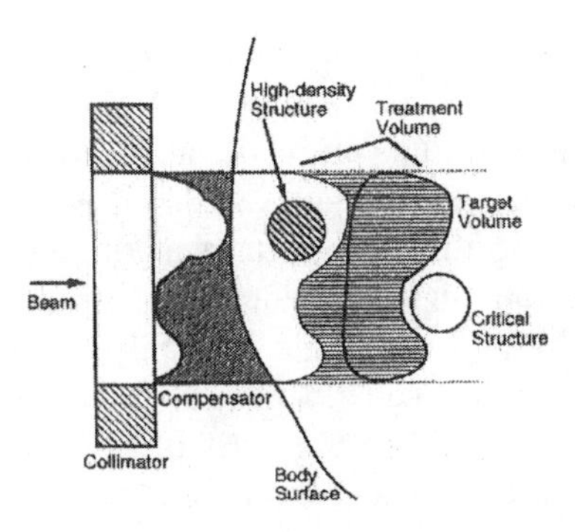

Figure 7.5. Schematic representation of the compensator; after W.T. Chu et al. [4.1]

Compensator (bolus). While many beam line devices are generic in their use, a few are specific to a particular treatment prescription. A patient specific device for modifying the beam range according to the patient anatomy is a *compensator* or *bolus*. Its function is to adjust the range of the beam across the target to conform the distal edge of the Bragg peak to the distal geometry of the target volume. Fig. 7.5 shows schematically the function of a compensator in two dimensions, and Fig. 4.3 shows a photograph of an compensator made from wax block. The details in the bolus structure are computed by considering the effects of tissue inhomogeneities due to bones and air cavities. The entire process, from obtaining the compensator contours based on the treatment plan to cutting the material on a numerically controlled milling machine, can be automated [4.1].

Range verification. Verification of the actual residual range placed inside the patient body is the primary safety tissue. An off line measurement can be done before a patient is treated with the assumption that the range remains constant during the treatment. Radioactive beam imaging is an example of range verification methods. The beam range can also be monitored on-line, i.e., during a patient treatment.

Off-line beam-range measurements. A measurement of the range of a beam can be made using a water column or binary filter along with two dose detectors, such as transmission ionization chambers. One detector placed upstream of the degrader measures the incoming beam. The second detector placed immediately downstream of the degrader measures the ionization of the exiting beam. The ratio of the two measurements as a function of the degrader thickness yields the relative ionization of the exiting beam. A Bragg ionization curve (Sec. 2.3.3, Fig. 2.5) is measured if the beam is monoenergetic. The beam range can be deduced from the absorber thickness where the Bragg peak occurs. For a modulated beam, the range, usually in water, may be defined for clinical purposes as the depth of the distal line of 90% of the isoeffect contour. Here the term isoeffect contour means the contour on which the biological effects are constant. Alternatively, it may be defined operationally as the water depth of the position in the distal dose falloff, where the relative ionization is 50% of that of the peak, i.e., the peak position plus a portion of the thickness of the distal falloff due to the range straggling. If there are other absorbing material, including the air space, between the water column and the patient, their range-modifying characteristics must be included in the computation of the residual range in the patient.

An alternative method for obtaining the range is to measure a depth-dose distribution in a water phantom (Sec. 9.3). Here a small detector, either an ionization chamber or a diode (Sec. 9.2.3), is moved in a volume of water and its readings are normalized to the incoming beam. It is important that the size of the detector be small compared with the field so that the change in dose as a function of depth is not dominated by the effect of the beam divergence.

A plastic scintillator (Sec. 9.2.4.1) can also be used as a range detector. In a device developed at LLUMC, the beam is stopped in a scintillator block and the output of light as a function of depth is viewed by a CCD camera. The light output is related to the energy loss of the beam, but is not exactly proportional to dose. Therefore, a calibration procedure is required to map the light output into a dose distribution. Its main advantage is that the entire range measurement can be done at once; however, the drawback is that such a device can be large, since the scintillation in the block must be optically imaged.

Another device developed at LBL is a series of ionization chambers sandwiched between degraders of predetermined thicknesses, which can be used to verify the range and the shape of the Bragg ionization curve using only a few beam spills from the accelerator. This method is as fast as that of the scintillators camera, but can take measurements only at predetermined depths.

Film behind a wedge absorber with a complex wedge angle can also serve to locate the distal edge of the beam. The range measured in the wedge material is converted into the water-equivalent range by applying the ratio of the integrated stopping powers of the wedge and water [4.1].

On-line measurement of the beam range. Since patient treatment must not be perturbed by a range measurement, an on-line range verification is difficult. The periphery of the beam, i.e., the area outside the collimated radiation field, may be used for a range measurement, provided the range is uniform across the entire field. A device located on the outside and upstream of the patient collimator aperture can make such measurements. A series of detectors, such as diodes, downstream of a metal wedge can be used to verify the range. Each detector in effect measures the energy loss of the beam at a different depth. The number of detectors determines the number of data points and the wedge shape determines the spatial resolution of the measurement. A variation of this is a rotating wedge with a detector behind it. The response of the detector and the position of the wedge can be correlated to give a depth-dose measurement. Both these methods require a normalization detector upstream of the wedge [4.1].

7.5.2 Range Modulation

Modulation of the residual range in order to spread out the thickness of the Bragg-peak dose region should be accomplished in two ways: *passive modulation* (Sec. 4.1.5) and *active modulation.* Use of range-modulating propellers or rotating spiral ridge filters falls in between these two methods.

The aim of the radiotherapy is to deliver the uniform dose to whole tumor volume (Sec. 5.1.5). The uniform physical dose distribution does not mean the uniform dose distribution from the biological point of view. In both beam modification methods, passive and active, the number of protons delivered to the target volume is given as a function of the proton range in tissue. In practice, each of the beams is characterized by the individual weight. These weights are defined on the ground of: (1) measurements of the proton's stopping power, (2) calculations or measurements of proton's LET as a function of depth, (3) existing survival curves and depth dose distributions and proton's LET [4.1].

Range-modulating propellers. The function of range modulation may also be accomplished by designing a propeller or fan-shaped stepped absorber, which is made to rotate rapidly in the beam so that the appropriate thickness of the propeller blades intercept the beam. The blades are made of Lucite, or Plexiglass, which are water-like in their absorbing and scattering characteristics. For balancing of the rapidly rotating blades, the propeller is usually made of two or four blades.

The stepped rotating absorber was first proposed by R. Wilson. This type of propeller has been used at the HCL with proton beams, for helium ions at LBL, and recently at LLUMC for proton beams. A small propeller for beams less than 2 cm in diameter was produced and used for ocular treatments.

Actual construction of the propellers is done by making a set of layers cut to the desired shape and then glued together. A water equivalent thickness of 4 mm of water is a practical Lucite thickness. Again Lucite is chosen because it is a low Z material thereby reducing multiple scattering and the associated increase in beam divergence. At LLUMC the propellers are machined out of plastic blocks. The useful radial size of the blades must be approximately three times the beam diameter [4.1].

Ridge filters. The above-mentioned plastic propellers have been used for proton or helium-ion beams. The main objections to them are their bulky size and the high rotational speeds. To overcome these shortcomings, several types of ridge filters have been made: stationary ridge filters, linearly oscillating bar ridge filters, and rotating spiral ridge filters (Fig. 4.3).

Bar ridge filters. The cross-Sec. al shape is calculated in such a way, that when the ridges are uniformly illuminated by a particle beam, the transmitted beam exhibits a relative abundance of penetrating particles of appropriate residual ranges. If they are mixed appropriately, an acceptable SOBP results throughout the target volume. The mixing is accomplished by the multiple scattering of the particles including in the ridge filter itself. Another practical way of designing the ridge filter is developing an analytical form of the ridge shape.

Because the particles traversing the thickest part of the proximal part of the SOBP and those the thinnest part (zero metal thickness) reach the distal part, the height of the ridges, in water-equivalent thickness, represents the width of the resulting SOBP. The half-base distance of the ridges, usually about 5 mm wide, is made comparable to but smaller than the transverse scattered distance suffered by the heavy charged particles at the target volume. This ensures that the particles of different ranges mix thoroughly so that particles arriving in the treatment volume produce the proper SOBP everywhere without imaging the ridges. If the ridge bases were too large so that the mixing by scattering is not complete, the ridges may be linearly oscillated in an orthogonal direction to the lines of the ridges and the beam axis to ensure the mixing.

When expressed in terms of water-equivalent thicknesses, the cross-Sec. al shape of the ridges is similar to that of the plastic propeller. For example, if plastics are t used to make a ridge filter for a SOBP of a 14 cm width, the cross Sec. of the ridges will have a base of 1 cm and a height of 14 cm. Such ridges are too sharp to machine out of plastics, but by using a high-density material the heights of the ridges are made manageable.

Metal bar ridge filters have been used with the proton beams at TSL, Uppsala, Sweden (tungsten - alloy), ITEP, Moscow, Russia (aluminum), PARMS, Japan (aluminum) [4.1, 7.5].

Spiral ridge filters. Another method to move the ridges across the beam is rotating a spiral ridge. Many types of spiral ridges have been developed. As an example we will present the filter developed at LBL. The spiral ridges are cut on a brass plate using a specifically shaped tool bit. One difficulty of machining this type of spiral ridges is encountered near the central axis, where the radius of the spiral becomes very small and the machining necessarily produces an imperfection, a dead spot. This problem was circumvented for the double scattering system by placing the central post of the occluding ring assembly at the axis, which covers the imperfection [4.1].

7.5.3 Transverse (Lateral) Spreading of Particle Beams

Many patient treatments require large uniform radiation fields, often as large as 30×30 cm^2, and occasionally even larger. This does not imply that such large areas are routinely irradiated in the clinic; the large-area capacity allows accommodation of target areas of various irregular shapes, sizes, and orientations. To obtain such large fields, the beam must be laterally spread out, in the direction perpendicular to its central ray, since the beam transported into the treatment area has a relatively small spot size. The aim is to produce a large field that covers the target area with a uniform dose with a variation of less than ±2% (Sec. 5.1.9). The tolerable dose variation depends on the particular clinical application for which the beam is used. Other important considerations are: optimization of such beam characteristics as the sharpness of the lateral dose falloff, the sharp falloff of the distal-peak dose, the beam utilization efficiency, dose rate, neutron production, beam fragmentation in case of heavier ions, the ease of beam tuning, repeatability, stability of the delivered dose distributions, and patient safety.

Many different methods for lateral spreading of the beam have been investigated. In the following discussions, these methods are divide d into static (passive) and dynamic (active) beam delivery systems (Sec. 4.1.5). A passive system (Sec. 7.5.3.1), such as the scattering system, spreads the beam into a uniform dose across the entire treatment volume at all times. On the other hand, a dynamic system (Sec. 7.5.3.2), such as the raster-scanning beam delivery system, moves a beam spot in a predetermined way across the treatment area and delivers the dose to only a part of the treatment volume at a time. A desired dose distribution results with such a system only after completing a preplanned course of irradiation.

7.5.3.1 Passive Beam Delivery Systems

Single-foil scattering method. The simplest beam delivery method uses a single scattering foil or plate. When a particle beam traverses a medium, each particle is deflected by many small-angle elastic scatterings mainly due to elastic Coulomb scattering from the nuclei within the medium, and its profile is broadened to a two-dimensional Gaussian-like distribution for small scattering angles. A narrow pencil beam scattered by a thin scatterer produces an approximately two-dimensional Gaussian dose distribution at isocenter.

An excessively thick scatterer is required to produce such a wide beam; hence only small radiation fields can be generated in this way. Small beams with radii < 2 cm prepared in this way are used in clinical applications in treating small targets, such as tumors in the eye or some small size intracranial disorders [4.1].

Double-scattering beam delivery method. As described above, the single-foil scattering method can provide the dose distributions of acceptable variations only in relatively narrow fields. If one tries to widen the circle of utilization by using a larger portion of the scattered beam, an excess dose would result around the central ray. To obtain a broader uniform dose distribution at the isocenter, some of these excess particles near the central ray must be removed. One of the methods developed for this purpose is the double-scattering method. Such scattering systems were developed for proton beams at HCL, and at LBL.

The example shown in Fig. 7.6a uses an occluding post of sufficient thickness to stop the beam particles which is placed in such a way that it blocks the central portion of the Gaussian distribution. Past the occluder, the transmitted beam intensity distribution is shaped as an annulus with a null in the middle. Its profile in a plane through the central ray exhibits two peaks as shown in the figure. The second scatterer, of an appropriate thickness and placed strategically, diffuses the particles in these two peaks filling the dose void in the middle, and produces at the isocenter a larger flat-dose area.

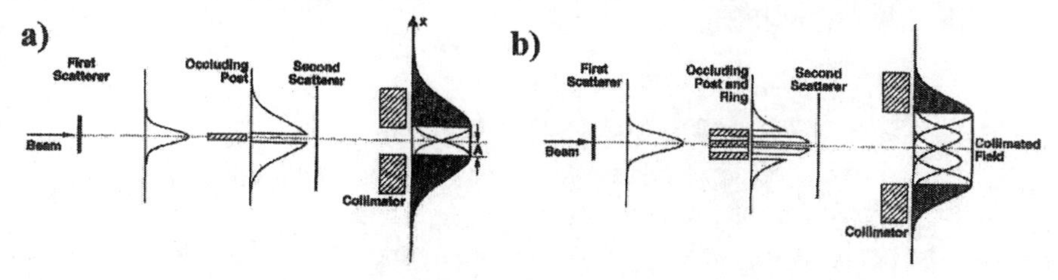

Figure 7.6. Schematics of a double-scattering method using a) central post occluder; b) annular ring plus occluder assembly; after W.T. Chu et al. [4.1].

A flat dose of even larger area can be obtained by using of annuli and / or post-occluder system as shown in Fig. 7.6b.

Another practical point to note is that the double scattering method requires thick scattering foils, which produces secondary particles for beam particles with $Z > 1$, which lowers the peak-to-plateau ratio and raises the dose beyond the Bragg peak. These fragments also lower the RBE and raise the OER values, thereby lowering the biological advantage. The beam utilization efficiency is low, typically 20%. The low efficiency implies that a large portion of radiation is absorbed in the occluder, as well as in collimators and scatterers, resulting in increased background radiation in the treatment room. This becomes a serious problem when a double-scattering system must be placed near the patient, such as in the case of mounting it on a rotating gantry. Shielding needed to block unwanted radiation may become unacceptably heavy [4.1].

Bimaterial scattering. The different scattering characteristics of heavy charged particles for different atomic mass scatterers may be exploited in the preparation of therapy beams. A high atomic-mass material scatters more with little range loss; whereas, a comparable low atomic mass material scatters little while modulating the range more. A pencil beam is laterally spread out to a Gaussian like beam spot and is made to impinge upon the second scatterer. In order to flatten the field, the rays near the central ray must be scattered out more than the rays further away from it. This differential scattering must be achieved while keeping the range modulation of the beam constant at all radial distances of the second scatterer. An elegant solution is a bimaterial (e.g., beryllium and lead, or copper and plastic) scatterers, which have successfully been produced and used at HCL and adapted at LLUMC. The particles suffer the same energy loss throughout the scatterer, but their scattering characteristics are a function of radial distance from the axis to scatterer [4.1].

7.5.3.2 Dynamic Beam Delivery Systems

A dynamic beam delivery system produces a desired radiation field when a controlled extraction of the beam from an accelerator is coupled with strictly prescribed patterns of the motion of the beam spot. For example, two dipole magnets, placed in tandem so that their magnetic fields and the incident beam form three orthogonal directions, can move a beam spot in a predetermined way to produce a desired dose distribution. It is also possible to devise a magnet with compound coils, a rotating set of permanent magnets, or electrostatic deflectors to accomplish similar functions. Discussed below are many different schemes devised for clinical use. These dynamic systems have a distinct advantage over the scattering systems in minimizing the material in the beam path, maintaining the beam range, reducing fragmentation of the beam particles, and decreasing the background radiation for the patients.

Rotating dipole. The simplest form of the dynamic beam delivery systems uses one rotating dipole. A beam incident along the central axis of a dipole is deflected in a certain angle, and if the dipole is rotated around the central axis of the incident beam, the exiting beam will wobble around the initial direction and produce an annular-shaped dose distribution at the isocenter. The resulting dose field is similar to that produced by the double-scattering method using a post occluder. A HCL system uses a 160 MeV proton beam, focused down to a 4 mm spot at the entrance of the magnet, and a dipole magnet with a 0.16 T magnetic field. The emerging beam is scattered by a 0.75 mm lead foil to produce a scattering angle of about 18 mrad. It results in a uniform dose distribution within ± 5% over a 15 cm diameter field [4.1].

Wobbler systems using two electromagnetic dipoles. A wobbler system consists of two dipole magnets placed in tandem with their magnetic field directions orthogonal to one another and to the beam direction. The magnets are energized sinusoidally with the same frequency but with a 90° phase shift between them. If the amplitudes of the magnetic fields are properly controlled, a beam entering the system along its axis emerges from it with the beam direction wobbling around the original beam direction, and "paints" an annular-shaped dose distribution. The diameter of the dose annulus is changed by adjusting the amplitudes of the currents in the wobbler magnets. A large area of uniform dose is obtained by painting the treatment area in several concentric annuli with different diameters, each with a certain predetermined particle-number fraction. The wobbler magnet power supplies, which provide sinusoidal currents, are easy to construct and reliable. The pertinent parameters to obtain flat fields are constant spill intensity, precise wobble radii and correct particle number fractions at various radii. Wobbler systems were selected as the beam-delivery system at HIMAC, and for NIRS.

There are tradeoffs for the wobbler beam delivery system compared to the double-scattering system. First, no scatterers are needed in the beam path and the same available beam range can be achieved at lower accelerator energy than that required for scattered beams. No beam is lost in scatterers and occluders. As the size of the flat field is readily varied according to the port size, less beam is lost in collimation, and therefore, the beam utilization is higher than that for a scattered system. Beam alignment is not as critical as in the case of the scattering system. However, the high stability of the beam entering the wobble magnets throughout the treatment time is critical to provide concentric annular dose distributions. Since the effective source size, i.e., the beam spot size is small, the wobbler produces sharper lateral dose falloffs than those attainable with the scattering system. The neutron production in absorbers and collimators is curtailed, and, therefore, the shielding requirement is lower than that for the double scattering method. This becomes an important consideration when one tries to place a beam delivery system on a rotating gantry. As a dynamic mode of beam delivery, the wobbler system requires active monitoring of the wobbler magnetic fields.

In the following two Sections, two scanning methods for producing large fields of uniform dose distribution are discussed. Scanning beam delivery systems are classified according to the ways in which the beam spot is moved, lamely, raster scanning, spot scanning, and pixel scanning, which are discussed in Sec. 4.1.5. Raster scanning employs a smooth motion of the beam spot while keeping a constant beam extraction. The last two scanning methods involve a more discrete spot motion. The spot is moved to the next position when the prescribed dose is deposited at a given position. In spot scanning, the spots overlap at least a half-width of the spot, whereas, in the pixel scanning, each pixel is individually irradiated and he adjacent spots overlaps only at their edges [4.1].

Specifications for scanning. A large field of a specified dose distribution may be produced by scanning a beam across a treatment area. In principle, scan speed and beam intensity can be varied as a function of the spot location in the treatment volume to generate the desired dose distribution. A schematic diagram of scanning is shown in Fig. 7.7. In general, a scanner consists of one or two dipole magnets, one for the fast scan in the x direction and the other for the slow scan in the y direction. Range modulation moves the stopping region of the beam spot, the Bragg peak, in the z direction.

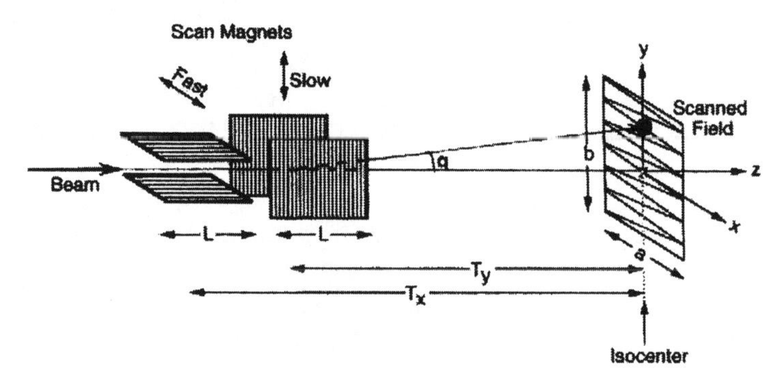

Figure 7.7. A geometry of beam scanning system. The length of each raster-scanning magnet is L, the magnets are placed at distances T_x and T_y from the iso-center, and the scanned field is a rectangle with sides a and b; after W.T. Chu et al. [4.1].

Lissajous pattern maker. If two dipole magnets are placed in tandem with their magnetic field directions orthogonal to one another and to the beam direction, and energized sinusoidally with two different frequencies, which are related to each other in a ratio of integers, a beam going through the system will draw a rectilinear Lissajous pattern at the isocenter. If the beam extraction level is held constant, a limited region at the middle part of the Lissajous pattern would exhibit a dose distribution of an acceptable uniformity since the rectilinear speed of the beam spot is approximately constant there. A system fabricated at Uppsala for 185 MeV proton beams produced a 2×2 cm^2 field. In this system, the beam-spot structures were smeared out by a thin scatterer. Two dimensional beam spreading using a toroidal magnet was also considered [4.1].

Raster scanning. A raster-scanning beam delivery system has been developed at LBL to broaden light-ion beams into large flat radiation fields. This system has several advantages over the scattering system as well as the wobbler system. The raster-scanning system uses no absorbing material in the beam path. Rectangular fields of various aspect ratios produced by a raster scanner conform better with many irregularly shaped ports than circular fields produced by a wobbler. Going beyond the simple rectangular scans by varying the extents of each fast scan, irregularly shaped fields may be produced, and provide better conformations of the radiation fields to irregular target volumes.

A large flat-dose field may be produced by moving the beam spot in rasters with a constant sweep speed while holding the beam extraction level constant. This can be achieved in two ways: in parallel rasters as shown in Fig. 7.8(a) or in zigzag rasters as in Fig. 7.8(b).

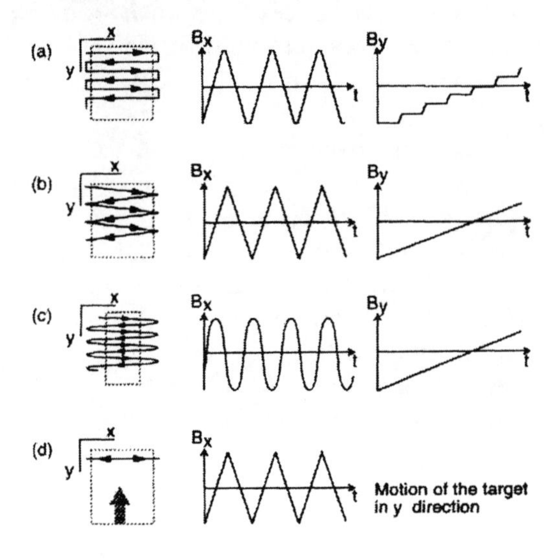

Figure 7.8. Schematics of various raster-scanning techniques. For each method, the scan pattern is shown in the left diagram, and the magnetic effective distance from the magnets to the field wave forms are shown in the right two diagrams. Shown are (a) parallel rasters, (h) zigzag rasters, (c) sinusoidal scan, and (d) fast line scan plus the target motion; after W.T. Chu et al. [4.1].

The latter can be considered as two sets of parallel rasters, one set moving in the + x direction and the other in the - x direction. These methods require a fast scan in x direction, and a slow scan in y direction. If the scans are accomplished with two magnets, their magnetic fields are varied as shown in the figure. The LBL raster scanner described above falls into this category. Sinusoidal variation of the fast-scan field can be considered as in Fig. 7.8(c). Only a small central part of the scanned field would exhibit an approximately flat-dose area; large portion of the beam outside of the area will be discarded. A scheme may be devised to enlarge the useful flat-dose area by modulating the beam-spill intensity according to the scan speed. Clamping the beam off while it is outside the constant scan-speed area could also increase the beam utilization efficiency. A two-dimensional scan can be accomplished with only a fast-scan magnet performing line scans while moving the target in an orthogonal direction as shown in Fig. 7.8(d) [4.1, 7.3].

Pixel scanning. A large uniform field can be made subdividing the treatment area into a large number of pixels and delivering a predetermined dose to each pixel (Fig. 4.4e). To achieve efficient pixel scanning, very fast magnets and fast monitoring systems are required. Considerations of multiple scattering of protons in the treatment volume and the edge matching of neighboring pixels make the smallest acceptable size of the pixel bigger than 1×1 cm^2.

The first two-dimensional pixel scanning system for proton radiotherapy was developed at NIRS to spread the 70 MeV proton beams. The system was called *spot scanning* but under our classification scheme, here it will be called a *pixel scanning system.* It employs two dipole magnets, which deflect the beam either horizontally or vertically, four sets of beam defining slits, a quick beam shutter, and three monitoring chambers. The power supplies of the dipole magnets are digitally controlled by a computer, which direct the 1 cm^2 beam spot to any desired position in the treatment field. A highly collimated proton beam is prepared through four slits to produce a 1 cm^2 beam spot at the isocenter and the grid size used for the spot movement is the same as the spot size. Using the maximum field of the scanning magnet of 0.1 T, they achieved a flat dose within 18 cm^2 fields with dose deviation of $\pm 2.5\%$. A parallel-plate transmission ionization chamber is used to monitor the delivered dose at each spot location. The spot speed at the isocenter is ~ 1 cm/ms. It is possible to irradiate a field of any irregular shape with an arbitrary dose distribution, and to correct any fluctuation in the beam extraction intensity. The beam-spot position stability is achieved through severe collimation paying a great penalty in the beam utilization efficiency. The dose rate within the 1 cm^2 beam spot is ~ 1 Gy/s and the system takes about 3 min to scan a 10 cm^2 field, i.e., 100 pixels. The system is also used to deliver a large radiation field of a complex dose distribution by varying the dose at each spot [4.1].

A large field of flat dose can also be obtained by moving a beam spot across the field in discrete steps; after positioning a beam spot at a given location in the field, a predetermined amount of radiation is deposited. The spot is then moved to the next position, and the process is repeated. This approach adopted at PSI to perform spot scanning of high-energy protons (Fig. 7.9). The target volume is scanned with a 250 MeV proton pencil beam slice by slice, like in computed tomography, by magnetic deflection, concomitant range modulation and patient translation. In the first stage of the project, the beam used is horizontal and stationary (fixed). In order to adjust the irradiated volume to the irregular and complex target volume a large number of pixels, or voxels, are necessary. If the target volume is, for example, 1000 cm^3, or 1 liter, the number of voxels is about 10 000 for a 5 mm distance between successive pixels. Since the treatment time should be limited to several minutes only, the irradiation times per voxel must be about 15 ms. Between two successive spots the beam is turned off and on by a fast switching magnet. Another similar magnet ensures that the beam travels to the desired coordinates. Both the magnet response times and those of the beam monitoring system should be of the order of 0.1 ms to achieve dose uniformity better than 1%.

In the system shown in Fig. 7.9, energy or range modulation is carried out by injecting small 5 mm thick polyethylene plates whose purpose is to reduce the beam range. The variable beam degrader shown schematically in Fig. 7.9 is intended for degrading the accelerator output energy down to the treatment energy of 250 MeV.

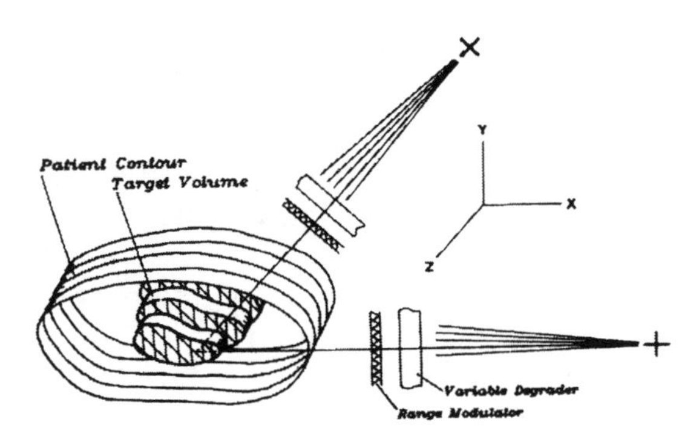

Figure 7.9. A system for a 250 MeV pencil proton beam scanning and modulation, developed at PSI, Villigen; after M. Blattmann et al. [7.6]

7.5.3.3 Conformal Therapy Delivery Using Variable Modulation

Fixed versus variable modulation. Heavy charged particle beams with Bragg peaks and their favorable peak-to-plateau ratios have been successfully used to treat tumors at a number of different sites in the body with less morbidity than with conventional radiation. In the systems described so far, passive range modulation was used to create fixed-width SOBPs as depicted in Fig. 7.10(a). The width of SOBP is determined by the thickest part of the target volume along the direction of the beam, and the Bragg peak is spread out to the same width for all rays in the beam (fixed modulation). The treatment volume, the region so irradiated with the fixed SOBP, is a cylinder, whose length is equal to the width of the SOBP. The distal envelope of the treatment volume is usually shaped by a compensator to conform it with the distal surface of the target. Generally, the longitudinal thickness of the target volume is not uniform and this method delivers a Bragg peak dose to tissues upstream of the target, where some critical normal tissues, such as the skin as depicted in the figure, receive the same high dose as the target. In spite of this limitation, successes in treatments using charged particle beams are the result of reductions of the high-dose volume and the dose to structures in the remaining irradiated volume when compared to external photon therapy. In realizing the full clinical advantage of heavy charged-particle beams, one of the important future developments in dose localization is to develop a beam delivery system that allows modulation of the spread Bragg peak over the target volume (*variable modulation*).

The shape of the *high-dose volume*, in other words the *treatment volume*, can be made to conform more closely to that of the target volume, as shown in Fig. 7.10(b). Such a system improve the therapeutic efficacy of the delivered dose and also increase the versatility of the beam spreading system for varied clinical situations [4.1].

Conformal therapy delivery in three dimensions. Several methods are proposed to reduce the dose in the surrounding healthy tissues (Sec. 2.1). Fig. 7.10(b) schematically depicts one of these methods. The treatment volume may be divided into many layers, as shown in the figure, and scanned layer by layer by changing the residual range of the beam. In order to reduce the number of layers, the Bragg peak is spread out to a minipeak of approximately 1 cm width, and these minipeaks are axially stacked by changing the range. The accuracy of obtaining correct SOBP using the range stacking depends on the slope of the minipeak. Analogous to the edge-matching problem in beam scanning, the gradual rise and fall of the minipeak facilitates correct stacking when the pulse-to-pulse energy variation of the beam is taken into consideration. The distal dose falloff of the SOBP is decided by the most distal minipeak. A pristine Bragg peak with the steepest possible distal dose falloff should be used at the deepest penetration.

The contours of the proximal and distal surfaces of the target volume are usually smaller than the widest lateral extent and, as the axial stacking proceeds, the aperture of the variable collimator is reduced in such a way that healthy tissues adjacent to the target volume, especially those upstream of it, are protected from the unwanted Bragg-peak radiation. It is especially important to spare the skin and critical organs from unnecessary radiation. For example, the skin surface in Fig. 7.10(a) would receive the full Bragg-peak dose in a fixed modulation method, whereas it will receive only the plateau dose, lower in magnitude and LET, in a variable modulation method as indicated in Fig. 7.10(b).

A 3D irradiation system using scatterers for lateral spreading was proposed at NIRS. The treatment volume is layered according to the shape of the distal surface of the target, and the dose is built up using a variable range modulator and a fixed bolus. As the range is shortened, the irradiation field is changed using a variable aperture, such as a multileaf collimator [4.1].

One way of delivering conformal therapy with variable-modulation SOBPs is to employ a raster scanner with controllable scan speed and a variable collimator. Requirements for accuracy and stability of the scan magnets are stringent. To ensure patient safety, the spatial distribution of the delivered dose must be monitored in real time using a large-area high-resolution dose detector. The system requires an automated control along with a means of very fast and fail-safe irradiation termination. Fast on-line monitoring is required of various system parameters, such as scanning magnet currents, extraction levels of the particle beams, aperture of the multileaf collimator, widths of SOBPs, and the residual range of the beam. Constancy of the beam extraction level is required over a large dynamic range in the extracted beam intensity.

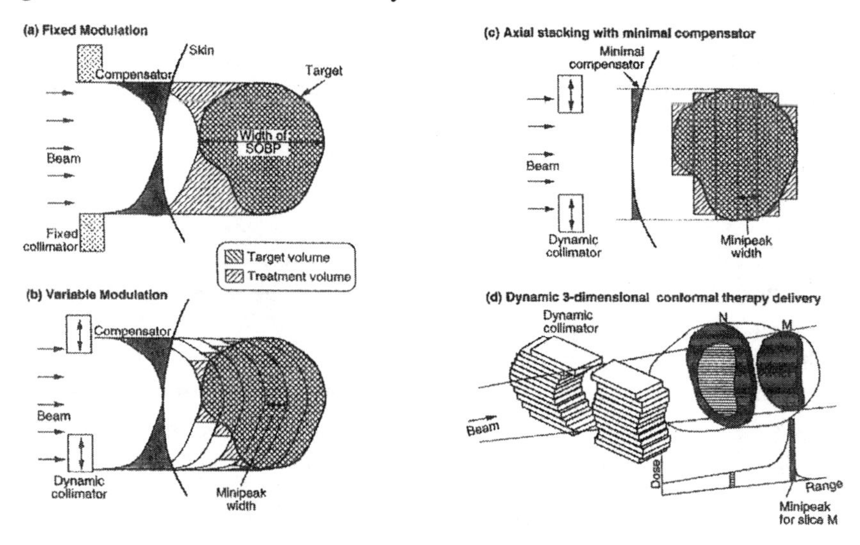

Figure 7.10. (a) The fixed-modulation method using a compensator produces a cylindrical treatment volume whose length is equal to the thickest part of the target volume. Normal tissues upstream of the target volume are irradiated unnecessarily. (b) The unwanted radiation may be trimmed by the use of the variable-modulation method. The treatment volume is made to conform more closely to the target volume than in (a). (c) Axial stacking with a variable-speed raster-scanning technique. (d) Schematic illustration of a 3D conformal therapy delivery; after W.T. Chu et al. [4.1].

7.5.4 Collimators

Shaping the radiation field to conform to the irregular volume, while conceptually simple, is one of the more difficult tasks in preparing the beam for a patient treatment. The various methods that have been investigated vary in their complexity and the degree to which they conform the radiation to the target volume. The simplest and earliest method uses a block of material with a fixed aperture, which conform to the projected area of the target volume as seen from the beam direction. A more refined method involves shaping a mechanically variable aperture to achieve the desired shape at various depths of the target volume. Another method, yet to be implemented clinically, requires no physical collimation of the beam, but relies on precise control of the position and profile of very small beam spots.

Collimator construction. The thickness of the collimator must be sufficient to stop the beam in the case of protons, but significantly thicker than the range of the primary beam in the case of heavier ions because fragments produced in nuclear interactions of the beam with the collimator material can travel further than the range of the primary particles. The tail dose for a heavier ion beam is significant. The LET of the fragments delivering this dose, however, is less than that of the radiation in the SOBP. The lower LET here implies a lower RBE, and therefore a decrease in the actual damage done by the particles penetrating the collimator. The collimator thickness is usually chosen to reduce the dose delivered by the penetrating secondaries to ~ 1% of the SOBP dose of the uncollimated beam.

Collimators can be made from any number of materials. The less dense the material the thicker the collimator will be in the beam direction; however, the weight will be approximately the same (± 20%) regardless of the material since the stopping power of a material is approximately proportional to its physical density (± 20%). One of the major concerns in the choice of collimator materials is the induced radioactivity from the beam stopping in the collimator. Generally the residual activity created from the exposure to a single treatment must be low enough for clinical personnel to work in the vicinity immediately following the exposure. Gamma rays coming from the induced activities are the most important component of the radiation because of their long range An absorbed dose of 1 Gy in a Cerrobend collimator results in an induced activity of 50 mSv at the surfaces with an initial half-life of about 5 min. Heavier materials such as tungsten have higher induced activity levels with longer half-lives and emit photons with higher energies compared with lighter materials such as aluminum [4.1].

Fixed aperture collimators. Fixed aperture collimators (Fig. 7.11) have an opening designed by clinical personnel based on the projected target shape. While this method cannot avoid delivering unwanted radiation to some normal tissue surrounding the target volume, the simplicity of these collimators makes them widely used. One method of making such fixed-shape collimators involves casting a low melting-point alloy in a styrofoam mold. Another method of making collimators is machining an aperture out of a metal such as brass. By automating the design and milling processes, these types of collimators may be fabricated routinely.

The major disadvantage of the fixed-aperture collimator is that a unique collimator must be cast for each patient port, which is a labor-intensive process. A clear need exists for a collimator with a variable aperture, but their mechanical complication has daunted their construction until recently.

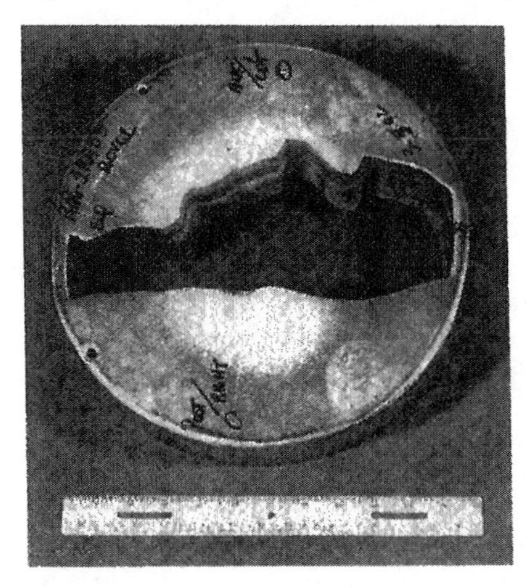

Figure 7.11. Collimator made of heavy metal used to shape the beam's profile to match the across section of the patient's tumor from a beam's eye view; after W.T. Chu et al. [4.1].

Variable aperture collimator. A variable aperture collimator is a device whose aperture can be changed mechanically to a desired profile. The first example to be discussed is a slit collimator consisting of a slit with moveable end blocks. Such a collimator can be used with line scanning beam delivery systems. During a line scan, the slit is aligned with the scanning beam, and the radiation field is collimated to a rectangular shape of a specified length by the slit and the end blocks which are moved to the desired positions. At the completion of a given line scan, the patient is moved to a position where the next line-scanned beam is aligned with the doses already delivered in preceding line scans. Repetition of this process produces an irregularly shaped radiation field in two dimensions. If the end blocks are made to rotate around pivots in such a way that their collimating edges align with the curved boundaries of the irregular port, the resulting irregular port shape conforms more closely to the target shape. This type of dynamic collimator has three controls: one for the patient motion and two for the linear motions of the end blocks. The pivoted version would require two additional controls for the angular displacements of the end blocks. The system can be adapted to 2D scanning, in which the patient remains stationary and the line scans are moved across the patient.

An example of a variable aperture collimator is the *multileaf collimator* (Fig. 7.12), which defines irregular shapes by of many movable absorber bars, called *leaves* or *fingers*. A multileaf collimator, whose leaves are manually moved, has been developed and used at NIRS. An automated multileaf collimator was developed at LBL, and whose leaves are moved by actuators under computer control. The leaves are stacked together to form what is called a jaw, and two jaws are placed in opposing positions. The leaves are driven by dc motors and a feedback loop is required for positioning the leaves and keeping them in place. The motors drive the leaves directly via a hall screw A sliding potentiometer attached to a leaf is used for measuring its absolute position. The generated signal is compared to a reference and the difference used to drive the motor. The leaf position is monitored by a magnetic encoder attached directly to the motor. Thus the position monitoring is independent of its control. The multileaf collimator is mounted in such a way that it can be rotated 90° around the central ray of the beam, allowing collimation of almost any field shapes encountered in radiotherapy applications. The leaves were made of steel as the best compromise for satisfying the following parameters: high density, high rigidity, ease of fabrication, durability, low induced radioactivity, and reasonable cost [4.1, 7.4].

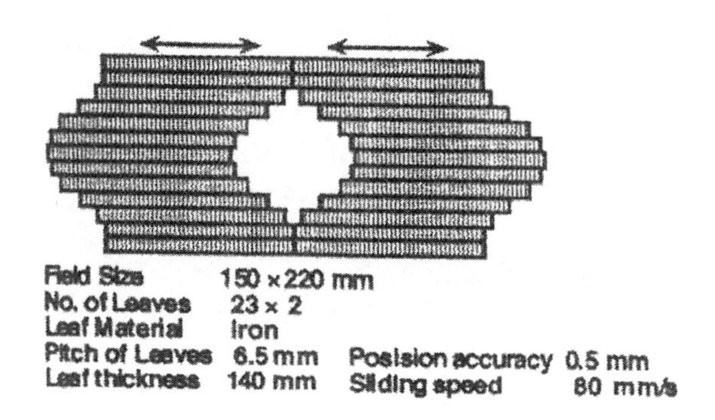

Figure 7.12. Specification on multileaf collimator proposed for HIMAC facility, after Y. Futami [7.4]

7.5.5 Dynamic Beam Shaping

Shaping of the treatment port by magnetic control of the beam without resorting to the use of a collimator has not been clinically implemented. Pixel scanning with pencil beams without collimation has been done on an experimental basis with proton beams (Sec. 7.5.3.2). This method requires the control of the beam profile, position, and intensity The lateral dose falloff of the collimated beam is defined by the profile of the beam spot. Because no physical means of stopping the beam at critical positions is used in this method, proper control of the beam must be guaranteed and an independent means of verifying the beams position and the dose being delivered must he provided [4.1].

Chapter 8

PROTON GANTRIES

It is an aim of any radiotherapy to deliver the highest possible dose to the tumor while keeping the dose delivered to surrounding healthy tissue under a tolerable level (Sec. 2.1). This aim can be best achieved if the dose is delivered in a tumor-shape conformed way (3D conformal radiotherapy, Sec. 7.5.3.3). Beams of light ions offer high physical selectivity (Bragg peak) combined with enhanced biological effectiveness, which makes them an ideal tool for 3D conformal tumor targeting. The highest degree of conformity could be achieved by combination of a rotating gantry with an advanced beam delivery technique (for example raster- or spot-scanning, Sec. 7.5.3.2). This trend can be recognized in recently built or currently proposed proton cancer therapy facilities, especially in dedicated hospital-based facilities where a high level of flexibility of the beam delivery system is required to treat a large spectrum of tumor sizes and sites (Secs. 4.2.3 and 4.3) [8.1].

The gantry includes the structure, which carries the necessary devices so as to allow the therapy beam to be directed from virtually any angle in a plane. One eventual aim of a gantry system, and the most demanding, is to be capable of *dynamic therapy*. This technique requires the beam angle to be adjusted during a treatment, thus minimizing the beam exposure at the surface while concentrating on the target volume. One of the key advantages of protons is the precision with which one can potentially deliver a treatment. The treatment area can be selectively treated while minimizing the dose to a critical structure, which can be a fraction of millimeter distant. To realize this possible precision, the hardware, for protons in particular, must be appropriately designed [8.2].

There are presently several high-energy gantries in operation or under construction. Consideration of gantry geometry possibilities leads to several paths (Sec. 4.1.4). It is interesting to note that practically all these paths have been or are being pursued. There has not developed a consensus about a preferred scheme, other than finding one, which results in the least expensive overall system and yet meets all the clinical requirements.

One question to ask is whether the accelerator and the gantry should be decoupled or physically integrated. Of course the latter requires an accelerator capable of being attached to a gantry structure and rotated through 360°. There is no proven method to date of miniaturizing an accelerator for high-energy protontherapy although a couple have been proposed. Another advantage of coupling the accelerator with the gantry is that the accelerator typically produces a non-symmetric phase space, which can be used if it is integrated in the gantry A main reason to separate the gantry is to feed a beam delivery system with multiple beam lines. Another issue is the relative location of the gantry and patient positioner relative to the isocenter. In the case where the isocenter is a point along the axis of rotation of the gantry, patient positioner to move as the gantry rotates, is not necessary but that requires the gantry to be large.

Last in this non-exhaustive list is the question of how the beam will be shaped (spread) to match the target. The use of passive scattering vs. active scanning may affect the gantry geometry. In particular, the location of the scattering or scanning devices may be an important factor. It is desirable to have a large effective source to isocenter distance. This can be achieved by a large physical separation between the beam spreading system and the isocenter after the last gantry dipole, or by incorporating these systems within the gantry optics and using close to point to parallel focusing from the spreaders to the gantry output (as at PSI). This latter technique tends require large apertures and therefore heavy dipoles [8.1].

The basic parameters of the existing proton gantries are discussed in this Chapter.

8.1 Specifications for the Proton Gantries

The gantry should accommodate a scattering as well as a scanning system. Its performance must meet the specifications listed in Table 8.1.

A number of parameters determine the features and layout of the gantry without an integrated accelerator. These include:

- **The distance between the last bending magnet and the patient.** The beam must be uniformly spread out to the desired treatment cross section. With the spreading devices after the last dipole, the effective source to isocenter axis distance (SAD) will be smaller than the distance from the last magnet to isocenter. The spreading can be done before the last magnet; however that will increase the magnet size significantly. This distance must be large enough to reduce to an acceptable level, the ratio of the skin dose to the target dose, to provide the desired field size while not losing too much energy in the spreading.

- **The space required for patient movement about the isocenter.** This will determine the minimum gantry stay clear envelope for the patient support system and alignment devices.
- **The magneto-optical properties and beam trajectory requirements.** This includes the beam particle and energy. It is related to beam spreading methods to be used and optimization of the gantry size and weight. Beam spreading methods can include scattering and scanning or combinations of both.
- **Input beam requirements.** Input beam should be achromatic, rotationally invariant (at least for scanning) at the coupling point to the gantry to achieve invariant beam properties independent of gantry angle.
- **Mechanical properties / isocenter requirements.** This category includes mechanical deflections and reproducible, as well as non-reproducible, positioning errors, particularly of beam sensing devices and resulting in beam pointing errors at the isocenter, during gantry rotation. The vacuum requirements are modest at the 10^{-2} Tr level.
- **Magnetic field switching speed.** For some applications it is required to change the beam energy transported by the gantry while treating a patient, either for depth modulation, or for changing treatment portals. The magnets on the gantry must be capable of this.

Description	Specifications
Size (diameter)	< 13 m
Rotation range	± 370° (overlap at the bottom)
Rotation accuracy	± 0.3°
Rotation step size	± 0.3°
Rotation speed	1 min. for full rotation
Braking	1° to complete stop
Source to isocenter distance	< 3m

Table 8.1. Physical specification of the proton gantry; after W.T. Chu [5.1].

Comparison of proton gantries parameters are presented in Table 8.2.

	LLUMC, USA	PSI, Villigen, Switzerland	NPTC, Boston, USA	NCC, Kashiwa, Japan	HIBMC, Hyogo, Japan
Date	1991 (1st), 1994 (2nd and 3rd)	Nov. 1996	1998	1998	2001
Number of gantries	3	1	2	2	2
Beam	250 MeV protons from weak focussing synchrotron	225 MeV protons degrades from 590 MeV beam cyclotron	70 - 235 MeV protons from fixed energy isochronous cyc-lotron	70 - 235 MeV protons from fixed energy isochronous cyc-lotron	70-230 MeV protons from proton / Carbon synchrotron
Gantry type	corkscrew	single-plane eccentric compact	single-plane large-throw	single-plane large-throw	single-plane large-throw
Weight [tons]	96	120	~100	~100	120
Overall diameter [m]	13	4	~9 (length ~9m)	~9 (lengh ~9m)	10.4
Rotation [deg]		370	370	370	360 on ball bearings
SAD	2.75	$\propto$	2.25	2.25	
Field size [cm]	40 diameter	20 × ↔	30 × 40	30 × 40	5 – 15
Isocentricity diameter [mm]	0.7 × 1.6	2	2	2	
Beam delivery	double scattering system		telescopic nozzle containing scatterers scanning magnets, range modulators and beam monitoring elements		
Accuracy and precision measurements	beam position: < 1.6 mm	alignment precision: < ±0.2 mm isocenter mechanical precision: < ± 1 mm		accuracy of the isocenter: ± 1 mm accuracy of the stop angle: ± 0.5°	precision: 1 mm
Manufacturer	Science Applications Interna-tional Corporation	PSI & Schar Engineering AG	Ion Beam Applications & Ge-neral Atomics	Sumitomo Heavy Industry, Ltd	Mitsubishi Electric

Table 8.2. Protontherapy gantries worldwide.

8.2 Proton Gantries Solutions

8.2.1 Corkscrew Gantry

In this gantry geometry, introduced by Enge and Koehler (Sec. 4.1.4, Fig. 4.2(2)), the major bending occurs in the plane of rotation so that these magnets do not sweep out a large volume of space as the gantry is rotated as shown in Fig. 8.1. The three dimensional aspect of the beam trajectory led to the name *corkscrew gantry*. The result is that the building to house the gantry can be smaller.

Disadvantages are the reduced volume available for the patient positioner motion and for access to patient within the gantry open space, and the greater amount of bend.

The corkscrew gantry has one set of two 45° dipoles (B1 and B2 in Fig. 4.2(2)) with a quadrupole triplet in between and another set of 135° dipoles and triplet. The first set bends the beam out of the plane of the beam line. The second set bends the beam in the orthogonal plane toward the isocenter. Since this system involves two orthogonal bends, and the whole system is required to be achromatic, then each set must be achromatic. In order to achieve *achromaticity*, four parameters are required, leaving two more parameters in addition to the pole edge rotations. This is adequate to ensure a reasonable beam size to be transmitted, but does not leave much flexibility for subsequent optics changes. The corkscrew gantry implemented at LLUMC uses four quadrupoles per bend plane.

The optics tune produces a waist at 50 cm from the exit of the last dipole. The input is presumed to be a symmetric beam with up to 24 mm mrad phase space area and ± 0.5%.

In the LLUMC implementation the gantry is primarily supported by large rings. The bearing is supported from the floor and one wall by support struts. The gantry is a cone shaped structure made up a of 7 foot circular ring at one end and a 16 foot ring at the larger end. The plates are connected by struts. The assembly is fabricated in Secs. small enough to transport into the gantry room. The magnets at LLUMC (Fig. 7.3) are aligned to 0.2 mm individually and 0.4 mm gantry overall. Measurements of beam pointing accuracy to isocenter result in an isoshape of less than 1.6 mm diameter [8.2, 8.3]. The main parametters of LLUMC gantry are presented in the Table 8.2.

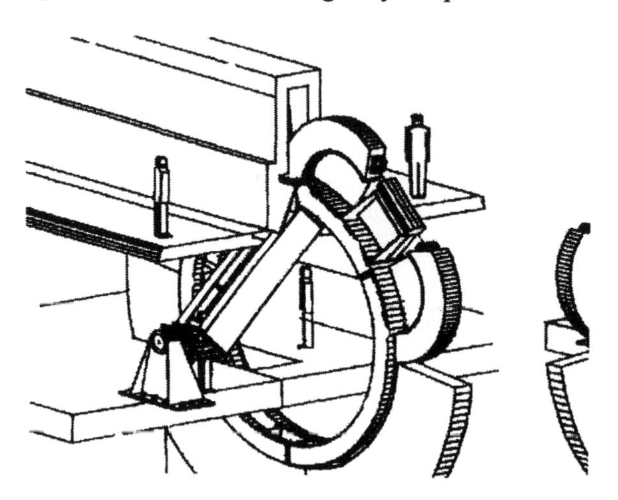

Figure 8.1. Schematic of Koehler corkscrew gantry; after J.B. Flanz [8.2].

A variation of the corkscrew gantry is the *supertwist gantry* suggested by Francis Farley (Sec. 4.2.1, Fig. 4.2(3)). This starts with the corkscrew physical layout concept and departs from the two orthogonal bend solution by stretching out the corkscrew in such a way that the total gantry length is longer and each magnet twists the beam in a trajectory through this path. Much of the focusing is done with pole edge rotations and the resulting system is achromatic at the end [8.2]. There is no information about practical realization of such a gantry.

8.2.2 Large Throw Gantry

A conventional gantry was built for MGH / NPTC by IBA and GA (Fig. 4.2(4)). The rotating elements begin with 4 quadrupoles in the plane of the beam switchyard which rotate with the gantry. They match the beam from the symmetric waist produced by the beam line to the gantry optics with an emittance of up to 32 mm mrad.

The beam is deflected through 45°, focused by five quadrupoles before it is bent through 135° and directed towards the isocenter. The distance from the output of the 135° dipole to the isocenter is 3.0 m. The gantry quadrupoles can be adjusted to produce a waist with a diameter of 12 mm at the isocenter. For scattering and wobbling, the quadrupoles can be tuned to produce a 10 mm radius waist at the center of the range modulator which is about 20 cm from the last dipole.

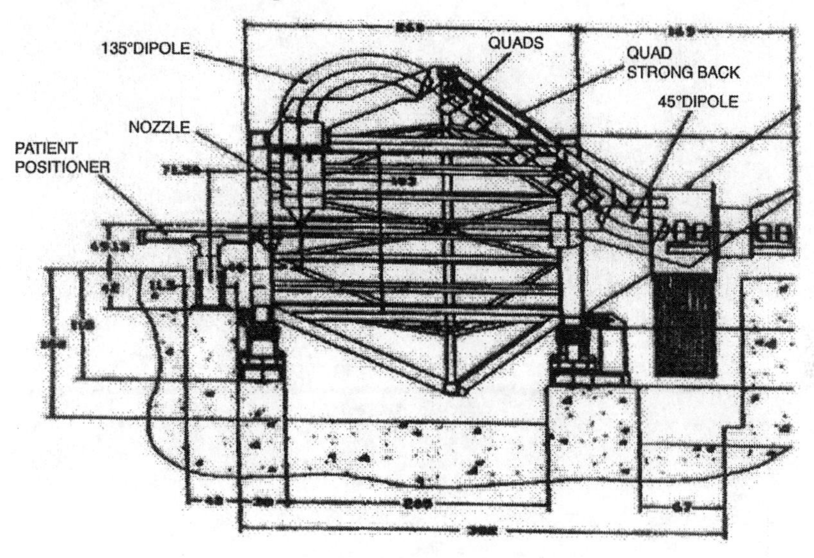

Figure 8.2. NPTC gantry layout; after J.B. Flanz [8.2].

There is sufficient space in the gantry to include beam position and profile monitors, which are capable of determining and correcting the beam trajectory angle and position to the required tolerance within the beam modification elements of the nozzle. This tolerance is basically submillimeter precision at the location of the scatters for scattering, or at the isocenter for scanning. The magnets accept a momentum spread of $\pm$ 5% $\Delta P/P$.

The rotating structure utilizes a configuration of rings, truss, and shell elements to support the magnets in a space frame. The structure is stiff and engineered to minimize its weight. The front ring is axially constrained. The truss elements are removable for ease of transport, assembly and possible repair. The structure is supported on both rings using a wiffle-tree assembly. Fig. 8.2 shows a schematic view of "large throw" gantry [8.2]. Fig 8.3 shows a picture of NCC gantry [8.11, 8.12].

Figure 8.3. NCC gantry; after T. Nijishio [8.11].

8.2.3 Compact Gantry

The PSI gantry design results from the special requirements of the beam at PSI and a general intention to design a gantry for potential users with inadequate space for a large throw gantry. A large beam phase space results from degrading the beam energy significantly (from 590 MeV to between 270 and 85 MeV just before the gantry. The gantry is a compact style, which spans a diameter of only 4 m and is designed to accommodate the spot scanning technique to used at PSI [8.2].

Fig. 8.4 shows the layout of the PSI gantry. The beam is first parallely displaced by two 35° magnets to a distance of 1.27 m from the gantry axis. The beam is then bent through 90° to be incident perpendicular to the supine patient, Quadrupoles and steering magnets arc used in the beam line to control position and focusing of the beam.

The supporting structure of the gantry consists of two large wheels connected with a welded box frame. The counterweight for the magnets of the beam line is included as part of the support itself and contributes therefore to the stability of the system. The structure is very rigid and massive. The magnetic elements of the beam line on the gantry are expected to maintain their alignment with a precision of about ± 0.2 mm during rotation of the gantry. The mechanical precision of the isocenter is expected to be better than ± 1 mm. The full weight of the gantry, including all rotating structures (beam line, patient table, supporting frame and counterweight), is about 120 tons. This great weight is mainly the consequence of degrading the beam from 590 down to below 250 MeV.

The patient table is mounted directly on the front wheel of the gantry, inside a rotating ring (1.4 m diameter) placed eccentrically (by 1.1 m) with respect to the gantry axis. When the gantry rotates, the patient positioned at the nominal isocenter moves on a circle of 1.1 m radius. The extra counter-rotation of the ring supporting the patient table is included in the mechanics of the gantry in order to maintain the patient table in the horizontal position during gantry rotation. The motors for the two rotations, of the whole gantry structure itself and of the ring on the frong wheel of the gantry, are synchronized in the software of the steering system of the gantry mechanics. The horizontal orientation of the table is controlled additionally by safety switches activated by gravitation. The *nominal isocenter*, i.e. the point where the beam axes intersect, is defined here in the reference frame of the patient table and is therefore the center of the ring and not the point where the beams cross the gantry axis.

The patient couch on the gantry allows for three translations and up to two rotations to be applied to the supine patient (normal 360° gantry rotation for all treatments and additional 120° rotation in the horizontal plane using a special couch for head treatments). Thus, for head treatments, the proton beam incidence can be chosen to be at almost any angle. Retractable mechanical holders are mounted in the sides of the yoke of the 90° magnet inside the anti-collision shield of the gantry head. These are used to move the instrumentation needed for the control of the positioning of the patient on the gantry (x-ray tube and film support) from the gantry head down to the isocenter level. This motorized motion also permits the insertion into the beam of the detectors used for proton radiography and tomography All the gantry motors are computer-controlled and work under remote control. The safety of the patient against mechanical injury is secured with an extended anti-collision system. The devices for the beam scanning are an integral part of the gantry design and are combined into the beam optics [8.4, 8.5].

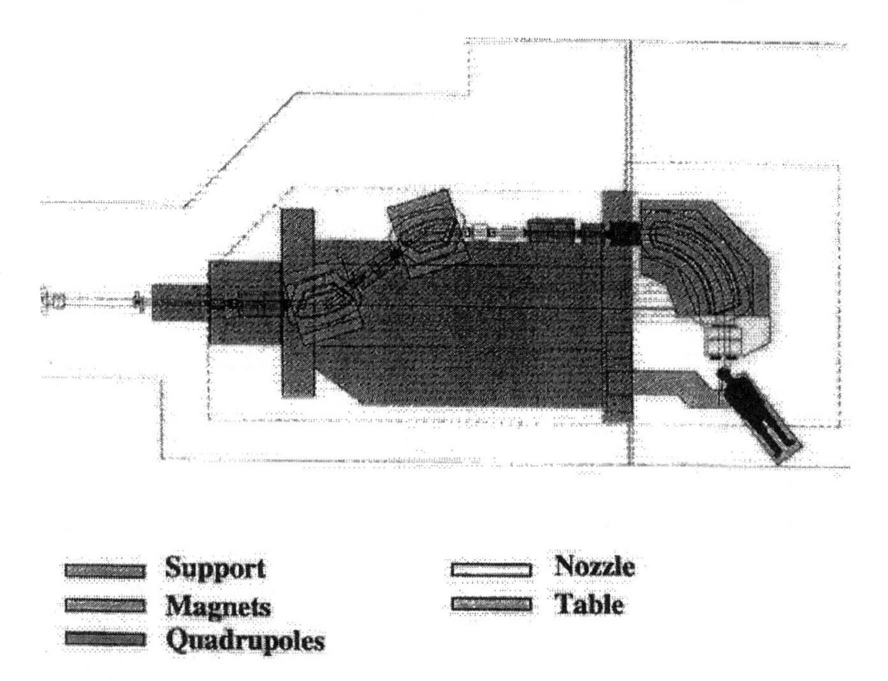

Figure 8.4. Schematic layout of the PSI proton gantry; after E. Pedroni et al. [8.4].

Spot scanning technique on the PSI gantry. The main advantage of spot scanning technique is that it allows for multiple beam port irradiations to be executed automatically on a compact gantry, without the need to modify or realign patient-specific hardware, such as range shifter wheels, collimators and compensators. With the PSI discrete spot scanning method, the dose distribution is constructed as the superposition of single dose spots (calculated in three dimensions) deposited at different positions in the patient body (Sec. 4.1.5, Fig. 4.4e). The spot sequence is executed according to a precalculated list of commands delivered individually for each patient by the treatment planning system developed at PSI. The position and the dosage can be chosen individually for each spot. Each spot is deposited like a static dose irradiation, by measuring the dose while the spot is being applied and by switching off the beam with the fast kicker magnet when the end of the single spot irradiation is indicated by the fast monitoring system in the gantry nozzle. The position of the beam is then adjusted, with beam switched off to the next spot position in the spot list stored in the computer memory. When all scanning devices are ready again for the new position, the beam is switched on, and the cycle repeats until the sequence of all spots in the treatment plan is completed.

The discretization of the spot applications has been chosen because this system is immune to beam intensity fluctuations of the split beam. The main disadvantage is the amount of dead time lost while the beam is being switched off during spot changes.

The position of the beam is first controlled by a sweeper magnet that displaces the beam along the gantry axis. This motion is the most rapid (about 30 ms is the time for a full sweep, around 3 ms dead time for a motion between neighboring spots) and will be the one most often executed. The sweeper magnet is mounted on the gantry immediately in front of thc 90° bending magnet. The beam is swept to the dispersive plane of thc gantry by up to ± 10 cm. This limit can be further extended by combining the magnetic scanning with an axial displacement of the patient table.

The second most rapid motion is performed with a range shifter system consisting of 36 polyethylene plates, each 5 mm thick, that completely cover the swept beam. This motion is used to control the position of the Bragg peak in depth. Additional plates are provided, one with half thickness, and the others with scatter foils used to modify the transverse size of the spot. The plates can be moved into the beam by pneumatic valves with a motion time of 30 ms and a dead time of about 60 ms per plate. The motion of the spot in depth is slower than for the sweeper direction, but, as this motion is used less frequently (middle loop of the 3D scan sequence), the dead-time contribution of the range shifter is similar to that for the sweeper magnet. Optical sensors provide the additional information on the individual setting of the plates, needed for the patients' safety system.

The third motion is performed by moving the patient table in the tangential direction (perpendicular to both the beam and gantry axis). This motion is the slowest and has the largest dead-time, but as it is normally carried out in a single sweep (the outer loop of the scan sequence), its contribution to the overall dead time is similar to that of the other devices.

They have chosen at PSI not to use a double magnetic scanning because they want to maintain the gap of the magnet so that it just covers the size of the unscanned beam. They therefore perform the magnetic scanning of the beam only in the dispersive plane of the gantry.

By choosing a proper shape for the entrance and exit faces of the poles of the 90° magnet, it is possible to displace the beam, by the action of the sweeper, such that all the resulting individual swept pencil beams are parallel to each other. The scanning of the beam can be so designed to be completely Cartesian in all three scanning directions.

The beam monitoring is performed with two plane parallel ionisation chambers. Position-sensitive ionization chambers with x and y strips are used dynamically to control the position of the scanned beam in front of the patient.

All scanning devices are monitored by at least two independent measuring systems and are checked with separate computer systems. A 3D treatment planning system based on complete sets of sequential CT images was developed at PSI specifically for spot scanning technique.

The devices necessary for the spot scanning technique have been designed for the deposition of typically 10 000 spots in 2 min in a 1 l target volume. This number was chosen to provide enough overlap between adjacent spot depositions in order to obtain a dose homogeneity over the whole target volume better than a few %.

The scanning system can profit from the very convenient time structure of the beam delivered by the PSI sector cyclotron (for medical application purposes a dc beam). The precision in controlling the dose was found in the set up for the horizontal beam to be better than 2%, with fluctuations of the split beam by as much as 100% [8.4, 8.5].

8.2.4 ITEP Proposal of Anti-Gantry (AG)

As a gantry project and realization still remains extremely expensive, so is seems necessary to search the way for lower investment costs. For this purpose the ITEP scientists tender a new anti-gantry system of an irradiation of an upright posed patient by a horizontal beam of protons or multicharge ions. They fulfilling a fixation of patient in a thin-wall shell, which is lookin-like cylinder, cut up on two halves along (Fig. 8.5a). A hermetic bag (2) filled on by small sized plastic pellets is put in a horizontally posed inferior half (1) of the shell. The patient (3) is put down on a bag. Then pump (4) suck out an air from the bag and the bag becomes solid. Further they covered the patient over by the second bag (5), and the second half (7) of shell. After, they connect both halves of shell in unit. Such hardened bags are stored in a "library" and used for fixing given patient daily. The MED.-TEC Inc. guarantees a safety of the relief of the surface of bags within 6-7 weeks.

The shell is established on the mobile stand (Fig. 8.5b). The upper and lower backs (8) of the shell are put in grider (9). Turning the gridder about horizontal axis (10), they turn up the shell in vertical position. The gridder is moved up- and downward along a vertical mount (11) of the stand. During the rotational-scanning irradiation, the shell with patient is rotated around of a vertical directions under a fixed or scanning horizontal narrow beam of particles (12). The design of the stand allows to place the shell with patient not only vertically, but also in horizontal or in intermediate position. In these cases, an irradiation with a method of linear scanning is carried out.

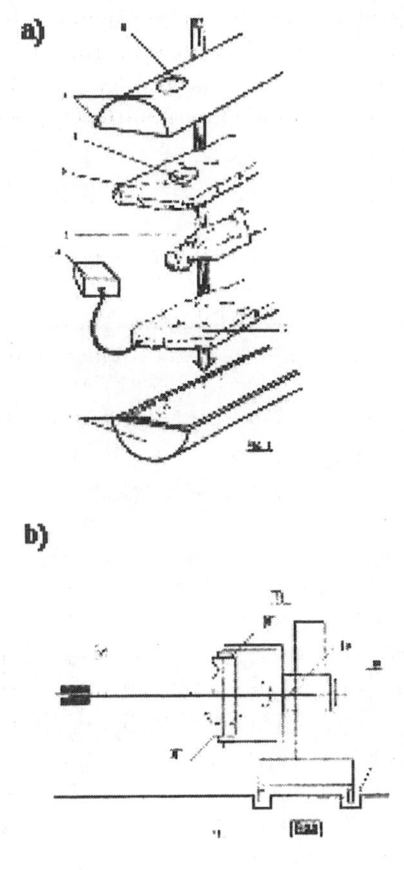

Figure 8.5. A ITEP proposal of anti-gantry: a) fixation of the patient, b) project of the shell and mobile stand, after B. Astrakhan [8.6].

The patients are prepared for an irradiation (including fixation, centration, serial horizontal CT) simultaneously and separately in six preparatory procedural rooms, placed outside of a zone of irradiation. The computer aided transport system automatically delivers the next patient from preparatory procedural room to radiative one and back together withmobile stand, without infrigement of quality of fixation and centratioin. Besides, through radiative room, the stands with patients are transported always two by two, in one direction, by the method, which one they have termed as the pendulum conveyer.

Such organization of preparation and irradiation of the patients allows to take the most of throughput of the accelerator (up to 1000-1200 patients per year) and carry oout the most conformal method – rotational-scanning or linear-scanning irradiation of tumors of any forms, sizes and locations. Anti-gantry can work with any medical accelerator of protons / ions needlessly either constructive alteration or adaptation. The area of 300 m^2 is necessary for accomodation of anti-gantry system (without rooms for the accelerator stuff). The cost price of a serial copy of the anti-gantry system is estimated for about 1 mln $, including cost of horizontal computer tomograps and an original construction of any procedural rooms. This is tens time cheaper than systems applying gantry to protons and in hundreds time cheaper than similar systems to ions of Carbon C^{12} [8.6, 8.8].

Prof. M.F. Lomanov from ITEP has commented the anti-gantry system proposition [8.9]. According his report, on the fixed beam there are the restrictions in selection of the patients. The sphere of diseases accessible for treatment was limited up to nineties by 5% of structure of oncological morbidity. All this has passed the analysis of the leading experts and contains in the IAEA document, Report on the Utilisation of Particle Accelerators for Protontherapy (7-10 July, 1998, F1-AG-1010, Vienna) [8.10]. There is shown that using gantry all modern means of topometry, Protontherapy applicability could extend up to 24-29% of all oncological morbidity structures. That is, exluding brachytherapy, practically up to a complete set of localizations accessible to external therapy. Certainly, it is possible to carry out revision of almost 50-year's experience of protontherapy. But before introduction in clinical practice of this offer, the cycle of preclinical researches confirming an opportunity of prospective technology to carry out the objective description of a target is absolutely necessary, to check up the preservation of a position of the patient and syntopy of a target and organs in a zone of interest at topometry in lying position and irradiation in sitting position. It is necessary also to determine a spectrum of localisations (probably, it will appear again 5%), by which it is possible to make a dose by the fixed beam, and to solve many other questions [8.9].

Chapter 9

RADIATION DETECTORS

9.1 Dosimetry Principles

The basic sequence of dosimetry procedures implemented during accelerator beam therapy is depicted in Fig. 9.1. The first stage is *absolute calibration of a dosimeter*, which most often is an ionization chamber or, less frequently, a semiconductor, thermoluminescent detector, or dosimetric film (Sec. 9.2). The second stage consists in *determining the absorbed dose at a well determined reference point* in a homogeneous standard medium, such as water or, sometimes, a plastic phantom (Sec. 9.3). In the third stage, the *radiotherapy field distribution* is determined, that is the absorbed dose distribution is measured at an arbitrary point in the phantom.

Before stage four is started, *data on patient topography* must be taken into account. These are obtained from such diagnostic techniques as radiography, x-ray computer tomography, MRI tomography, etc. The main point in stage four is to determine the *absorbed dose distribution in a particular model of a patient.* All the procedures implemented in this stage, including the biological effectiveness of the type of the radiotherapy beam used, are generally called *radiotherapy planning* (Chapter 10).

The fifth stage in the sequence of dosimetry procedures is to perform patient irradiation proper, that is to *deliver a dose to the target (tumor) volume*, monitored or measured by an internal accelerator dosimetry.

Direct patient dose monitoring is of particular importance, since it makes it possible to detect systematic errors made radiotherapy planning and in some irradiation techniques used. In addition, random errors in dose calculation, patient positioning and accelerator operation can also be exposed. What is important is to monitor the dose level inside critical organs or close to them. However, direct measurement of the dose in the tumor volume, except for some special cases, is not feasible. Instead, in practice, *surface doses on the patient skin* are determined with ionization chambers, semiconductor or thermoluminescent detectors.

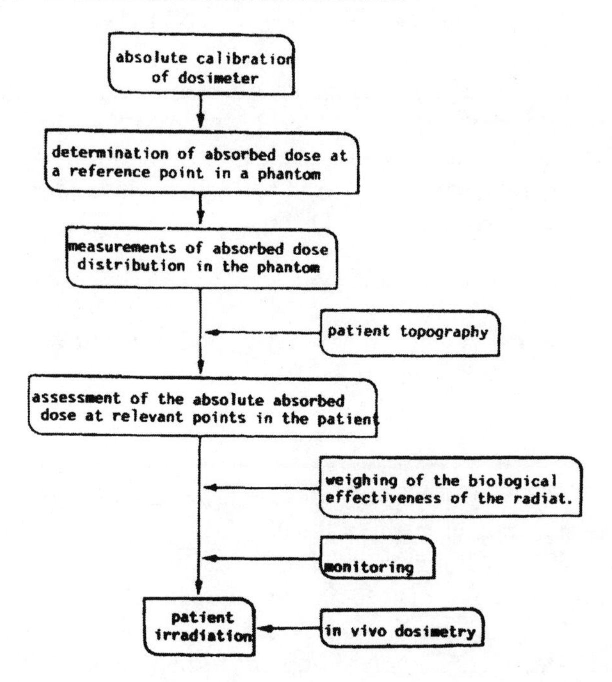

Figure 9.1. Sequence of dosimetry procedures in radiotherapy; after J. Mijnheer [9.1].

According to the recommendations included in some dosimetry protocols (NACP 1980), the dosimetry of patient doses should be carried out at two safety levels. At the first level, systematic errors in radiation therapy planning and irradiation procedures should be detected. This includes measurements for each radiation field during first irradiation sessions as well as those made when the patient anatomy is changed.

The second level incorporates measurements carried out during the whole series of irradiations, with the aim of detecting any errors in setting up irradiation parameters and in accelerator output values.

The main trend in the development of the present-day radiation therapy dosimetry is to make a detailed analysis of errors at all stages of measurements and, consequently, to minimize them. For this purpose, *clinical dosimetry procedures* are being standardized with the aid of dosimetry protocols and codes of practice described (Sec. 9.4)

In view of the well-known dependence of field parameters on the field size, accelerator dosimetry techniques have witnessed a division into *small field dosimetry* and *large field dosimetry*. The term *small field* has not been universally defined; it is, however, usually taken to describe square fields whose side is between 5 and 8 cm. It is clear that small field dosimetry must also account for measurements of the smallest possible fields with diameters or sides sometimes as small as 1 cm [1.22].

9.2 Beam Monitoring, Dosimetry and Microdosimetry

9.2.1 Monitoring Chain

In general, dosimetry systems for radiotherapy and radiosurgery have three main functions: (1) measurement of the dose being delivered to the patient in real time in order to terminate the irradiation at the prescribed dose, (2) measurement of the lateral distribution of the radiation delivered in order to insure that the patient prescription is satisfied, and (3) measurement of radiation field parameters for controlling the beam delivery system.

The purpose of the dosimetry system is to measure precisely the delivered dose to the patient. The detectors used in such systems must measure a quantity proportional to the dose imparted to the patient, must not significantly perturb the radiation field, and must measure the dose independent of the LET distribution in the beam. This is especially important for light-ion beams, which contain particles with a wide range of LET values due to fragmentation of the projectile particles. In addition, large radiation detectors with fine spatial resolution are important for comparing the prescribed and delivered radiation distributions over the entire treatment area. A required dose accuracy for clinical purposes of $\pm$ 2.5% (Sec. 5.1.5), measured over a large dynamic range, adds to the difficulty of monitoring the radiation.

The components of the monitoring chain strongly depend on the choice of the beam delivery systems. While for a passive system the techniques used in conventional radiotherapy can be the basis of the monitoring chain, when an active system is used most of the monitoring system should have to be of novel design.

9.2.1.1 Ionization Chambers

Ionization chambers (Fig. 9.2a) are the primary dose measuring devices. The ionization of the gas contained in the detector is proportional to the energy loss, which in turn is approximately proportional to the dose absorbed by the detector. This dose can then be related to the absorbed dose in another medium, such as human tissue, by the ratio of the stopping power of the gas to that of the medium. The ionization Q (measured in Coulombs) produced by the beam passing through such a chamber is proportional to the *absorbed dose D* (measured in Gray), by what is known as the Bragg-Gray equation:

$$Q = \frac{D\rho V}{W} \tag{9.1}$$

where ρ is the mass density (kg m^{-3}), V is the volume (m^3), and W is the ionization energy (eV) per ion pair. The amount of ionization per unit of deposited energy varies for different types of radiation. However, for proton and light-ion beams at clinically relevant energies there is only a weak dependency of the W value on particle type and velocity. A value of 34.3 eV for protons in air and 33.7 eV for heavier ions in air are standard W values. Values of W for other gases are recommended by the *International Commission on Radiation Units and Measurements* and more recent measurements have been made for protons in nitrogen gas [4.1].

In principle any type of reliable ionization chamber used for photon dosimetry can be employed in proton dosimetry. However, most centers involved in proton dosimetry use *tissue-equivalent ionization chambers*, made of A-150 plastic. If other types of ionization chambers are used, as listed for example in IAEA Technical Report Series No. 277 [9.2], the formalism can easily be adopted. The choice of the cavity gas filling of the ionization chamber is strongly dependent on the knowledge of the energy required to create an ion pair in the gas $(W_{gas}/e)_p$ for proton beams [9.3].

Transmission ionization chambers are the most widely employed systems to measure the total beam charge, current and uniformity in the case of conventional radiotherapy beams.

Homogeneity has to be controlled in a broad beam arrangement, so that recombination problems do not arise. To this purpose a set of transmission chambers, usually two, subdivided into sectors is used [1.4].

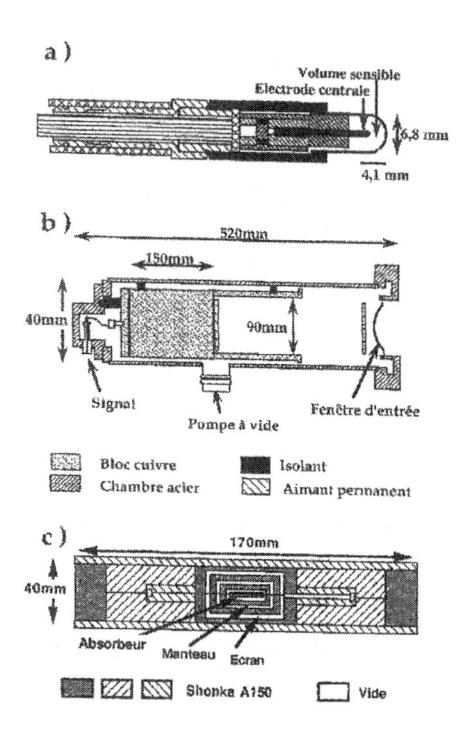

Figure 9.2. Schematics of some currently used proton beam dosimeters: a) ionization chamber, b) Faraday cup, c) calorimeter; after A. Mazal et al. [9.4].

Longitudinal ionization chamber has been constructed at TSL for Uppsala facility. This parallel-plate ionization chamber is used to measure a narrow proton beam, which is introduced between the plates and parallel to them. In this design, the plates do not have to be thin as the particles do not penetrate them, and the electrodes consist of thin layers of silver on plexiglass plates. The chamber measures the total number of particles.

A *position sensitive ionization detector* has been developed at PSI, which consists of a high-voltage foil tilted with respect to two parallel signal planes. The ratio $Q_1 / (Q_1 + Q_2)$ of the charges collected on both sides of the tilted foils measures the mean position of the beam. Resolutions of better than 0.5 mm have been achieved. Since this detector is used in conjunction with a spot scanning beam delivery system, the collection time is required to be less than the dwell time of the beam to obtain a true position measurement. Measurements show this to be under 1 ms with charge measurement times on the order of 10 μs. To control the dose being delivered the response time of the detector system must be even faster than this or on the order of the minimum dose divided by the dose rate and the required dose accuracy. Less than 100 μs is required for the PSI detector [4.1].

Calibration methods for ionization chambers. The dose absorbed by an ionization chamber and averaged over the sensitive area can be calculated from the charge collected during the irradiation, the charge collecting volume, pressure, and temperature of the gas in the active volume, and the W value. By correcting the ionization chamber reading with the ratio of the stopping power in the gas and that of another material, for example tissue, the dose, which would have been received by the other material in place of the ionization chamber can be deduced.

Mainly because of uncertainties in the W value, it is desirable to calibrate an ionization chamber against some other method before using it in a dosimetry system. For irradiation with monoenergetic particles a *Faraday cup* can be used as a calibration device (Sec. 9.2.1.3). In practice, however, Faraday cup measurements exhibit errors of up to ± 5% mainly because of uncertainties in the charge collection and, second, because light-ion beams often contain an added mixture of secondary particles. The most basic and absolute measurement of absorbed dose is done with a calorimeter which measures the deposited energy independently of particle type and energy. Water calorimeters have successfully been used and can serve as a calibration standard (Sec. 9.2.2.1).

In order to be able to measure on-line the dose received by a patient, a second type of calibration is needed. The ionization chamber and other dosimetric measurements of the radiation field have to be related to the dose absorbed in the target volume at isocenter. This is best done by placing a small calibration detector, usually a calibrated *thimble ionization chamber*, in a tissue equivalent phantom simulating the treatment geometry and calibrating the dosimeters against it [4.1].

9.2.1.2 Wire Chambers

In radiotherapy with heavy charged particles, wire chambers are primarily used for beam monitoring during beam-line tuning. The beam characteristics of interest are: locating the centroid of the beam spot within 1 mm, and monitoring the beam shape, which is generally Gaussian varying between 1 and 15 cm. The fact that good spatial resolution is obtained with a small number of signals and variable sensitivity make these devices extremely practical for this purpose. In the simplest form, a wire chamber consists of two planes of evenly spaced parallel wires separated by a given distance, and the intervening volume filled with gas. One wire plane is used as the signal plane, and the other as the high-voltage plane. The high-voltage plane can also be a solid foil conductor instead of parallel wires. While the common application of wire chambers in particle physics is to count traversing particles on a particle-by-particle basis, in medical applications wire chambers are more often used as integrating devices that integrate spatially along a wire as well as temporally during an irradiation procedure. Beam intensities from 0.1 pA to 0.1 mA are used for experiments ranging from microdosimetry to radioactive beam imaging [4.1].

On line monitoring of beam profile and centroid position. Two planes of orthogonal sense wires can show the horizontal and vertical projections of the beam profile. The centroid and the width of the beam can be computed in each plane for beam-stability information. A comparison between the beam profile and a desired profile is a useful guide to the operator for tuning the beam delivery system. Monitoring beam profiles is important for efficiently transporting the beam and obtaining the correct dose distributions with the beam delivery system. Typically a wire chamber is positioned upstream of the beam delivery system and a second wire chamber is located downstream of the delivery system, near the patient. The active area of the second chamber is typically large to accommodate the large radiation fields [4.1].

Dose profile monitoring. At NIRS. a multi-wire ionization chamber (MWIC) was developed to monitor the beam profile of the dose delivered by spot-scanning methods. The chamber is operated in an ionization mode. The chamber has 100 of 0.18 mm diameter tin-plated copper collector wires, 12 mm long and spaced 1 mm apart. Two high voltage planes are placed on either side of the collector wire plane at a distance of 2.5 mm. Each wire is surrounded by a sensitive volume of 1 mm by 12 mm by 5 mm. The chamber has 1.5 mm thick Lucite windows. A similar multi-wire ionization profile monitor has been used at PARMS and LLUMC [4.1].

9.2.1.3 Faraday Cups

These devices can be used to measure the *number of particles*, i.e *beam current* and, therefore, of *particle fluence*, by measuring the charges collected on an electrically isolated and evacuated container, which stops the beam (destructive measurement). Typical Faraday cup designs (Fig. 9.2b) maximize their collection efficiency by considering the thickness of the cup, the guard ring bias voltage, the uniformity of the bias field, the vacuum, and the effects of nuclear interactions. When monoenergetic heavy charged particles impinge upon the Faraday cup, the dose is calculated from the measured number of particles per unit area and the mass stopping power of the particle at the given energy in tissue. In practice Faraday cup measurements are combined with an ionization chamber, exposed in air, in order to calibrate the ionization chamber for dose determination in a phantom [4.1, 9.5].

The disadvantage of the use of Faraday cup is that some precautions should be taken with measured clinical proton beam. Firstly, the whole beam has to interact with the cup, otherwise the measured current will be restricted to that of the particles transmitted by the cup diaphragm. Secondly, the net charge collected must be limited to that of the beam particles and, therefore, effects such as back-scattering from the absorber, lack of total particle absorption, background current, electron production in the entrance window, etc. must be taken into account. At present, no Faraday cup is working as a standard for proton beams, but this fact is only due to lack of suitable proton sources and not to the existence of basic problems [1.4, 9.5].

9.2.1.4 Scintillators

Scintillators have been used extensively in high-energy nuclear physics experiments for single-particle detection. In medical accelerators they have been used primarily for beam control because of their fast response time and large dynamic range of operation when used in conjunction with phototubes. When used to measure large beam currents, radiation damage to the solid scintillator can limit their lifetime and effect their response. A Xe gas scintillator has been used to control the beam at LLUMC. Because light production in the scintillator is not simply proportional to the energy loss of the particle due to quenching, their use as a dose detector requires special calibration. Recently plastic scintillators have been used in conjunction with charged coupled devices (CCD cameras) to measure the lateral uniformity of a radiation field and to measure the scintillation light as a function of depth for range verification [4.1].

9.2.2 Reference Dosimetry

Any relative dosimeter has to be calibrated against an absolute one. To this end it should first of all be pointed out that published material on reference dosimetry of proton beams is not as abundant as for photons and electrons. Instruments intended for the measurement of absorbed dose need to be calibrated in the frame of a proper calibration system. This metrological system must be based on a *primary standard* which provides the reference absorbed dose unit without any previous calibration, i.e. by an absolute measurement. While for photons and electrons, primary standards are well established, no similar system is so far available for protons. The difficulty in realizing an absolute dosimetry for protons with energies of interest in radiotherapy stems from both, the poor knowledge of some physical parameters and the many effects to be accounted for in the data analysis. Two methods, based on *calorimeters* and *Bragg-Gray cavity ion chambers*, respectively, are worth considering for absolute dosimetry of proton beams. Each of them presents its own advantages but also some problems.

9.2.2.1 Calorimeter

Absolute dosimetry by calorimeter (Fig. 9.2c and 9.3) is the most direct and potentially the most accurate method, provided that the kinetic energy of the particles interacting with the calorimetric medium is completely converted into heat. They yield the smallest uncertainty in the determination of the absorbed dose for proton beams, as the measured quantity is the temperature rise caused by the energy absorbed. The calorimeter should be made of a tissue equivalent material for protons, with no heat defect when irradiated and with good thermal properties, in particular as far as the thermal capacity and the thermal diffusivity are concerned. As no material with all these properties is today available, some compromises must be accepted: the thermal properties of water have been well known since long, but it should be demonstrated that water heat defect for protons is the same as for photons and electrons. For other nearly tissue equivalent materials, the problem of thermal properties is even more open than for water. In some laboratories water calorimeters have been built for photons and electrons, but they have not yet been recognized as primary standards because of some problems in measurement reproducibility. A standard of the same type for protons can be constructed, but additional research work is needed to overcome the problem of the heat defect in water when irradiated with proton beams. The sensitive elements of the calorimeters used for proton dosimetry can be water, graphite or A-150 tissue-equivalent plastic [1.4, 9.4].

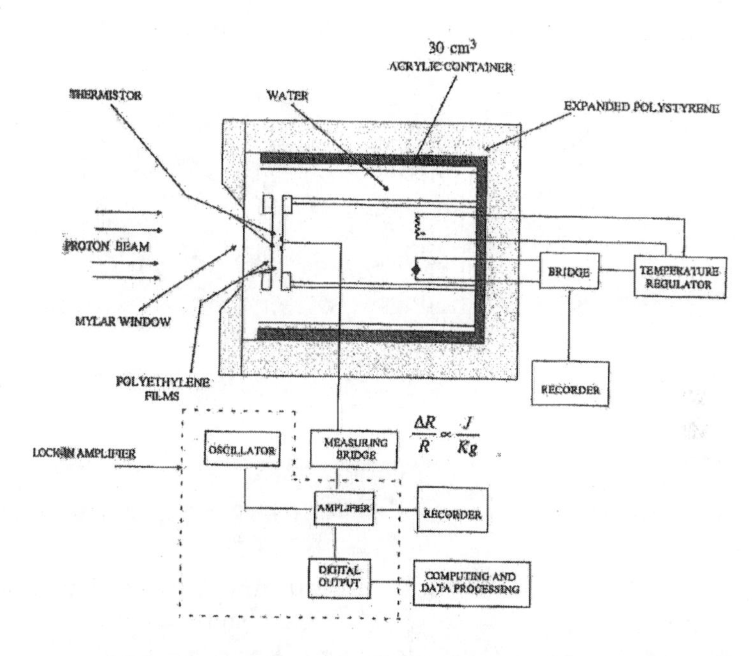

Figure 9.3. Schematic of the water calorimeter for horizontal proton beams. Only the main thermistors and the immersion heaters are shown. Other components as the gas inlet in water, the mechanical for varying the position of the measuring point, etc. are not shown; after U. Amaldi et al. [1.4].

9.2.2.2 Bragg-Gray Cavity Ion Chamber

The Bragg-Gray cavity theory widely employed for absolute dosimetry of photon and electron beams, also applicable to proton beams, as long as the basic requirements of the theory are satisfied. The production of γ-rays and the perturbation of the particle fluence and energy spectrum may require smaller corrections than for photons, provided that a proper chamber design and a homogeneous chamber-phantom system is realized. As the effective collecting must be accurately known and a plane-parallel chamber is the most appropriate for protons (Fig. 9.4), particular care is required in chamber design to avoid electric field distortions. The chamber (Fig. 9.4b) provided with a suitable guard ring is embedded in a polystyrene block to realize a homogeneous system within the polystyrene phantom (Fig. 9.4a). To allow measurements at variable depths in the phantom and to avoid charge-storage effects, the phantom is made of thin movable polystyrene slabs [1.4].

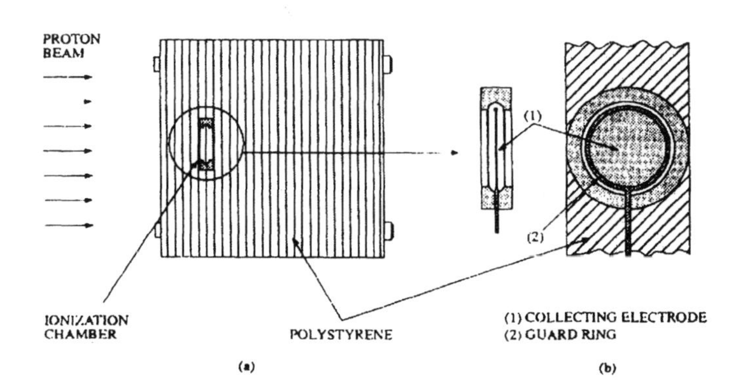

Figure 9.4. Schematic of the plane-parallel ionization chamber and related phantom for absolute measurement of absorbed dose in proton beams; after U. Amaldi et al. [1.4].

The aspect, which calls for further investigations, is the choice of the stopping power entering the Bragg-Gray expression: at proton energies larger than some of MeV nuclear interactions, are not negligible. Therefore the stopping power must be calculated at an energy lower then occurring in absence of nuclear interactions. To this end, the energy spectrum of secondary protons should be known and this requires adequate theoretical calculations. Finally, it should be emphasized the lack of direct information about the mean energy require to form an ion pair. Accurate experimental values are only available for very low energy protons; an approximate value can be obtained by extrapolation to higher energies, with an uncertainty of at lest 4%. Therefore improvements are also needed on this issue [1.4].

9.2.3 Systems for Relative Dosimetry

Relative dosimetry includes all those measurements that are not performed with reference systems and in reference conditions. Relative dosimetry may be carried out either in a phantom or in the patient (in-vivo dosimetry) and different detectors and techniques are more or less suitable for different purposes. It is important to use detectors whose response does not vary in different conditions, as for example at different depths on the beam axis or in fields of different dimensions. The response of detectors can also vary in a well known way, as is the case when one wants to calculate a depth-dose curve from a depth-ionization curve using an ionization chamber and applying thc stopping power ratio correction [1.4].

9.2.3.1 Ionization Chambers

Ionization chambers can be employed as relative systems in many situations with reproducibility better than 2%. The main advantages of this kind of detectors are the high stability of response and the very good knowledge of the effects of the ambient conditions. On the other hand, ionization chambers are in general delicate devices that must be handled with great care. Moreover the sensitive volume of commercially available chambers does not allow to reach a very good spatial resolution. However some technical improvements can be studied, in order to make ioniazation chambers competitive also from the point of view of resolution [1.4].

9.2.3.2 Thermoluminescent Dosimeters

Thermoluminescent dosimeters (TLDs) are inorganic materials, such as LiF of CaF_2, that do not scintillate promptly, but rather trap electron-hole pairs created from the exposure to ionizing radiation. Upon heating the exposed material, the trapped electrons can escape leading to light emission which can be detected. The light output is a function of temperature, often called a *glow curve*; the total light output is then related to radiation exposure. The light output is a function of the LET and the particle species composition of the heavy charged particle beams. Thermoluminescent LiF dosimetry has been used to characterize the therapeutic heavy charged particle beams. These methods are not amenable to realtime analysis, but are useful for treatment verification purposes. In general, thermoluminescent dosimeters have precision better than 2%. The accuracy of TLD's is critically dependent on the standardization techniques used. However, the accuracy better than 5% is easily attained for radiotherapy beams [1.4].

A TLD system based on LIF: Mg, Cu, P (Gr-200A) has been tested in CAL for dosimetric purpose of TERA Project.

9.2.3.3 Diodes

Semiconductors work as radiation detectors because ionizing radiation liberates enough free electrons in the silicon crystal to generate a small current across the junction. *Diode detectors*, connected to appropriate circuity, can be used for relative dosimetry. The high sensitivity to radiation, small size and absence of external bias make their use advantageous. Moreover, the signal coming from the semiconductor during dosimetry is very fast and independent of atmospheric pressure variations. Moreover, the usefulness of diodes as dose detectors derives from their small size. Measuring dose distributions with fine spatial resolution is difficult with ionization chambers, but simple with diodes, which intrinsically have small active areas. As a solid, they absorb more dose in a small volume. A calibration of the diode must be done before use since they suffer radiation damage over extended exposure. They are, however, excellent for relative measurements over short time periods. Arrays of diodes have successfully been used for measuring profiles of radiation fields and Bragg ionization curves [1.4, 4.1].

9.2.3.4 Films

Films have been used since the discovery of X rays and is still the most economical method for measuring spatial dose distributions with fine spatial resolution. Clinical dosimetry of complex dose distribution associated with charged-particle radiotherapy requires a rapid and convenient technique with good spatial resolution. The spatial resolution achievable with film is better than that of any other method. It is limited by the resolution of the digitizing procedure which, in turn, is limited by the time required for digitization and data reduction. The darkening of the film, i.e. optical density, after exposure to a heavy charged particle beam depends more on the particle fluence than the dose deposited. In other words, it depends on the LET of the particles. For this reason film is primarily a relative dose detector and is used for measuring lateral dose distributions rather than depth-dose distributions. The use of *cellulose triacetate (CTA) film* dosimeters have been also investigated with proton beams. The optical density change per unit absorbed dose is measured. The CTA dosimeter was found to be useful to obtain dose distributions with high spatial resolution in samples exposed to proton beams [4.1].

9.2.4 Microdosimetry

Microdosimetry measurements utilize gas proportional counters to study energy losses on the microscopic scale of a cell. Gas proportional counters are used with sizes and at gas pressures that replicates the energy deposition in a cell and which have enough gain for reasonable signal processing. Experimental microdosimetry is performed with proportional counters made on tissue-equivalent conductive plastic and employing tissue-equivalent mixtures as counting gases. Dose, being defined as a macroscopic quantity, does not account for the fluctuations, inhomogeneities, and stochastic nature of the effect of radiation on matter, even though these effects are important in understanding radiation biology of cells and smaller structures. To understand the effects of the inherently statistical nature of the energy deposition in a cell, a detector comparable in size to a cell is required [1.4, 4.1].

Microdosimetry is based on a conceptual and mathematical framework coming from the probability theory. For the following discussion it is sufficient to recall the definition of specific energy per critical volume, z^*. The critical volume is a sensitive volume, which has experienced at least one event of energy deposition. The specific energy z is the quotient of ε by m, where ε is the energy imparted by an ionizing radiation to a mass m. When the volume is small and the total absorbed dose is low, a large amount of sensitive volumes have experienced no events and therefore for them $z = 0$. The average values of z and z^* are different, being $z^* > z$. For high doses, obviously $z^* = z = D$. In such case the macroscopic absorbed dose D is meaningful even for microscopic volumes.

Microdosimetric investigations of therapeutic hadron beams are scarce and concentrated on fast neutrons. The only data available to assess the microscopic energy distribution in therapeutic ton beams are the microdosimetric measurements collected at the 160 MeV proton beam at HCL [9.2]. Using those data, z^* has been calculated at different steps along the beam axis and at different absorbed doses.

In Fig. 9.5 the thick line represents the mean specific energy per critical volume, which has been chosen to have 1 μm of mean diameter. The quantity $\overline{z^*}$ is plotted against the absorbed dose, for different positions in the depth dose profile. The two thin lines are drawn at a distance form the average value $\overline{z^*}$ equal to the standard deviation of z^*. The shadowed area between the lines is therefore the area of the main fluctuation of z^*. The dashed line is the percentage number of volumes struck by at least one event [1.4].

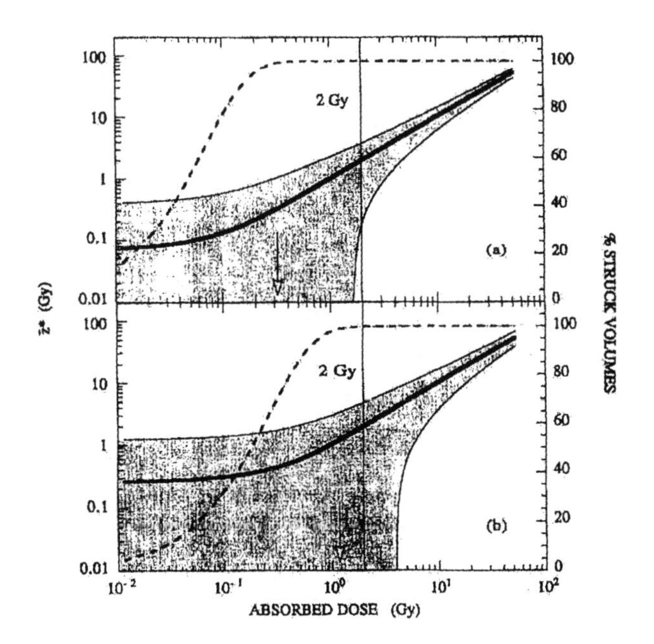

Figure 9.5. Microdosimetric parameters versus macroscoppic dose: a) in the middle part of plateau region, b) in the near part of Bragg peak region of 160 MeV proton beam; after U. Amaldi et al. [1.4].

9.3 Phantoms

In clinical dosimetry, phantom materials should be closely matched to tissue, i.e. they should have very similar absorption and scattering properties for the radiation beam considered. They are used for two purposes: firstly to simulate the presence of tissue when determining the absorbed dose under reference conditions (reference field size, reference depth, etc.) and secondly to obtain basic dose distributions, i.e. percentage depth doses, off-axis data and isodose curves.

In most clinical proton beams a SOBP is produced, which can be of different widths, depending on the clinical requirements. The center of this SOBP is a region of dose uniformity and calibration of the beam should therefore be done in the middle of the SOBP. Absorbed dose for proton beams is specified in water and calibration measurements in the middle of the SOBP are more easily performed in water. However, measurements can he made in a *solid phantom* (e.g. acrylic or solid water) provided the depth of measurement is scaled to the depth in water. The scaling factor can be obtained from the ratio of the CSDA (continuous slowing down apporoximation) ranges in the different materials, and can be derived from the data given in ICRU Report [9.3].

9.4 Dosimetry Intercomparisons and Protocol for Protontherapy Beams

When efficacy of new radiation therapy modality is being investigated, it is important that dosimetry be compatible among the various institutions involved in the clinical trials and that the dosimetry be accurate. This is necessary for the comparison of radiobiological and clinical results from the different institutions and for the pooling and analysis of data.

Institutions currently follow one of two dosimetry protocols, either the protocol established by the *American Association of Physicists in Medicine* (*AAPM*), or that provided by the *European Heavy Particle Dosimetry Group* (*ECHED*). Both protocols permit the use of calorimeter, a Faraday cup, or the most common device used to measure proton dose, the ionization chamber. Measuring proton dose with a calorimeter is considered the most accurate method, but because it is an extremely delicate and difficult procedure it is not a candidate for standardization. Ionization chambers, on the other hand, are easy to use and can be calibrated with a cobalt beam ^{60}Co [9.6, 9.7].

Methods and equations for converting the charge measured by an ionization chamber into dose absorbed by muscle or water are similar in both the *AAPM* and the *ECHED* protocols. However, the factors employed in equations, such as the proton stopping power and *w*-value, are different in the protocols. In addition, some institution report dose as absorbed dose to water, others as absorbed dose to muscle. The goal of the intercomparison was not only to determine the consistency of dose calibration between the institutions but also to evaluate what effect the different protocols have on any observed differences.

The examples of such a comparison are presented in [9.8, 9.9].

9.5 Examples of Monitoring and Dosimetry Systems

The components of the monitoring system strongly depend on the choice of the beam delivery systems discussed in Chapter 7. While for a passive system the techniques used in conventional radiotherapy can be the basis of monitoring chain, when an active system is used most of the monitoring system will have to be of novel design. At present stage the attention is mainly devoted to passive systems, since it is not possible to give detail information on the monitoring of active systems, which is still under study in the centers, where such techniques are going to be soon employed. The monitoring system should control the delivered absorbed dose and, therefore, the beam integrated charge, current and energy, the beam position and its uniformity in the transverse plane.

The correspondence between prescribed and delivered dose at a reference point, within a given overall uncertainty, which has been fixed to be $\pm 2.5\%$, is the primary purpose of a monitoring system. Dose delivery must be promptly stopped as soon as the prescribed dose is reached. The dose per pulse, according to the beam data and the different beam delivery conditions, can be estimated to be of the order of a few cGy at the Bragg peak. The capability of turning off the beam during an extracted pulse is essential and, therefore, the reaction time of the control system must be much shorter than the pulse duration.

Dose rate has always to be controlled and kept within the accepted limits; the control has to be more stringent in the case of an active system. In this case the requirements on homogeneity on a transverse cross Sec. of the beam could fail if the beam current is not strictly controlled. In principle, it is possible to modulate the scanning velocity on the basis of the beam intensity, or to modulate the beam intensity on the basis of constant velocity [1.4].

GSI concept of the dose-verification system. In a heavy-ion therapy facility, it is particularly important to perform a fast routine verification of the good operation of the whole-beam delivery system. For this purpose at GSI facility they have started to realize a telescopic radiation-detector system, in order to compare the prescribed and the delivered three-dimensional dose distributions over the entire treatment volume. With the active beam-delivery system developed at GSI, a pencil beam is scanned pixel-wise over the transverse treatment area slice by slice, each slice corresponding to a different beam entry energy, until the whole target volume is treated. In this case, the beam profile and its position, apart from the uncertainty due to the multiple Coulomb scattering, are not depending on the penetration depth at every instant; thus one position-sensitive device is sufficient to identify the transversal beam profile and its locus in the complete target volume. The therapeutic beams will use $^{12}C^{6}$ with energy varying from 90 to 430 MeV/u and fluences from 10^6 to 10^8 ions per spill (2 s). The measurement of the dose therefore requires a detector with a large dynamic range in the order of 10^4 and a good linearity for high intensity radiation. On the other hand it is necessary to get a spatial resolution at least comparable to the dimensions of the voxels as from a typical CT scanner, i.e., of 1 mm.

All these requirements can be met by coupling parallel-plate ionization chambers measuring the dose integrated over a relatively large area, together with a single *multiwire proportional chamber* (*MWPC*) that can provide the beam profile and its position with the requested spatial resolution. Along the beam path, the ionization chambers (Fig. 9.6) with a sensitive area of 25×25 cm^2 sandwiched between 1 or 5 mm thick removable plastic plates, will provide the depth-dose profile measurement. In front of the ionization chamber stack, a single MWPC will provide the bidimensional position information with a spatial resolution of 1 mm over the transverse treatment area. Combining finally the readout of the MWPC with the dose measured along the beam axis, it will be possible to reconstruct a three-dimensional dose profille.

In order to accommodate the detector setup according to the different treatment plans. a range shifter with a range from 19 to 300 mm of water will center the enlarged Bragg peak on the ionization chamber stack, providing a longitudinal spatial resolution of 1 mm for the enlarged Bragg peaks and up to 50 μm for single peak measurements [9.10].

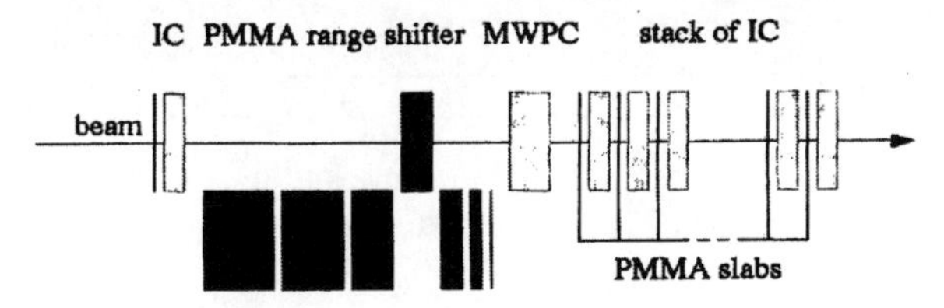

Figure 9.6. Sketch of the three-dimensional dose verification system, after C. Brusasco et al. [9.10].

9.6 Cost Estimations

Monitoring systems and reference dosimetry are essential components of the hadrontherapy center. The cost estimations concerning monitoring and dosimetry system were performed by U. Amaldi et al. [1.4]. Since at present no dose and beam position monitor is commercially available, the cost estimate can only be based on the costs of the chosen dosemeters or detectors; design and development are not included. Monitoring for passive beam delivery can be similar to the conventional systems and the cost estimate given in Table 9.1 would include a dual ionization chamber dosimetry and its logic system. On the other hand for an active beam measuring device the total cost estimate is determined mainly by the number of electronic channels needed for an intensity x-y position decoding system. Estimate is given for three passive systems and three systems employing magnetic scanning [1.4].

Monitoring systems	
Passive systems	
Hardware (3 units)	150 000
Realization and testing (software and controls)	500 000
Active systems	
Hardware (3 units)	1 500 000
Realization and testing (software and controls)	500 000
Reference dosimetry	
Development and material	400 000
Relative dosimetry (in-vivo dosimetric systems)	
Realization and testing	250 000
Phantoms	150 000
Microdosimetry	
Development and materials	350 000
Phantoms	100 000
TOTAL (kLit)	3 900 000

Table 9.1. Cost estimate for monitoring and dosimetry system for Italian Hadrontherapy Center (for 1994); after U. Amaldi et al. [1.4].

Chapter 10

TREATMENT ANCILLARY FACILITIES

This chapter contains the following cathegories: (1) pre-treatment equipment such as diagnostic tools, (2) treatment planning software and hardware, (3) patient positioning and alignment devices.

10.1 Performance Specifications for Treatment Ancillary Facilities

10.1.1 Performance Specifications for Pre-Treatment Equipment and Diagnostic Tools

10.1.1.1 Patient Accrual

A computer facility must be networked to nationwide protocol control, picture archiving and computer systems, therapy planning computers and treatment control computers.

A typical patient flow in the Department of Radiation Oncology would be that a patient should be referred to the Department for consultation by a radiation oncologist. When the patient is determined to be a proton beam therapy candidate, the patient information will be entered into database. A protocol coordinator will be notified to determine whether or not the patient can be entered into the existing or the immediate future protocols. The patient protocols as well as protocol accrual must be connected by PACS (Picture Archiving and Computer Systems) network for nationwide information transmittal. The computer, which will be used for these purposes must be able to communicate with any computers within the proton medical facility. When all these data collection are done or in place, the patient will go to diagnostic work-up [5.1].

10.1.1.2 Diagnostic Equipment

The diagnostic imaging equipment must be provided to furnish following capabilities

- ability to obtain image data from CT, MRI, SPECT (single-photon emission computed tomography), angiogram, and gamma camera through PACS network systems for acquisition by the treatment planning computer,
- the use of a flat table top insert for CT and MRI .the ability to incorporate all the immobilization devices which will be used for treatment,
- long term storage capability of image data from all imaging devices, such as CT or MRI, must be provided by optical disks in the Radiology Department, otherwise an optical drive must be provided in the treatment planning area [5.1].

10.1.2 Performance Specifications for Treatment Planning Software and Hardware

10.1.2.1 Treatment Planning

The therapy planning system must provide following capabilities:

- *dose calculation for non-coplanar beams*
- *3D dose calculations incorporating multiple scattering*
  True 3D dose calculations incorporating multiple scattering effects using, for example, the differential pencil beam (DPB) algorithm must be available. The dose calculation algorithm must be tested against, and the results must be comparable to, Monte Carlo calculations in realistic patient geometries.
- *modeling the beam delivery system*
  The therapy planning software must be able to model realistically the beam delivery system, be it a scattered or scanned beam. This includes geometrical considerations such as non-zero source diameter and beam divergence. In addition, if a scanning beam-delivery system is chosen, an interface must exist between the Treatment Planning System and the

Integrated Treatment and Accelerator Control System (ITACS) to execute the therapy plan accurately by the treatment delivery system.

- *designing 3D smeared compensators*
  The capability to design 3D smeared compensators must exist. A smeared compensator is a device designed to minimize the possibility of underdosing the target volume. This is generally done by replacing the nominal compensator thickness (calculated based on patient CT data) at each point with the minimum compensator thickness within some specified radius of the calculation point. This radius is selected by the treatment planner and represents the maximum distance a patient is expected to move during his/her treatment. For skull-base tumors, this distance is generally between 0.2 cm and 0.3 cm.
- *computing "worst case" dose distributions*
  The therapy planning software must be able to produce "worst case" dose distributions which illustrate the effects of patient motion during treatment as well as the effects of compensator misalignment. The program must also be capable of producing dose distributions reflecting the uncertainties in the input quantities for the dose calculation, namely the CT numbers, beam range data and beam fluence.
- *specifying the resolution of the calculation*
  The user must be able to specify both the resolution of the calculation (i.e., the points in the CT matrix at which a calculation is desired), and the CT planes on which the calculation is to be performed.
- *fast dose calculations*
  The speed of the dose calculations must be such that the time required to calculate a 3D dose distribution is not the time-limiting factor in the treatment planning process.
- *dose-volume histogram calculations*
  The software must be provided to compute and display the dose-volume histograms of selected targets and critical tissues
- *user friendly quality of the system*
  The user interface must be "friendly" in that a person with greater than one year of treatment planning experience must not have difficulties learning and using the program. A manual must be provided.
- *various dose display options*
  Isodose lines and color-wash display options must be available. On-line annotation of an isodose distribution displayed on a graphics terminal must be allowed so that the user may enter descriptive comments or

notes. A hardcopy device must be available so that these annotated isodose or color-wash displays may be included as documentation in the patient's chart [5.1].

10.1.2.2 Requirements for Image-Manipulation for the Therapy Planning Code

Following image manipulation capabilities are required for the therapy planning system:

- *editing CT numbers*
  The capability to edit CT numbers must exist. This includes the option to set CT densities outside of the patient equal to zero, as well as to alter CT numbers within a specified region.
- *window and leveling capabilities*
  Window and leveling capabilities must exist. The number of gray levels available must be such that none of the resolution inherent in the CT scan is lost.
- *image magnification options*
  A capability of magnifying selected parts of images must be provided. This is especially important in contouring mode.
- *contouring modes*
  The following contouring modes must be available: (1) an option which draws a curved line between points entered with a mouse or track ball; (2) a option which connects points with a straight line, or smoothly interpolate between the points with splines; and an automatic thresholding mode. In addition, must a reliable automatic contouring method become available, this method must be incorporated in the treatment-planning system. (3) A contour modification option must exist, especially for physicians reviewing contours). It must be possible to interpolate a contour on a given CT slice from other contours in the same structure on adjacent slices.
- *digitally reconstructed radiographs (DRRs)*
  The program must be able to produce DRRs. It must be possible to use any specified part of the patient's CT scans to make the DRRs.
- *projecting contoured structures on DRRs*
  It must be possible to project contoured structures on DRRs.
- *the collimator design options*

Options must be provided to add automatically a specified fixed margin to a collimator and to add manually or modify a variable collimator margin.

- *image correlation of CT images with other images*
  The code must be able to display and manipulate other image data sets besides CT and MRI, i.e., SPECT, angiogram, and gamma camera. Image correlation capabilities must exist whereby volumes-of-interest defined on one imaging modality can be transferred to another imaging modality. This must include both the correlation of two 3D data sets (e.g., CT and MRI scans), as well as the correlation of 2D projection information with 3D data sets. The transfer of dose distributions calculated on CT to other imaging modalities must be possible.
- *beam's-eye-view alignment aids*
  The treatment planning system must be capable of producing beam's-eye-view alignment aids. These must include both DRRs, as well as portal overlays. To produce a portal overlay, it is necessary to project the patient's bony anatomy, which has been contoured by a dosimetrist, in the beam's-eye-view and magnify it to the same magnification as the portal image used for patient setup and alignment. The overlay is then superimposed on the portal image to evaluate how w ell the patient is aligned. This may be done manually, with a portal film and a transparent portal overlay, with the advent of new technology, this process could be done electronically [5.1].

10.1.3 Performance Specifications for Patient Positioning and Alignment Devices

Patients treated with proton beam must be well immobilized and precisely aligned with the treatment beam to take full advantage of the dose localization potential of this treatment modality. A patient support system for the gantry beam must hold the patient securely in the supine or prone position, and permit proton beam entry from any oblique direction, without danger of collision with the gantry or beam shaping hardware. Although the immobilization of the patient relative to the couch is the responsibility of the user, the patient support system must be able to support at least 98% of all potential patients in such a way, that all points in any conceivable target can be accurately and reproducibly aligned to the beam to within $\pm$ 0.5 mm of their intended position. The couch and gantry together should have 6 degrees of freedom: 3 translations and 3 rotations and should be compatible with remote patient positioning followed by transport to the treatment room, in order to minimize the setup time in the room.

Immobilization of the patient in radiotherapy has two functions: the first is simply to hold the patient still during the treatment; and the second is to allow the repositioning of the patient into the same position many times, for example, for the CT and MRI scans and for every fraction of the treatment [5.1].

10.1.3.1 Immobilization

Immobilization material in the beam. The immobilization must place as little material in the beam as possible. Such extraneous material, if thick, would degrade the beam penumbra unnecessarily and unacceptably. The immobilizing material, including the support structures such as chair, table, or pod, must not present to the beam any rapid variations in thickness, such as sharp edges, because adequately compensating for these would be impossible. Thus the acceptable material within a port must be uniform or smoothly varying with a thickness less than 5 mm water equivalent. This requirement applies to any possible direction of treatment.

The immobilization devices must not cause artifacts in the CT or distortions in the MRI. No ferromagnetic materials must be used in the immobilization for MRI imaging. No metal may be present in the immobilization for CT imaging, except for small fiducial markers, pieces 2 mm^3 or less in volume. The immobilization must not attenuate the x-ray beam so much that the quality of the images obtained is compromized [5.1].

Ease of immobilization and releasing of the patient. The immobilization must allow the patient to be put in place and removed quickly, easily, and safely The patient must be completely freed from immobilization within 30 s from the time of distress.

The quick removal is especially important in the case of any emergency, e.g., in the case of fire or earthquake. Thus the immobilization device must be of the "quick disconnect" design, so that the patient can be completely freed up in short time from the time of distress, including the time required for staff to travel from the control area to the patient.

Immobilization for non-ambulatory patients. For non-ambulatory patients in the case of a building emergency it must be possible for the technologists to remove the entire immobilization assembly with chair couch from the patient positioner, placing it on a dolly, and roll it out of the treatment room to a safe place [5.1].

Immobilization and patient emergency. Immediate access (< 30 s) must be provided for the technologists to attend the patient in case of emergencies [5.1].

10.1.3.2 Patient Alignment Accuracy

In general the requirements for the immobilization may be stated in terms of the relation of the target to the collimator and compensator. The collimator and compensator are designed to allow some misalignment of the patient and still adequately treat the target and at the same time spare the nearby critical organ. This means that there is a margin of error all around the target as seen from the beam's-eye-view and the immobilization must hold the patient so the target remains inside that margin of error.

The design of the compensator allows for a margin of error in the depth of the distal surface of the target or the proximal and lateral surfaces of a critical organ, should one be just distal to the target. Change of depth (water equivalent) can occur if there is a body rotation so as to give a different entrance point of the beam or a change in the intervening tissue between the patient's surface and the distal target surface. The immobilization must control the rotation of the patient and the skin surface. Of course, it cannot control changes in internal anatomy, such as stomach-intestinal-bowel contents, breathing, or heartbeat. Thus, the immobilization must not allow rotations or skin contour changes of a magnitude that would place the target outside its margin of error in depth.

The margins of error are specific to the different regions of the body. The basic quantities which will be specified are the translation as a rigid body, rotation as a rigid body, and the flexure of the body. The first two are straightforward but the last, which refers to any twisting, bending, or stretching of the body must be broken down into specific quantities applicable to each part of the body.

The translation and rotation requirements are summarized in the Table 10.1.

Stability. Stability is defined as the amount of movement during the time of one treatment procedure, which should not last longer than 45 min.

Reproducibility. At the start of any procedure the patient is placed in the immobilization and aligned to laser lines or previously determined couch coordinates. This initial alignment must bring the patient to the correct position within the numbers given here.

	Stability		Reproducibility	
Region of body	Translation	Rotation	Translation	Rotation
Head and neck	< 1 mm	< 1°	< 5 mm	< 2°
Thoracic spine	< 2 mm	< 2°	< 5 mm	< 3°
Lumber spine, sacrum	< 2 mm	< 2°	< 5 mm	< 3°
Other parts of the body	to be specified			

Table 10.1. The translation and rotation requirements; after W.T. Chu et al. [5.1].

Control of body flexure. The immobilization must prevent the flexing of the body. Flexure for this purpose includes all other movements of the body, which can be specified and controlled to some extent with immobilization. Each part of the body can in general flex in different ways and thus will be described separately. In the head the primary moving part is the jaw. It can rotate about a lateral axis through the temporomandibular joints. Then the requirement is that rotation be limited to 1°. This corresponds to approximately 2 mm of movement at the tip of the chin. The spinal column from foramen magnum to coccix is much like a rope. It can bend in two directions, differently for each adjacent pair of vertebral bodies. It can stretch or compress along its own axis, and it can twist about its own axis. These quantities must be specified locally, that is, on the scale of one vertebral body to the next. Thus the requirement for bending of the column is that it must be kept to 1° of change from the desired value per 10 cm of length along the cord. This angle is measured as the angle between tangent lines to the axis of the column [5.1].

10.1.3.3 Patient Couch

The couch which will hold the patient in horizontal position (e.g., supine, prone, or decubitus) for gantry treatment room must be specially designed to satisfy listed specifications. Couch motion ranges are presented in Table 10.2.

Type of motion	Specifications
Lateral motion	± 30 cm from table center
The motion along the body axis	> 120 cm
Vertical motion	from the beam center to 30 cm above the floor
Table rotation around the vertical axis	± 95°
Table rotation around the horizontal axes	±5°
Flexing under weight	< 1 mm
Absolute table position accuracy	± 1 mm and ± 1°
Relative table position accuracy	± 0.5 mm and ± 0.2°

Table 10.2. Couch motion ranges; after W.T. Chu et al. [5.1].

Couch motion control. Automatic and manual control methods of the couch motion must be provided. The control computer must monitor couch position all the time. The speed of the couch motion must be variable from 1 mm s^{-1} to 10 cm s^{-1} so that the requirements of precise, small movements and large repositioning between ports, can both be accomplished with the required accuracy and with minimum lost time. The couch must have collision detectors to protect the patient in the event of an incorrect positioning command. There must be fixed detectors at the corners of the couch and at least two detectors on wires, which can be fastened anywhere to the couch or immobilization. Tripping any detector must stop all motion in the room, including the couch, gantry, and everything on the gantry. A reset procedure must be provided to allow the correction of the unsafe situation and the continuation of operations.

All computer commands must receive permission from a person in the treatment room before they can be carried out (except for the case of dynamic treatment mode); this may be accomplished through a request displayed on a terminal in the treatment room and with a affirmative response or by a permit switch at the couch. In non-patient mode, i.e. any irradiation in which no patient is present, this requirement is not needed. The manual controls must include a "dead-man" switch configuration and a "stop" button that overrides a computer command and stops all couch motion in 0.1 s. These controls and a position readout must be operable by a person standing at the couch itself [5.1].

Couch movement for dynamic treatment. In this case, the patient would be moved in one direction dynamically under computer control during treatment. Thus lateral beam spreading would be needed in only one dimension; the effect of spreading in the other dimension would be created by the couch motion. The two horizontal translations must be available for use during treatment in the event that very large field sizes are needed. Couch speed would be 2 cm s^{-1} or less in this mode [5.1].

Patient couch / chair for fixed beam room. For treatment in the horizontal fixed beam room, it is expected that patients would need to be positioned in a chair in addition to a couch. For this purpose, there must be either a chair attachment to a couch, or a stand-alone chair. Either way, the chair must have all the motions required for the couch above, namely, 3 translations and 3 rotations. The horizontal translations must allow excursions of ± 30 cm about the body center and motion along the patient axis (height change above floor) must be adequate to allow treatment from near the top head to the top of the pelvis. This requires 120 cm of vertical motion. To accomplish treatment to the head requires a beam height above the floor of at least 150 cm. Full rotation of ± 185° about the vertical required in order to use any beam angle in the axial plane.

All controls of the motion of the chair must be the same except that for dynamic mode it is the vertical and the motion perpendicular to the beam direction which must be available movement during treatment [5.1].

10.1.3.4 Patient Alignment

In each treatment room there should be a coordinate system with its origin at a point along the beam axis, at the center of a gantry's circle of rotation or of a couch rotation for a horizontal beam line (isocenter). For both rooms the first axis is vertical. The second is horizontal in the plane of gantry rotation or perpendicular to the beam line in the horizontal beam line room. The third is perpendicular to the plane of gantry rotation or along the beam line in the horizontal room. The directions of the axes must be so as to form a right handed coordinate system [5.1].

The equipment for alignment. The process of alignment should first make use of lasers for the rough alignment of external anatomy or marks on the immobilization to laser lines marking the three axes of the coordinate system. But final alignment is performed with x-rays imaging internal anatomy. These images are compared with images from the treatment planning showing the desired relation of the collimator, compensator, and isocenter to internal anatomy. Then the positioner should move the patient to the correct position.

For many patients this must be done in the treatment room. The precision needed for many treatments requires that the time between alignment and the beginning of treatment be kept to a minimum. Initially the alignment will likely be done with film, but eventually a digital imaging system is required so that the time between the start of the imaging to the presentation of the image to the person who must make the decision about moving be no more than 10 s. Then for other patients whose precision requirements are not so great, there must be a separate area outside the treatment room itself, where the alignment procedure can take place. This area must have the same lasers, x-ray tubes, film holders, and patient positioner in the same geometry as the treatment room itself. After alignment in this room the patient in the immobilization is moved to the treatment room, positioned to the determined couch or chair coordinates, and treated.

Both areas must have the following equipment:

- a set of *orthogonal lasers* which have a reproducible accuracy of ± 0.5 mm at isocenter. The laser lines must be clearly visible in normal bright room light.
- *x-ray tubes* to give standard orthogonal views of patient anatomy and beam's-eye-view for all directions in which that's possible. If possible they must have the same SAD (source to axis distance) as the effective SAD of the proton beam. The x-ray sources must be able to be positioned on the room axes to an accuracy of 0.5 mm. The tube which gives the beam's-eye-view must be positionable to within 0.5° of the particle beam direction.
- *x-ray detection system.* Although the initial detection system is likely to be film, there must be the capability for a digital imaging system. If an adequate digital system is available at the time the room is built, then that is what must be installed. The detection system must have a spatial resolution to give a pixel size of 0.5 mm or less for a plane through isocenter [1.22, 5.1].

Patient position verification. Verification of the patient target relative to the beam requires the use of isocentric lasers, isocentric diagnostic x-ray tubes and radiographic detectors. These detectors could be x-ray film cassettes or digital portal imaging devices. Manufacturers must show compatibility of any facility designs with these devices. In each treatment room, the isocenter, and associated coordinate system, must be defined and all positioning and movement must be referred to the isocenter. The precision aligning of the patient to the collimator, compensator, and proton beam is the final critical step before actual treatment. In each treatment room there will be a coordinate system with its origin at a point along the beam axis, at the center of a gantry's circle of rotation or of a couch rotation for a horizontal fixed-beam line. This point is called *isocenter*. All movements of the patient will be related to this coordinate system [5.1].

Laser and x-ray localizers for alignment aid. For each treatment room, four each orthogonal alignment lasers and x-rays must be provided according to the specifications given in Table 10.3.

Parameter	Specifications
Reproducible accuracy of lasers	± 0.5 mm at isocenter
Positioning accuracy of x-ray sources	± 0.5 mm on the beam axis: ± 0.5° of the beam direction

Table 10.3. Specifications for lasers and x-rays for alignment; after W.T. Chu et al. [5.1].

The process of alignment should first make use of lasers for the rough alignment of external anatomy or marks on the immobilization to laser lines marking the three axes of the coordinate system. The laser lines must be clearly visible in normal bright room light. Final alignment should be performed with x-rays imaging internal anatomy. These images are compared with images from the treatment planning showing the desired relation of the collimator, compensator, and isocenter to internal anatomy. x-ray tubes to give standard orthogonal views of patient anatomy (i.e., AP and lateral) and beam's eye view for all directions in which that's possible. Then the positioner should move the patient to the correct position. To maintain smooth patient flow in the treatment rooms, the time between alignment and the beginning of treatment should be kept to a minimum. Initially the alignment can likely be done with film, but eventually a digital imaging system is required so that the time between the start of the imaging to the presentation of the image to the person who must make the decision about moving be no more than 10 s. If an adequate digital system is available at the time the room is built, then that is what must be installed. A digital imaging system must work in conjunction with the treatment planning system. There must be the capability of displaying the digital, possibly enhanced, images on monitors throughout the department within 10 seconds after capture. There must be the ability to display images from the treatment planning system on top of the digital setup image and for a person to manipulate one of the images in order to determine any movement needed to put the patient in the correct place. The system must then calculate the needed move and after receiving permission from the technologist in the treatment room perform the actual move [5.1].

Simulation room. Simulation room must be provided to pre-setup the patient outside of the treatment room. For certain patients whose alignment requirement is less stringent than those requiring precision alignment in the treatment room, there must be a separate area outside the treatment room itself where the alignment procedure can take place. This area must have the same lasers, x-ray tubes, film holders, and patient positioner in the same geometry as the treatment room itself. The pre-setup room (simulation room) must have the same imaging geometry and capabilities as the treatment room. This may be achieved by mounting x-ray units on a gantry in which they may rotate around the patient and keep the same radius as the x rays in the treatment room. If the source-to-patient distance for the gantry is more than about 2 to 2.5 m, it might be difficult to use such a distance. In this case, a different geometry would be required and divergence corrections would be required in the treatment planning program to provide a prediction for the appearance of the setup x-ray. There must be a positioner in this room, which has the same position readout coordinates as the one in the treatment room. After the correct position is achieved, it must be possible to transport the patient between the external setup area into the treatment room without significantly altering the patient position. This means that it must be possible to dock the pod with the gantry simply and reproducibly. The docking mechanism must have reproducibility of ± 1 mm when a pod setup is transferred from a pre-treatment room to a treatment room [1.22, 5.1].

10.2 Treatment Planning Software and Hardware

One of the fundamental points for a correct clinical use of proton beams is represented by the *treatment planning system.* Planning a radiation treatment is a multistep task which extends from the initial evaluation of patient anatomy, through the therapy simulation to the 3D display of dose distributions. The algorithms, which can be used to calculate the dose distribution include: (1) broad beam, (2) pencil beam and (3) Monte Carlo algorithms. Monte Carlo calculations are compared with experimental results. The eye treatment planning has been considered separately because it shows peculiar features with respect to all other anatomic sites [1.4].

10.2.1 Algorithms

The algorithms, which can be used to calculate the dose distribution include: (1) broad beam, (2) pencil beam and (3) Monte Carlo algorithms.

A *broad beam* is a non-divergent large enough beam so that the relative depth-dose curve, DD_∞, on the central axis does not depend on the field amplitude. In the broad beam model, or ray-tracing model, protons are followed along their path from the source to the point of interest. Given the DD_∞ curve in water, the dose at any point in the patient is evaluated by calculating the water-equivalent path length (integral of the water stopping power along the path) and using corrective factors for the beam intensity on the transverse plane, for the inverse of the square distance and for the penumbra. For a monoenergetic beam the D_∞ can be obtained from a depth-dose curve measured in correspondence to a certain *source to surface distance* (*SSD*). The dose distribution for modulated beams is obtained through the summation of all the single beam contributions. The main limit of a Broad Beam algorithm is that it cannot consider the scattering effects due to the inhomogeneities in thc patient, so that it cannot accurately predict the perturbations in the dose distributions (hot and cold spots) near the interfaces and the end of range in the shadow of these inhomogeneities [1.4, 10.1].

With a *pencil beam* algorithm the effects of multiple scattering on dose distributions can be considered. A pencil beam is a beam with an infinitesimal transverse Sec. . In large field conditions the dose can be predicted by the summation of single pencil beam distributions. The dose distribution of a single pencil beam in a water phantom can be calculated in various ways:

- its shape can be described through an analytical function including a central axis term and a gaussian off-axis term in which σ is the root mean square (rms) of the multiple scattering distance at the calculation depth. In the case of inhomogeneous structures the physical depth is replaced by the water-equivalent depth and, in any case, σ is a function of the calculation point. With this basic idea, several methods can be developed, by varying for instance the way of evaluating σ. The starting experimental datum is always DD_∞;
- Monte Carlo method is used to calculate the differential dose distribution, that is the dose deposited in a homogeneous water phantom in a cylindrical ring between the depths z and z + dz and the radii r and r + dr and normalized to the maximum dose D_0. In the case of a heterogeneous

medium, the depth is replaced by the water-equivalent depth. Even in this version of the pencil beam model the differences in the multiple scattering effects due to the different spatial position of the inhomogeneities are not considered [1.4, 10.1].

Monte Carlo algorithms, at present, cannot be routinely employed for the calculation of dose distributions because they require very long computing time. However their use is very useful to perform theoretical experiments, verify the algorithms used in the clinical practice and calculate the kernel in a pencil beam model [1.4, 10.1].

Several home-made codes exist which often resort to drastic approximations for proton transport such as to neglecting nuclear interactions, *general purpose codes* also exist which allow proton tracing: PTRAN and GEANT [10.2]. PTRAN, at present, allows evaluating the dose only in a homogeneous water phantom and it considers that charged particles produced by nuclear interactions arc locally absorbed. GEANT, on the contrary, allows proton tracing with any geometry and inhomogeneity, the secondary particles produced by nuclear interactions are followed through the hadronic shower generators FLUKA or GHEISHA [10.3]. Total cross sections are said to be known within 3%, but the accuracy of partial cross stion values, for which only fragmentary experimental data exist, is not well known. Fig. 10.1a) compares a percentage depth-dose curve measured with an ionization chamber at the HCL in free beam conditions, that is without any beam modifying devices in the beam line, and the result of a GEANT simulation, assuming a gaussian energy distribution with a mean value of 158.5 MeV and a standard deviation of 1.5 MeV. Even if the beam divergence has not been considered because at HCL the source-to-axis distance (SAD) is greater than 5 m, the depth-dose curve is well reproduced by the simulation. In Fig. 10.1b) a comparison is shown between the depth-dose curve for the HCL beam and the result of a PTRAN simulation obtained by Berger [1.4].

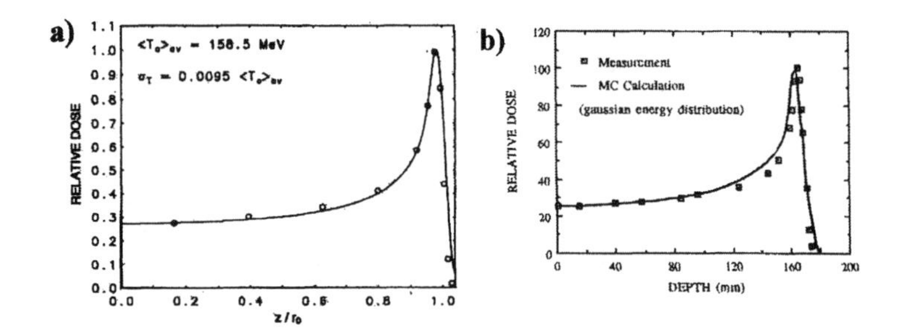

Figure 10.1. Comparison between a depth dose curve for HCL beams and the results of a) PTRAN, b) GEANT simulation; after U. Amaldi et al. [1.4].

10.2.2 Assessment of a Plan and Comparison of Competing Plans

The design of a treatment plan is an iterative process in which a set of beams are designed, each of their dose distributions is calculated, and the resulting combined dose distribution is assessed. In this Section we shall discuss methods by which this assessment is made and how competing treatment plans can be analyzed. On a more general note, we shall also discuss the use of comparative treatment planning as a tool to assess the relative merits of different radiation modalities and delivery methods.

Analyzing treatment plans. The analysis of a treatment plan consists of two tasks. Firstly, an assessment of the estimated dose distribution in relation to the clinical requirements and restrictions for the patient must be undertaken. Secondly, the practical aspects of the plan must be considered to ascertain whether the treatment can be delivered reliably and safely.

The most direct method of analyzing a dose distribution is through visual inspection and requires that a graphical tool is available which allows for the display of dose in relation to the defined target volume, critical structures and patient anatomy. The calculated dose can be overlaid over the patient model by the use of dose contours or by *color-wash techniques*. In the latter method, the patient and dose data at any point are represented by intensity and color, respectively, and can provide a visually striking representation of the estimated dose in relation to the patient's anatomy. For 3D dose analysis, orthogonal cuts through both the anatomy and dose-data sets can be useful, whilst the interactive selection of the cut planes is an important capability which can ease the problem of mentally reconstructing a three-dimensional distribution. In addition, for isodose contour and color-wash representation of dose, interactive selection of color bands is important, as is the ability to query or profile the distribution for a more detailed and quantitative analysis.

Visual inspection of a dose distribution remains the most important tool for the analysis of a treatment plan. However, when doses have been calculated three-dimensionally, the amount of information that has to be assimilated can make this analysis difficult. For this reason, it is now common practice to also compute *dose volume histograms* (*DVHs*) for any *volume of interest VOI* (Fig. 10.2). Two forms of DVH are in common use. The so-called *differential DVH* is in fact a conventional histogram, in which the frequencies occurrence of each of the defined dose bands within a VOI are plotted. From this can be derived the more useful *cumulative DVH.* In this form, the DVH directly provides clinically useful dose-volume information, such as the volume of a VOI that receives a dose equal to or greater than a specified level.

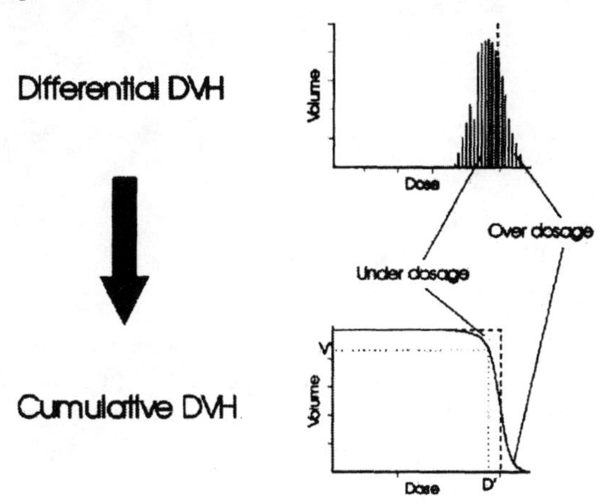

Figure 10.2. Two forms of dose volume histogram (DVH): differential and cumulative; after A. Lomax and M. Goitein [10.1].

In its most convenient form, the DVH provides a direct method of visualizing the dose-volume characteristics for the VOI, and provides a succinct method of representing three-dimensional data in a one-dimensional format. However, in this process of data reduction, spatial information is inevitably lost and it is important that DVHs and visual inspection of dose distributions are seen as complementary tools. Indeed, by combining information from these two modes of analysis, detailed information about the quality and deficiencies of particular plans can be extracted.

Comparative treatment planning. Assessment of an individual treatment plan is an important task in radiotherapy, especially in the case of a non-routine treatment in which the technique has been tailored to the patient's particular problem. It is also important to be able to assess which of two or more competing treatment plans is thought to provide the best solution, given the patient's indications.

When comparing treatment plans, it is valuable to display competing dose distributions side-by-side, so that direct visual comparison of the plans can be made. Similarly, the display of DVHs for the same VOIs, but derived from different plans (Fig. 10.3), should be overlaid on the same axis, providing a direct comparison of their dose-volume characteristics.

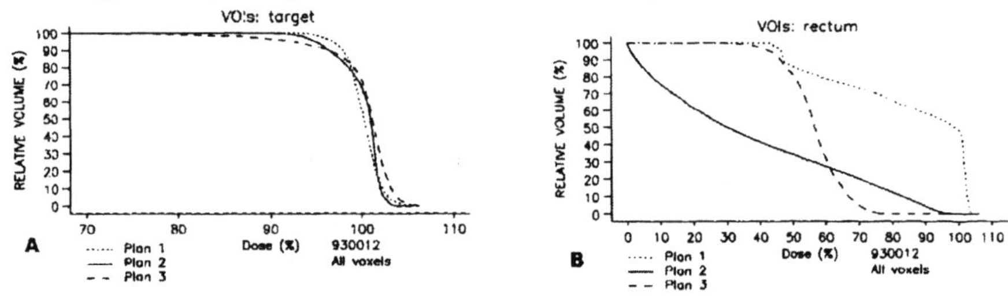

Figure 10.3. The display of dose-volume histograms from three competing treatment plans: a) for the target, b) for the rectum.

As another aid to assessing and comparing treatment plans, biological models have been developed which attempt to predict the *tumor control probability* (*TCP*) and *normal tissue complication probabilities* (*NTCPs*) given a particular distribution of dose within the target and organs at risk. They provide a mechanism by which a complex three-dimensional dose distribution can be reduced to a single medically meaningful parameter per VOI. However, such models should be used with caution. All are based on dose-response data for whole- and partial-volume irradiations, which are either derived from results in the literature or have been estimated from clinical experience. Unfortunately, the data are at present extremely sparse, and the validities of the parameter fits are therefore questionable.

Comparative treatment planning has been used extensively to explore the theoretical possibilities of new planning methods or treatment techniques, and has been used extensively in assessing the role of protontherapy. Such comparisons can, of course, in no way be seen as replacing properly designed clinical trials. These, however, require long follow-up times and are difficult to perform in radiotherapy. Comparative-treatment planning is currently one of the most important methods of gaining rapid insight into the possible efficacies of new treatment approaches.

Optimization of treatment plans. The planner has a large number of degrees of freedom available when designing a treatment plan. This gives great flexibility for shaping the dose to match the target volume or avoiding critical structures, but on the other hand, makes the selection of the set of parameters which can best solve a particular planning problem - an overwhelming task. In recognition of this problem, the application of mathematical optimization procedures to treatment planning has been pursued vigorously in recent years to the point where such techniques are entering clinical use. The main thrust of recent research has been towards optimizing individual beam weights and the calculation of optimal-intensity profiles for intensity-modulated beam delivery. A number of optimization algorithms have been applied to this problem, ranging from *iterative least squares minimization*, a *random search technique*, the use of *simulated annealing and genetic algorithms*. In addition, there is a considerable debate as to what should be optimized. Some methods optimize to prescribed dose levels, whereas a number of techniques attempt to optimize to the biological response of the target and critical structures. Both criteria can produce acceptable results, although there is some evidence to show that including some amount of biological knowledge into the optimization process can improve the resultant dose distribution in certain cases [10.1].

10.2.3 Eye Treatment Planning

Almost all centers where ocular treatments are performed use, as basic program, that elaborated by M. Goitein and T. Miller and successively modified in several centers (MGH, PSI, Clatterbridge). The problem of eye treatment planning must be dealt with separately from that concerning other anatomical sites because, with the exception of the Chiba Center, in the eye case the input data are not CT images, but a series of ophthalmological information relative to anatomic characteristics and to the position of reference radiopaque clips surgically implanted. In the treatment room couples of orthogonal RX projections are then acquired with the eye axis at several angles. Starting from this series of information and considering a spherical homogeneous model, the eye geometry of the patient is reconstructed, including its internal structures, the implanted clips and the tumor. The following treatment parameters are then optimized: the eye axis orientation, the maximum range and the modulation thickness and the field aperture profile. The therapy planning system must provide, as output, isodose curves on suitable planes intersecting the eye and dose-volume histograms for the most critical ocular structures [1.4].

10.2.4 Available Systems for Treatment Planning

At present, commercial therapy planning systems allowing dose distributions calculation for proton beams do not exist, but at least two companies, Siemens and Helax, are developing a new system (or improving an already existing one) which can deal with this kind of radiation. Some centers, where therapeutic proton beams are already employed, use home made systems in continuos development. The Table 10.4 summarizes the existing treatment planning systems used during the process of charged particle therapy planning [3.6].

Year	Created by	System name	Status
1979-93	LBL	LBL System	Not available
1980	MGH	Rx	Distributor MGH
1980	MGH	EYEPLAN	Distributor MGH – EYES only
1990-96	MGH/Siemens	V-Treat (AXIOM)	Not available
1991	PSI	PSI System / Pion	Disributor PSI
1995	DKFZ / Royal Marsden	Voxelplan / Proxelplan	Adapted by GSI, NAC, DKFZ
1996	Radionics / MGH / HCL	P – Knife	Not available
1997	LLUMC / Permedics	OptiRad 3D	Commercial pending FDA
1998	Tsukuba	Hitachi system	In-house system
1998	DKFZ	OCTOPUS	Used in Berlin – EYES only
1994	Orsay / Curie	ISIS	Distribution ?
1998	CMS / MGH	FOCUS	Commercial release 1999
1998	DKFZ	KonRad Plus Protons	Research only
199?	Uppsala / KVI	Helax (+ protons)	Distribution
	Render Plan		?
	Adac		?
	Michigan		?
	Varian		?

Table 10.4. Treatment planning systems used during the process of charged particle therapy planning; presented by S. Rosenthal, MGH at the Workshop on Treatment Planning Systems, PTCOOG XXXI; after J. Sisterson [3.6].

The Siemens Company developed Axiom, a system for conventional radiotherapy. The system includes the algorithms for the calculation of proton dose distributions developed in collaboration with M. Goitein's group at MGH. Programs using broad beam and pencil beam models.

The Helax Company (Sweden) is completing its system, TMS, including the software for proton beams. It is based on the pencil beam model. The program is in use in GWI in Uppsala, Sweden.

The ISIS treatment planning system was developed in France by C. Rosenwald (Institut Curie). The program is in use at the CPO facility.

At PSI a model based on the superposition of elementary beams is being developed. The model needs the measurement of the initial phase space of the beam (beam shape and divergence measured in air and estimation of the momentum band) which is described with a gaussian distribution. The algorithm is mainly studied for an active scanning system of a narrow beam [10.4].

At the MGH, at present, a therapy planning system developed by M. Goitein is employed. The ray-tracing algorithm that is implemented is rather crude, but it shows the important feature of evaluating the dose distribution variability range, related to the unavoidable uncertainty due to the patient motion [10.5]. The same system is in use at LLUMC.

In 1994 the NAC obtained a sophisticated three dimensional Protontherapy planning system, which is entirely based on the VOXELPLAN planning system, made available to the NAC by the DKFZ, Heidelberg, Germany together with a Protontherapy module which was provided by the Royal Marsden Hospital, UK. The NAC system is now called PROXELPLAN. The proton module has been modified and refined by the NAC's Division of Medical Radiation to make it suitable for clinical use. The main features of the PROXELPLAN system are: (1) fully 3D treatment planning (i.e. non-coplanar treatments) based on CT data; (2) digitally reconstructed radiographs can be obtained and compared with portal x-ray images; (3) dose volume histograms are calculated; (4) stereotactic co-ordinates of beam entry and exit points are calculated; (5) beam shaping is done according to a beams-eye view of the treatment volume; and (6) isodoses are displayed in the different scanned transversal planes as well as in user selected frontal and sagittal reconstructions. The proton module uses a ray line tracing algorithm which does not permit the accurate calculation of compensators. A pencil beam algorithm is currently under development [10.6].

In summary, there is at present no need to develop a treatment planning code. At the moment, the existing therapy planning systems are under continuous development and the same happens for the facilities which provide proton beams. There follows the necessity that the therapy planning system must be open and modifiable to take into account any possible new device inserted in the beam line and any upgrading both in the therapy planning system and in the machine (for instance, the chance from a passive scattering technique to an active scanning system). Such a requirement, that is fundamental mainly in the initial phase of the project, imposes the presence of staff trained in computer science and, treatment planning and the establishment of active cooperations with the centers where treatment planning systems are under development [1.4].

10.3 Patient Positioning and Alignment Devices

The immobilization requirements for the protontherapy patients are more stringent than those for conventional x-ray therapy. Most of these patients is selected for protontherapy because of the close proximity of one or more radiosensitive organs to the target volume or because of the necessity to deliver a high dose to the target volume sparing the surrounding healthy tissues the margins between the target and the treatment volume must 50 be kept as small as possible. Regardless of tumor location a poor patient immobilization could compromise tumor control probability; in the case of a radiosensitive organ lying adjacent to the target volume the dose must he kept within the known tolerance dose for the organ.

To minimize the volume of normal tissue irradiated it is necessary to reduce patient motion and positioning uncertainties. Three different items have to be considered to treat this matter and they can be summarized as follows: (1) immobilization techniques and devices, (2) data acquisition and treatment planning with the patient in the immobilization device treatment position, (3) verification of the correct position with coaxial diagnostic techniques [1.4].

10.3.1 Immobilization Techniques and Devices

Two different and important items are included in this Section and are represented by the immobilization devices themselves and by the patient positioning system or couch. The two subject are closely interrelated to provide correct patient immobilization and treatment technique. The position assumed by each patient during irradiation must be consistent with different conditions such as treatment protocols, tumor site and target volume and it is supposed that the supine and seated position can cover almost all the possible events in radiation oncology and in proton treatments in particular.

Supine position. The supine position may be used in radiation oncology to treat almost every site in the human body. A face mask or body cast is used to immobilize the patient and to obtain a stable configuration. The immobilizing device may he fixed to the treatment couch to allow for a reproducible easy positioning on the treatment table every day.

Seated position. The seated position is mainly used in the irradiation of ocular tumors. In this particular case both the patient and the eye must be immobilized. The patient is held in the prescribed position using a rigid bite block and face mask, while immobilization of the eye is obtained by voluntary fixation upon a flashing light for the duration of the treatment. Registration of the correct alignment of the proton beam to the tumor is obtained by translation of the patient along the perpendicular x, y and z axes. Rotational movements of the head around two perpendicular axes (x, y), intersecting at a point located on the central axis of the proton beam (x axis), provide adjustments that prevent or decrease irradiation of the ocular appendages [1.4].

Immobilization devices. For many cases a contoured mask or cast may be produced using radiolucent thermoplastic material (Acquaplast^{TM}) with or without multiple perforations. Such a material can easily be formed into masks for partial body casts after immersion in hot water (65°) for a few minutes. Acquaplast^{TM} call so be moulded directly around the body contours where it becomes rigid in a few minutes. The ohtained casts or masks do not change their shape over a long period of time and can he stored for additional therapies or diagnostic procedures.

The use oft the perforated thermoplastic material offers the following advantages: (1) the material is partly transparent so that the patient anatomy can be seen without cutting the mask and decreasing its strength; (2) air circulation is possible and the mould is not uncomfortable for the patient; (3) large areas of the body may be covered with this material; (4) breathing and vision are possible when it is used to produce face masks; (5) the same mask or cast may be used either during photon or proton irradiation.

The disadvantage of the perforated material is its relatively low rigidity, but the problem may be overcome by using a combination of the perforated and non-perforated material or different thickness of the same one. Non perforated plastic is stiffer and may he easily moulded and combined with the perforated one.

The thermoplastic masks are frequently used in combination with other devices (bite block) to obtain a complete fixation of the head and neck in both the seated and supine position. The *bite block* is composed of different parts:

- a costume dentature that fits into thc patient's upper teeth or palate for edentulous people;
- all arch, positioned not to interfere with the entering beam, is coupled to the bite block. Different angles and length adjustments are possible to fit different head positions. A locking system prevents changes from the original settings throughout the course of the treatment;
- a frame supports the neck allowing alignment of the patient's head to the couch. The frame may be provided with arms, which are used to lock the mask into its original position. Different materials are used to produce frames and many of them are commercially available. Carbon fibre boards, that may be formed into different shapes, are particularly interesting because of their stiffness and lightness;
- a ball joint allows fine adjustments of the bite block to the most comfortable position for the patient and treatment technique.

Effective immobilization of the head and neck or other parts of the body cannot be obtained by thermoplastic masks but require a more complex and sophisticated system. *Vacuum forming of plastic* can he used in such cases. The technique was developed for orthotics and prosthetics but may be applied to any situation in which a rigid, firm and light structure, is necessary to hold or constrain a limb or any other part of the body positive model, with an open cavity at one extremity, is obtained using plaster of Paris and holes of about 0.5 cm are made through the surface into the cavity. The plastic material (Plastic-Vac^{TM}) is heated to about 100°C and then positioned over the positive model. A control valve, inserted in the cavity of the positive model, is connected to a vacuum system so that the heated softened plastic is forced towards the model. The plastic cast can be cut to remove the inner positive plaster model and engrinded and polished to be applied to the patient [1.4].

10.3.2 Therapeutic Couch and Chair

Couch. The design and construction of the treatment couch is not a simple task and different solutions, with different precision results, may be possible. One possible solution is the use of a commercially available couch like those that are provided with linear accelerators; this solution was adopted at LLUMC. The main disadvantages of this solution are the limited number of movements with respect to the precision that is necessary to achieve with proton beams, fact that such a couch does not provide patient tilt and the high movement accuracy required by proton treatments.

Mechanisms and movements of the couch depend mainly on the available space inside the gantry. The design must take into account that the operators should be free to move around the positioned patient [1.4].

A specifically designed couch was built for the NPTC at MGH, Boston. The draft of the proposal presented by Ion Beam Applications (IBA, Belgium) to MGH is shown in Fig. 10.4.

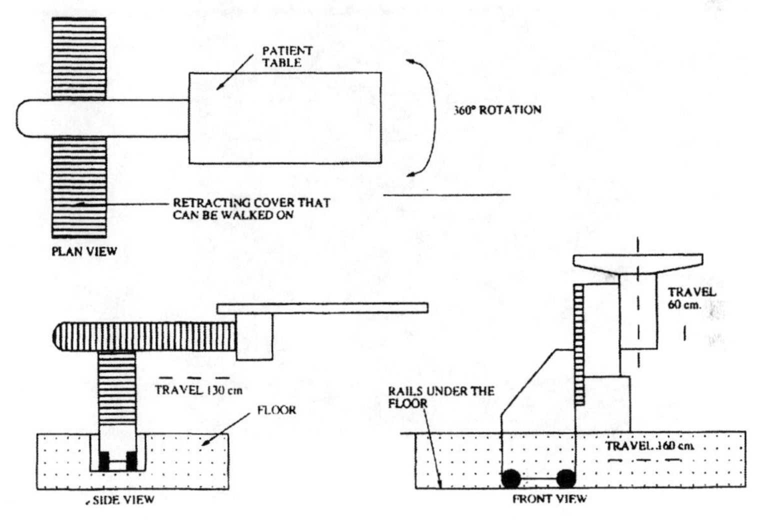

Figure 10.4. Three views of a schematic representation of the couch project prepared by IBA for MGH.

Chair. Different types of chair are used to treat patients in the seated position and in particular to treat ocular tumors. Patient positioning is obtained with movements (translation of the chair) along the x, y and z axes (three degrees of freedom). The patient head can be rotated around two perpendicular axes (y, z) intersecting at a point on the central axis of the proton beam (two additional degrees of freedom). The movements along the perpendicular x, y and z axis allow thc positioning of the tumor center or of the center of the target volume to the central axis of the proton beam. Rotations around the x and y axes will allow the removal of as much as possible of the ocular appendages from the beam. An additional degree of freedom is finally provided by setting a polar (α_x) and azimuthal (α_y) angle for the central axis of the eye. A schematic representation of the eye and of the directions along which movements can he done is shown in Fig. 10.5.

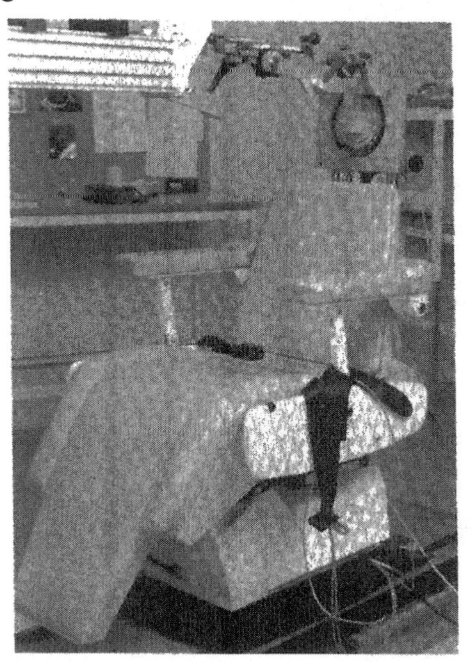

Figure 10.5. The chair in the medical treatment room in the The Svedberg Laboratory, Uppsala, Sweden; after E. Grusell [10.12].

The center of these axes is positioned in the center of the tumor or target volume. When the patient has been positioned with appropriate movements of the chair the eye is fixed according to a polar (α_x) and azimuthal (α_y) angle. The eye will remain in this position during all the treatment time. The reference point around which the latest movements are performed is the center of the eye. Additional movements of the head can be done. The head holder (mask and bite block) can rotate around a perpendicular (y) and a transverse (x) axes to remove ocular appendages from the treatment field (Fig. 10.6). Movements of the chair, eye and head provide the six required degrees of freedom [1.4]. Two different models are used; one is the chair in use at MGH / HCL while the other is in use at PSI.

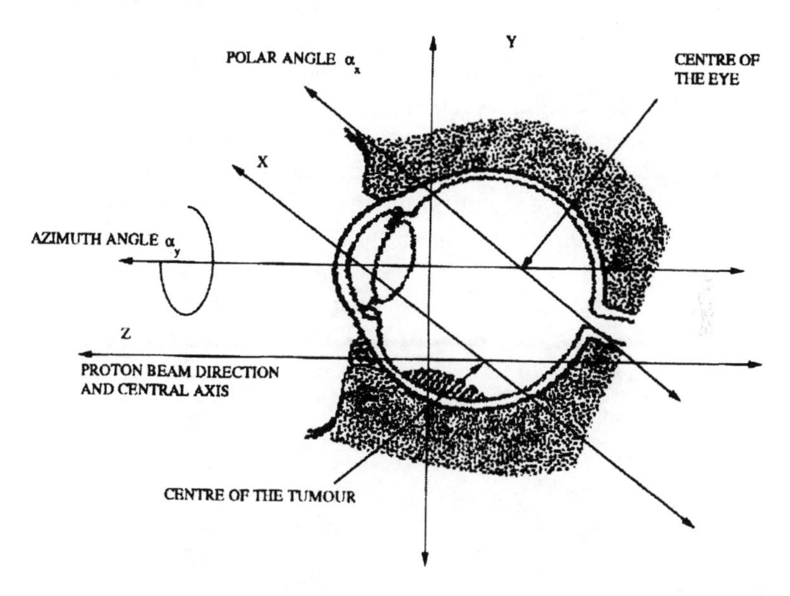

Figure 10.6. Schematic representation of the eye and the x, y and z reference system along which movements of the chair are performed; after U. Amaldi et al. [1.4].

10.3.3 Verification of Patient Positioning

Localization of the treatment volume begins with the acquisition of a series of thin (1 – 3 mm) CT slices to obtain an accurate model of the volume to be reconstructed. This 3D CT map provides the target position with respect to bony landmarks. In the case of head and neck tumors the entire cranium can, for instance, be reconstructed and used for orientation as opposed to a limited number of selected bony landmarks or fiducial markers. Digital radiographs are obtained from the CT data along directions that simulate the viewing directions of one or more independent x-ray tubes or treatment beams. Computerized modeling has demonstrated that target position can be determined with submillimetric accuracy.

The alignment of the patient to the proton beam is routinely obtained using lasers and verification films or a digital imaging system is used to visualize the fiducial markers (clips or skeletal landmarks). The position of the landmarks, in the three dimensional reference system, is compared with the position of the corresponding fiducial markers on the *digital reconstructed radiography* (*DRR*) and movements to reposition the patient directed to the patient positioner. The stereoscopic localization of the fiducial markers can be made on a conventional radiogram and with the use of a digitalization tablet. Another possibility is represented by the comparison, performed by a computer, between the original DRR and a digitized radiogram (Fig. 10.7) taken before each treatment session. In the last example the software comparing the images will automatically direct the movements of the couch to reposition the patient (precise movements along the x, y and z axes, as well as rotation and tilting). In both cases (digitalization tablet or comparison between digitized images) a second set of radiograms is necessary to verify that the correct setup has been achieved. The localization of implanted fiducial markers is the method in use at MGH. Approximately 15 min are necessary to setup and verify the position of the patient.

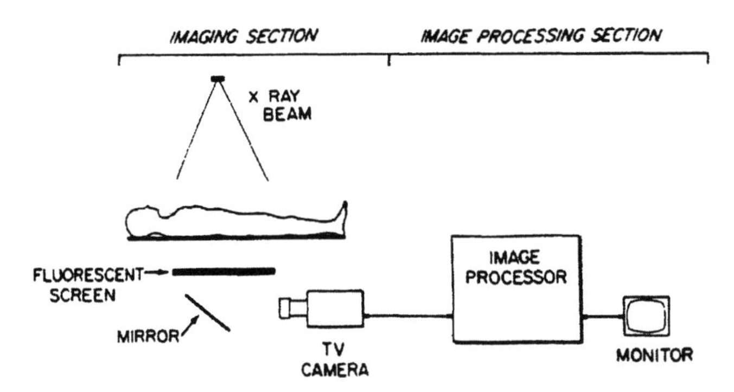

Figure 10.7. A TV camera is used to digitize the x-ray image. The processor determines the position with respect to the x, y and z axes, of the fiducial markers and make comparison with the DDR. Movements are directed automatically to the patient positioner; after U. Amaldi et al. [1.4].

The time necessary to set up and verify the correct position of the patient depends on many factors and technical conditions. The availability of a gantry should allow the treatment of about 4 patients per hour (mean estimated time at MGH).

10.3.4 Treatment Simulation

In many instances treatment with proton beams may be combined with photon irradiation. With non coplanar isocentric techniques it is necessary to shape the photon portals according to the proton apertures and to provide the user with an accurate reproduction of anatomic features from the view point of the treatment source (beam'-eye-view). To do this, digital radiographs are obtained from the CT slices along directions that simulate the viewing directions of the treatment beam. On the reconstructed radiograms the treatment volume or target will be delineated as it is seen by the beam and the radiation oncol0gist wil l shape the photon aperture according to this particular view. Personalized cerrobend apertures are needed; a simulator is necessary to reproduce this treatment conditions. The typical layout of a commercial simulator (VARIAN, Ximatron) is shown in Fig. 10.8 [1.4].

10.3.5 Examples of Patient Handling Systems in the Existing Proton Facilities

10.3.5.1 Patient Handling at PSI

The patient is prepared for the treatment in the preparation room. All operations are performed with the patient lying immobilized in the supine position in his individually moulded couch. The patient couch is transported using the special carriage system (patient transporter), as shown in Fig. 10.9.

The patient is then transported in the positioning room. The patient couch is coupled to the CT table using the same clamping mechanism as for the proton gantry. In this way a well-defined mechanical reproducibility of the position of the couch on the gantry and in the CT will be guaranteed for all phases of treatment. In order to achieve the required mechanical precision, the entire patient support system of the commercial CT has been replaced by a new table designed specifically for this task.

The routine positioning control is performed with a dedicated computer tomograph (CT). The same CT-unit is used for the data taking of the: (1) CT-slices used for treatment planning and (2) for the patient positioning verification before the delivery of a single fraction. Reference scout view images are taken at the time of collection of the CT slices for treatment planning The positioning of the patient in his couch is checked before each fraction by taking the same scout view images and comparing them on a computer workstation with the reference images. By checking the position of the patient using the same CT as used for treatment planning, we obtain the most direct validation of the treatment planning data before each fraction. After the completion of the patient positioning in the CT, the patient is moved into the proton treatment room.

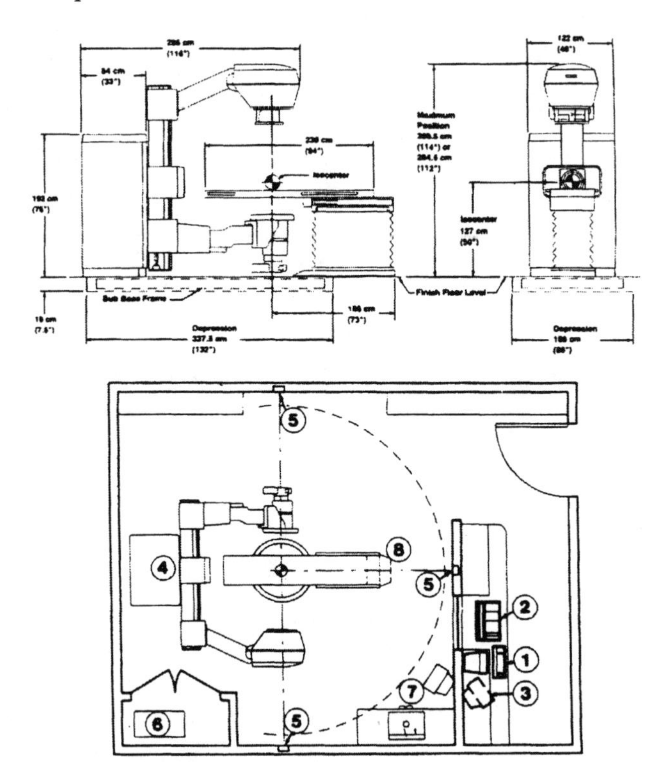

Figure 10.8. The VARIAN™ Ximatron: (1) workstation, (2) control console, (3) fluoroscopic monitor, (4) stand gantry and couch, (5) laser positioning lights, (6) high-frequency generator, (7) in-room monitor, (8) pendant; after U. Amaldi et al. [1.4].

In the gantry room the couch is coupled to the table of the proton gantry (Fig. 10.9). If necessary the position of the patient with respect to the couch is checked again with x-rays or proton radiography [10.7].

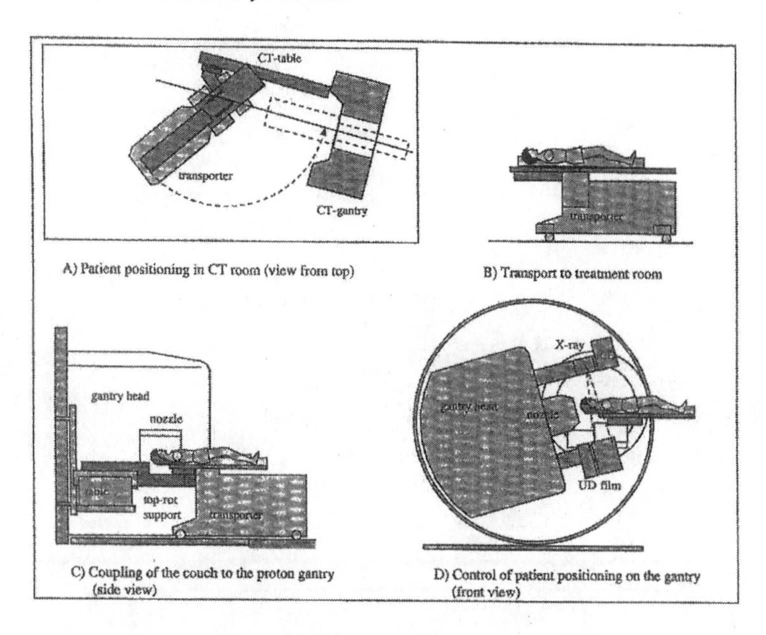

Figure 10.9. Overview of the patient handling system developed for the PSI Protontherapy facility; after E. Pedroni et al. [8.4].

10.3.5.2 Patient Positioning System at NPTC

Movements of the patient positioning system mechanisms are integrated in the patient positioning system joints. The couch is supported on a turntable (rotary axis) allowing the couch to rotate about a vertical axis. The resulting motion envelope that the system can sweep out is shown in Fig. 10.10. The overall patient positioning system operating envelope is a major driver of the gantry design in that it sizes the minimum inner-diameter of the front gantry bearing ring. Since the patient must be rotated 90° with respect to the gantry axis of rotation, the radius of the gantry ring must be substantially larger than the height of the patient. The patient couch assembly mounts directly to a load cell. The load cell provides a means of measuring a patient's weight and the center of gravity position of the couch during treatment. The load cell will acquire the data necessary for an algorithm to compensate for the deflection caused by the variable weight and location of the patient on the patient positioning system [10.8].

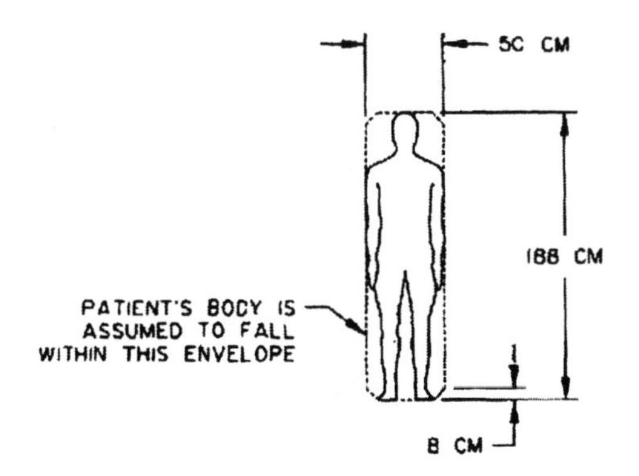

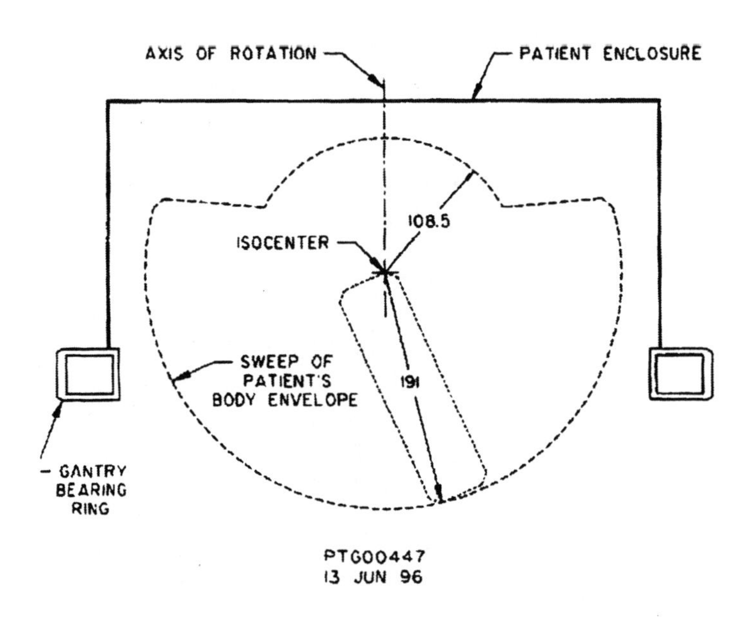

Figure 10.10. Working envelope of patient positioning system at NPTC; after J.B. Flanz [10.8].

10.4 Systems for Irradiation Gated by Respiration of the Patient

10.4.1 HIMAC Respirated-Gated Beam Control System

A beam control system for irradiation treatment gated by respiration of a patient has been developed at HIMAC (Sec. 4.2.3, Fig. 4.12) in order to minimize an unwanted irradiation to normal tissues around tumor. The system employs a position sensitive detector to sense a target movement, an rf-knockout extraction with FM and AM, an optimized operation pattern of HIMAC synchrotron ring, a beam deceleration as a residual beam aborting system, and an interlock system for a safety treatment. The irradiation treatment gated by respiration has begun since June 1996.

The system for the irradiation gated by respiration requires essential design considerations as follows; (1) a permitting irradiation signal should be accurately generated to permit irradiation only when target is at the designed position, (2) a start and stop of beam extraction should respond quickly to a trigger signal, (3) an operation pattern of the ring should be optimized to give maximum effective irradiation dose-rate, 4) an aborting system of residual beam should be provided to avoid undesired activation, 5) an interlock system for a safety treatment should be applied [10.9, 10.10]. An operation pattern of the system for the irradiation gated by respiration is schematically shown in Fig. 10.11.

10.4.2 NAC Stereophotogrammetric (SPG) System

The basic components of the patient support and positioning system used at the NAC (Sec. 4.2.2 and Fig. 4.9) are an adjustable patient support system (chair), eight video cameras and two personal computers. The patient positioning system makes use of real-time SPG techniques and is linked to a computerized chair. When a patient undergoes a CT scan to locate the treatment volume, small reference targets affixed to a custom-made plastic mask, which fits the patient's head precisely.

Several improvements to the SPG system have recently been implemented. The patient movement monitoring policy has been modified to monitor movements about the patient's final set-up position rather than the calculated position. Fig. 10.12 shows a diagrammatic two-dimensional representation of the new monitoring policy at NAC. The solid circle indicates a positioning tolerance of 0.5 mm. The dotted circle indicates the previous monitoring tolerance of 1 mm about the calculated target position. The dashed circle indicates the new monitoring tolerance of 0.8 mm about the final set-up position. The hatched area indicates the worst possible case where the new monitoring policy will allow movements outside the initial monitoring sphere [10.11].

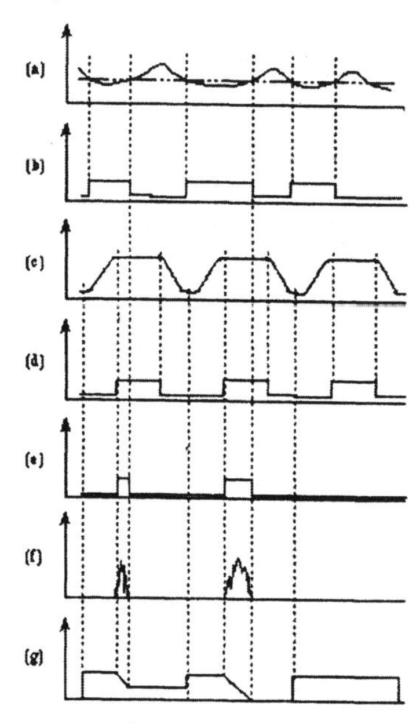

Figure 10.11. Schematic operation pattern of the system for irradiation gated by respiration: (a) respiration signal, (b) permitting-irradiation signal, (c) operation pattern of ring, (d) flat-top signal, (e) beam on / off signal, (f) beam spill, (g) circulating beam intensity; after K. Noda et al. [10.10].

It is now also possible to temporarily interrupt the beam in the case of a small movement. A temporary interruption of the beam comprises the insertion of only the low-energy Faraday cups, i.e., stopping the beam between the injector and main cyclotron. If the patient moves back into position within a certain preset time, the beam is restarted automatically. A small movement is defined as such a movement where there is a chance that the patient might move back into position during the specified time period. This is typically possible when the patient takes a deep breath or coughs. To facilitate this a watch period was implemented in the patient monitoring routine [10.11].

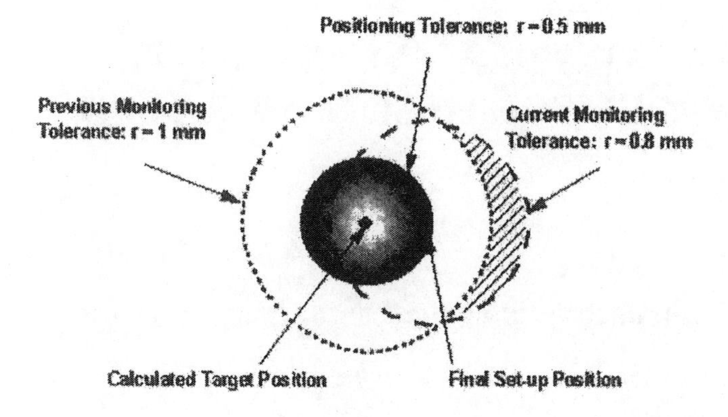

Figure 10.12. A diagrammatic two-dimensional representation of the new monitoring policy at NAC, after A.N. Schreuder et al. [10.11].

Chapter 11

CONTROL SYSTEM OF THE PROTONTHERAPY CENTER

11.1. Control System of the Protontherapy Center

The control system automates and centralizes all the activities needed for a correct management of the Protontherapy center. A short list of typical services is:

- functions tied to the performance of the accelerator and injectors complex;
- operations connected with thc management of the treatment rooms;
- connection with treatment planning systems, medical diagnostics and with the information system of the central hospital and the remote centers of the national oncological network;
- supervision of both conventional and radiological safety systems;
- sophisticated control service-like early fault detection;
- management of the data hank concerning the whole installation as well as the local network;
- acquisition, display and analysis of data collected in the experimental area.

The control system must also provide the tools for the integration of the medical diagnostic facilities with the treatment planning and beam delivery equipment. A wide-band local area network is a good solution that makes possible the rapid distribution of large amounts of information related to the medical digital images [1.4].

11.2 Control System Proposed for Italian Hadrontherapy Center

Description of the System. The proposed structure is based on a distributed processing, hierarchically organized on two physical levels as shown in Fig. 11.1. The links to the treatment planning, diagnostic equipment and national oncological network RITA (Sec. 4.3) are also sketched on the figure. The upper level consists of a computer cluster situated in a single control room, for the supervision, data collection and presentation of the information and machine status to the operators.

The lower level is made of a number of microcomputers, each one devoted to the control of a group of similar equipment, such as the magnet power supplies or the rf system. The control systems for the treatment rooms are also included in the second level. These systems can be further divided into two categories: 1) systems devoted to the control of complex components (gantries), 2) equipment directly used by the radiation technicians for routine treatment operation. The duties of these systems should be similar to those found in conventional radiotherapy. They are connected to the control room where the irradiation operations are coordinated.

The heart of the distributed control system of the center is an advanced local area network called *FDDI* (*Fibre Distributed Data Interface*).

Computer System. At present the machines most widely used for particle accelerator controls are represented by the so-called workstations computers based on RISC (Reduced Instruction Set Computing) microprocessors, having UNIX operating system An alternative solution, already competitive in terms of price to performance ratio, is represented by IBM-compatible personal computers (PC).

Treatment Rooms. Close to every treatment room there is the equipment for the treatment control. These systems are conceived to give support to the radiation technicians and to assist them in order to limit, as far as possible, human errors. A diskless workstation (or PC) linked to the LAN is foreseen for the treatment management, together with the conventional closed circuit television monitors devoted to patien's surveillance. The interface should be graphic and especially conceived for these users. A communication channel between the medical staff and the operators in the control room is also foreseen [1.4].

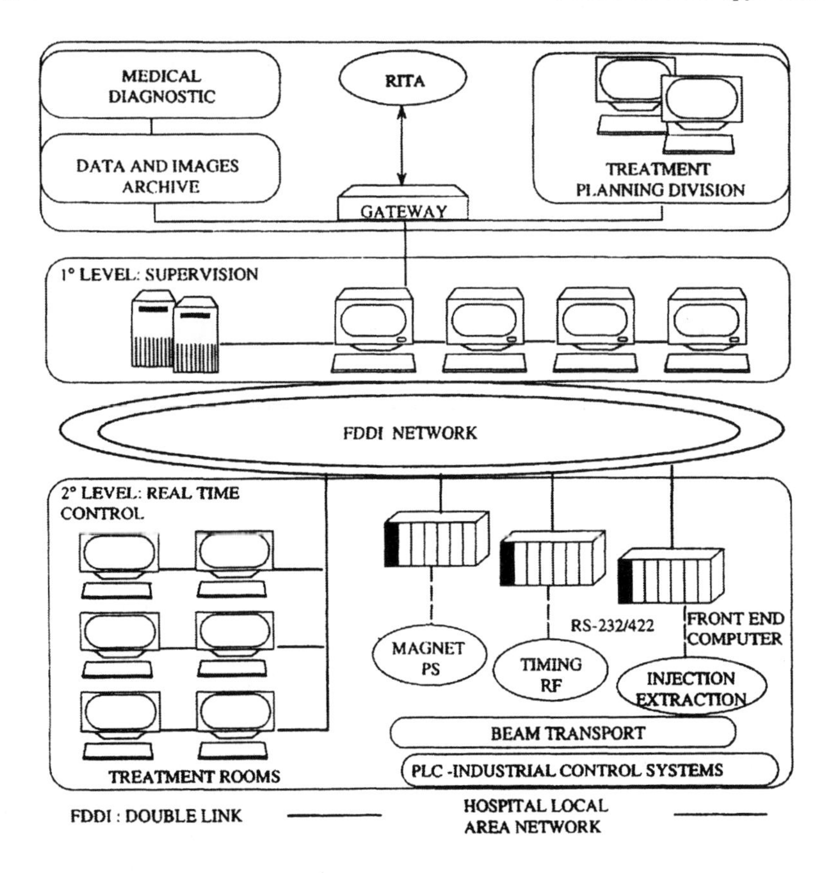

Figure 11.1. Architecture of control system for Italian Hadrontherapy Center; after U. Amaldi et al. [1.4].

11.3 Safety Systems

While safety is difficult to specify, a safe system will include redundant guards against any conceivable failure mode. Specific attention must be given to beam delivered outside the chosen target area, incorrect dose rate, incorrect total dose, collisions of patient with nozzle components, avoidable exposure to facility workers, and accidents involving fire, electrical or vacuum systems.

The control of the all operations connected with health and safety at the Protontherapy center should carried out by dedicated system interfaced with the control system, but completely independent on it. The safety system must have the following characteristics:

- high reliability: this translates into the use of good quality products and a high redundancy;
- fail safe: in case of a severe failure of the system, the accelerator has to be stopped and the main systems automatically turned off;
- complete automation: this is in order to avoid, as far as possible, human errors;
- high intervention speed;
- simplicity of operation and maintenance [1.4, 5.1].

11.3.1 Safety Requirements

Patient Monitoring. There shall be a minimum of two pairs of cameras and screen operable for continuing viewing of the patient during treatment. This continuous patient monitoring devices must have focus, iris adjust and zoom capability. There shall be a intercom between the treatment console area and the treatment room operable for verbal communication between the patient and staff during treatment [5.1].

Emergency Buttons. There should be three levels of power off clearly marked emergency buttons: 1) cut off the power to the entire facility, 2) cut off the power to the room and equipment a person is working, 3) divert or turn off the beam within 10 μsec to the beam dump after the button is pushed.

There shall be a minimum of one level two emergency button in each of treatment console area, at least one level two emergency button located in the treatment room and the accelerator control console must be equipped with one level one and one level three buttons. Any radiation hazard areas must be equipped with one level three emergency button. These area include beam switchyard and beam transport area [5.1].

Door Interlocks. There shall be at least two electronics switches located on the door or door jamb. These switches shall function as the level three emergency button and thus stops the radiation exposure within 10 μsec when the door was opened. Closing of door shall not cause the radiation to be resumed. The radiation shall resume only when door is closed and the operator initiates the exposure, assuming that all the other interlocks are properly set [5.1].

Radiation Monitor. A red and white light shall be located above the treatment room door for indicate whenever there is the beam in the accelerator and beam lines or not. The control console shall be equipped with an indicator indicating that the accelerator's main power is on.

There must be an alarm system which must sound "alarm" and initiate red light flashing should any one of radiation monitors placed in strategic places read more than the national limits. The alarm system must simultaneously trigger either to divert the beam or shut off the power to an accelerator so that the radiation readings in the entire facility accessible to all personnel in the facility including patients and visitors must instantly become less than the limits.

Neutron levels at 1 m from the isocenter must be less than 0.1% of the dose at the isocenter [5.1].

Mechanical Safety. Patient couch, gantry and any other treatment aids in contact with or adjacent to the patient shall be locked or "frozen" as soon as the power is cut off either by activating one of the emergency buttons, power failure or intentional power shutoff.

In case of power failure, there shall be emergency light turned on in the treatment room, possibly in the treatment maze, a hallway leading to the patient waiting area, and any other locations so that the patient can be easily and safely evacuated to safe locations [5.1].

11.3.2 Safety System Proposed for Italian Hadrontherapy Center

The complex has to check a great number of different equipment located in different areas. To satisfy these requirements it is worthwhile to choose a solution similar to those usually employed in high-risk installations. This consists of a distributed system based on PLCs (Programmable Logic Controller) connected through a local network to a main computer, which acts as a supervisor. The PLC is an electronic device that, by a set of instructions located in a programmable memory, carries out operations normally undertaken during controls of processes, such as sequential tests, timing, counting and optimizations.

A possible scheme of the architecture of the safety system is shown in Fig. 11.2.

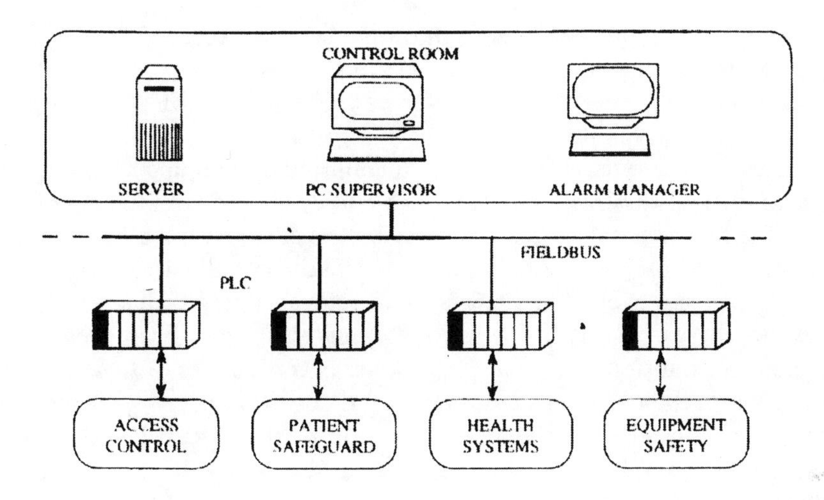

Figure 11.2. Architecture for the safety system for Italian Hadrontherapy Center; after U. Amaldi et al. [1.4].

Over-Irradiation Protection. The system dedicated to avoid *over-irradiation* or *accidental irradiation* of the patient includes: the sensors and diagnostics of the accelerator, the fast beam abort system, the treatment room interlocks and beam line shieldings, the ionization chambers for the monitoring of the secondary beams.

Beam monitoring (beam intensity, position and profile) is ensured by dedicated instrumentation in the magnetic transport system (Sec. 9.2). On the other hand the main functions of the monitoring / dosimetry system in radiotherapy are to provide a real-time measurement of the dose delivered to the patient and an information on the spatial distribution of the radiation delivered to ensure that the prescribed dose is given. This detector system should be independent of the beam transport and beam delivery systems and it must work with a response time sufficiently short to provide both thc data acquisition system and thc operator with on-line information on the irradiation status. To maximize safety, at least two independent devices for monitoring the dose and terminating the irradiation are needed. The local supervision of the monitors should be performed by hardwired control equipment [1.4].

Radiation Monitoring. Photon-neutron radiation detectors connected to the safety system, will overview the radiation level. Their tasks are: 1) to stop the machine whenever a dose threshold level is exceeded, 2) to indicate possible beam losses, 3) to prevent the access of personnel in u-le synchrotron vault, after the accelerator has been stopped, until the radiation level is below a given threshold.

Access and Operation. The underground building housing the accelerators and the treatment room is a surveyed area with a single access controlled by the medical physics service. The areas with the higher risk of irradiation are obviously the accelerator vault and the treatment rooms. These zones are well delimited by concrete shielding (Chapter 12). The access to these areas is via a maze closed by a door. Each area can be declared an exclusion area, an *area of supervised access* or of *limited access*, depending on the machine status and on the dose equivalent level measured in the area. Consequently the access to the area can be denied, supervised or overseen. The number of persons present in the area will be automatically deduced, in the phase in which the area is accessed, by a counting system installed at the entrance of the controlled zone [1.4].

Chapter 12

SHIELDING FOR PROTON FACILITY

The use of proton accelerators in the 60 – 250 MeV energy range for cancer radiation therapy has grown since the first trials in the mid-fifties and now about 20 hospital and non-hospital based facilities are in operation world wide (Chapter 4). These accelerators produce secondary radiation, mainly neutrons, from beam interaction with accelerator components (the magnets in particular), beam delivery devices (such as collimators) and the patient. Although the beam intensity is much lower than that typical of research accelerators (a few 10^{10} protons / s versus values larger than 10^{13} protons / s), yet the fact that such accelerators may be installed inside a hospital and / or in highly populated areas calls for shielding. The shielding should reduce the radiological impact on the external environment essentially to zero [12.1].

The design of the radiation shielding walls around the equipment has a significant impact to both the cost and functionality of the facility. Reducing the mass of the shielding walls, while maintaining acceptable personnel and patient exposure levels, has the potential to substantially reduce the cost of the facility. The design of the treatment room and beam line entrances also effects ingress / egress times, costs and construction techniques. Much of the time spent in this aspect of design is devoted to finding the proper trade-offs of costs, construction complexity and space [4.2].

12.1 Performance Specifications for Shielding

Most of the dedicated protontherapy accelerators have similar main parameters: proton beams average 20 nA and energies from 70 to 250 MeV. So the existing and planning Protontherapy facilities can be used as a paradigm for studying radiation protection, even though other facilities may include different types of accelerators and facility designs.

Neutrons are the dominant secondary radiation field for high-energy proton facilities. The shielding walls must: 1) stop the most energetic protons and 2) reduce to acceptable levels the secondary neutrons and gamma rays produced by the protons as they interact with the accelerator, degrader, beam analyzer, beam transport lines, gantries, nozzles, patients and shielding walls [4.2].

The technical concerns include: 1) identification of source terms for fast neutrons and gamma rays with respect to location, intensity directionality, and energy; 2) verification of the attenuation properties of shielding for the radiation of concern, principally neutrons, by experimental and theoretical methods; 3) determination of the reduction of dose equivalent by ducts and mazes needed in the radiation therapy facility; 4) evaluation of the radioactivation hazard, especially that which occurs in the treatment rooms and can affect clinical personnel [12.1, 12.2].

12.1.1 Radiation Exposure

Examples of existing annual limits on dose equivalent in USA and Italy are presented in Table 12.1. For example, in the United States the limit for the general public will be probably reduced to 0.1 rem, without changing the radiation worker limit. However, it is believed that the latter will also soon be reduced to 2 rem, under the urging of the ICRP and NCRP. The ICRP has also issued a recommendation for a change (generally, an increase) in the neutron quality factor.

Groups with annual limits on dose equivalent	**Limits**	
	California (USA)	Italy
Occupational radiation workers	5.0 rem (50 mSv)	
Facility personnel		0.2 rem (2 mSv)
General public	0.5 rem (5 mSv)	0.1 rem (1 mSv)

Table 12.1. Examples of existing annual limits on dose equivalent.

12.1.2 Attenuation in Shielding

In general, the shielding thickness d required to reduce the neutron dose equivalent rate to the maximum permissible value is given by:

$$\frac{d}{\lambda(\theta)} = \ln\left(\frac{H_0(\theta)\cdot I\cdot f_{loss}\cdot t_{loss}\cdot U\cdot T}{H_{\max}\cdot r^2}\right) = \ln\left(\frac{H_0(\theta)\cdot S}{H_{\max}\cdot r^2}\right) \tag{12.1}$$

where $\lambda(\theta)$ is the attenuation length in concrete in the direction θ with respect to the beam direction, r is the distance between the source and the point of interest ($r>d$), $H_0(\theta)$ is the source term along the direction θ, I is the average proton current, f_{loss} ($0 < f_{loss} \leq 1$) is the beam loss factor, t_{loss} is the macroscopic duty factor (i.e., how many hours per day the beam is lost at that particular point), T is the occupancy factor of the environment beyond the shield, U ($0 < U \leq 1$) is the use factor of the shield and $S = I \cdot f_{loss} \cdot t_{loss} \cdot U \cdot T$ [12.1].

The examples of the calculated dose-equivalent attenuation lengths are presented in Table 12.2. The calculated dose equivalent attenuation lengths in Table 12.2 are in very reasonable agreement with the well-established neutron attenuation lengths in concrete. It is well established that other shielding materials (earth, iron, big-density concrete) can replace concrete if substituted on the basis of areal density, except that a sufficient thickness of hydrogenous material must be on the personnel side of the shield.

Experimental			**Calculated**		
θ(°)	H_0r^2 (10^{-15} Sv m^2)	λ_{eff} (kg m^{-2})	θ(°)	H_0r^2 (10^{-15} Sv m^2)	λ_{eff} (kg m^{-2})
0	8.6±0.8	910±30	0-10	6.6±0.4	991±28
22	4.6±0.5	876±34	10-30	5.0±0.2	1040±21
45	2.1±0.2	746±24	40-50	2.3±0.1	894±21
90	6.9±0.8	519±21	85-95	1.0±0.2	534±26

Table 12.2. Zero depth source intensities and attenuation lengths from Siebers [12.4].

12.1.3 Neutron Sources

Neutrons are the dominant secondary radiation field for high-energy proton facilities. The highest energy protons will cause the greatest hazard. Neutrons are produced by protons interacting with the structural materials of the accelerator aparatus, with targets inside the treatments and experimental rooms together with the contribution from the accelerator through the primary walls. Dimensions and neutron yields for Fe, Cu, and H_2O targets are presented in Table 12.3. The angular distribution of the neutrons escaping from the three targets is shown in Fig. 12.1.

Target material	Radius (cm)	Length (cm)	Yield (neutrons per incident proton)
Fe	5.8	7.5	0.768
Cu	5.0	6.5	0.994
H_2O	20.0	20.0	0.20

Table 12.3. Dimensions and neutron yields for Fe, Cu, and H_2O targets; after U. Amaldi et al. [1.4].

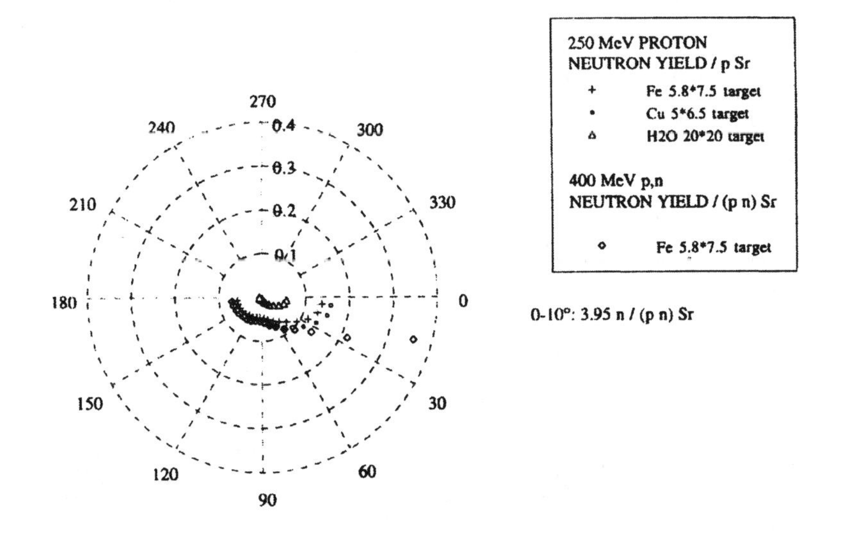

Figure 12.1. The angular distribution of the neutrons escaping from the reference targets; after U. Amaldi et al. [1.4].

12.1.4 Attenuation in Mazes and Ducts

While plug doors are commonly used in physics-oriented accelerator facilities, they are undesirable in a therapy facility because of possible adverse reactions of patients. It is also noted that the treatment room at the end of a maze (which must be large enough to pass a hospital gurney) is a copious source of radiation: the patient is in reality a beam dump and other losses occur in the beam transport system.

Most of the dose equivalent transmitted through maze is carried by low-energy neutrons. Thermal neutrons are generated by collisions with the walls and tend to build up as the maze is traversed [12.3].

Ducts, defined as shield penetrations with an average diameter of 30 cm or less, are needed for power, cooling, and other utilities. The principal vector of dose equivalent is low-energy neutrons, and their transport behavior in ducts has been understood for at least 40 years by the nuclear power industry[12.3].

12.1.5 Activation

Radionuclides are principally produced by high-energy protons and by low-energy neutrons, especially those that thermalize and capture. Number of isotopes produced in concrete per incident proton (vertical axis) versus isotope are presented in Fig. 12.2.

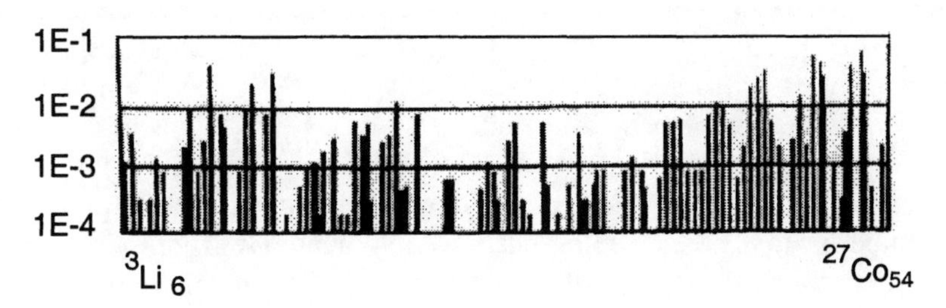

Figure 12.2. Number of isotopes produced in concrete per incident proton (vertical axis) versus isotope. The incident proton energy is 230 MeV and all cascade particles down to 20 MeV contribute to production; after H.B. Knowles [12.3].

There is special concern about the clinical personnel who must tend the patients after treatment. Typically, a brief 1 – 2 min. irradiation is followed by a 15 - 20 min. interpatient time. It can be shown that the clinical staff are likely to have more dose equivalent from radionuclides with half lives of the order of the interpatient time then from shorter- or longer lived activities, provided that the latter have not been allowed to build up over a long time. Components with such a long decay times would of course be routinely replaced, but nature has been unkind in the matter of radionuclides created in the body itself: there is a high production of the radionuclide ^{11}C, which decays by positron emisson with 20.4 min. half-life. When the total dose equivalent to clinical personnel during one interpatient period was determined, assuming an iron or copper component of the beam nozzle and a loss of 10 nA of protons in 20 g / cm^2 or materials, it was found that, at a distance of 1 m, the dose equivalent would be 0.57 mrem.

12.2 Bunker Project for Italian Hadrontherapy Center

12.2.1 Beam Loss Sources

Hadrontherapy Center is a proposed hospital based facility, which will be built in the North of Italy (Sec. 4.3, Fig. 4.16) [1.4]. The main accelerator is a synchrotron capable of accelerating H^- ions up to 250 MeV. Protons will be extracted by charge exchange. The required vacuum in the synchrotron doughnut is about 10^{-10} torr, to avoid the stripping of the second electron of the H^- ion in the atomic collisions with the residual gas molecules. Two separate linacs will inject H^- and light ions into the synchrotron at 11 MeV and 3 MeV/u, respectively. Since only a modest beam intensity is necessary for therapy (of order of a few 10^{10} protons s^{-1} and about 10^9 $^{16}O^{8+}$ ions s^{-1}), only a modest average beam current needs to be injected and accelerated in the synchrotron. The exceeding beam current available from the injectors, is planned to be used for alternative purposes, e.g., radionuclide production for PET diagostics and neutron production on a beryllium target for BNCT. The Center for Oncological Hadrontherapy will consist of two buildings: an underground, heavily shielded area (bunker) housing the accelerators and the treatment rooms, and a surface building above ground with conventional facilities and office space. The bunker has a surface area of about 3500 m^2. Two treatment rooms will be equipped with an isocentric gantry. This unit allows 360° rotation of the terminal tract of the beam line around the patient in order to vary the direction of irradiation, as is done in conventional radiotherapy.

The expected radiation sources due to beam losses in the accelerator, beam transport and delivery systems are identified by letters in Fig. 12.3.

12.2.2 Shielding Project

Fig. 12.3 show a plan view of the facility and the position of the radiation sources and the shielding thickness required to meet the annual dose equivalent.

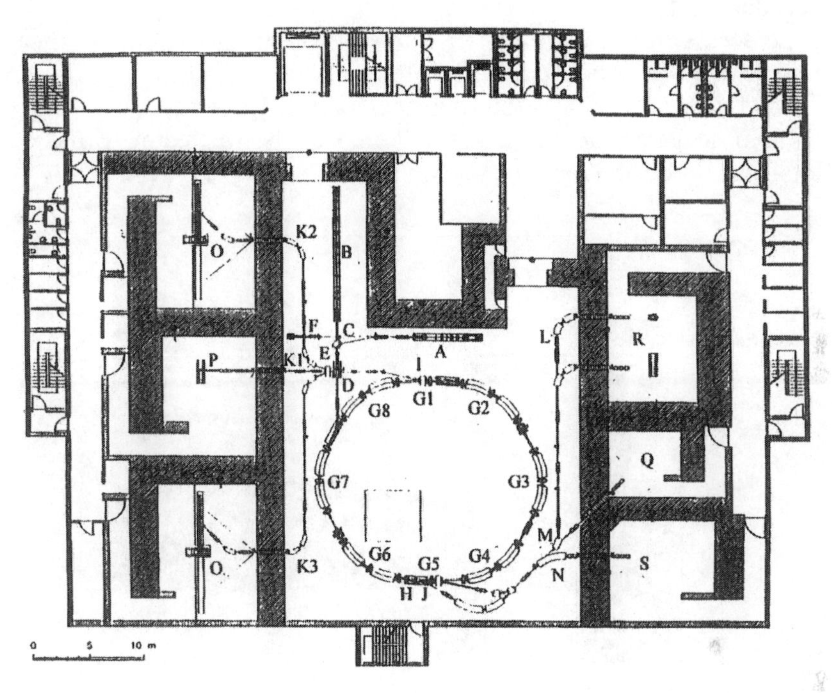

Figure 12.3. Layout of the bunker of A.2 floor of Italian Hadrontherapy Center; after U. Amaldi et al. [1.4].

The bunker includes:

1. two treatment rooms equipped with an isocentric gantry capable of transporting protons up to 250 MeV;
2. one treatment room equipped with one horizontal and one vertical beam, also for 250 MeV proton beams;
3. one room equipped with two horizontal beam lines: one for eye tumor irradiations and one mainly devoted to the treatment of head and neck tumors;
4. one room with one horizontal beam for experimental activities with both protons and light ions (dosimetry, radiobiology. calibrations. etc.);

5. one room devoted to future light ion treatments;
6. two smaller rooms served by the 11 MeV proton beam from the injector, one for the production of positron emitting radionuclides for PET diagnostics (^{11}C, ^{13}N, ^{15}O, ^{13}F), the other for BNCT thermal neutron production [12.2].

The access mazes were designed by optimizing the length and Sec. of their legs and their wall thickness with the dose equivalent limit of 2 mSv per year fixed in areas accessible for personnel. The mazes were designed for isocentric gantry rooms and for horizontal beam treatment room. An example mazes design for gantry room is presented in Fig. 12.4.

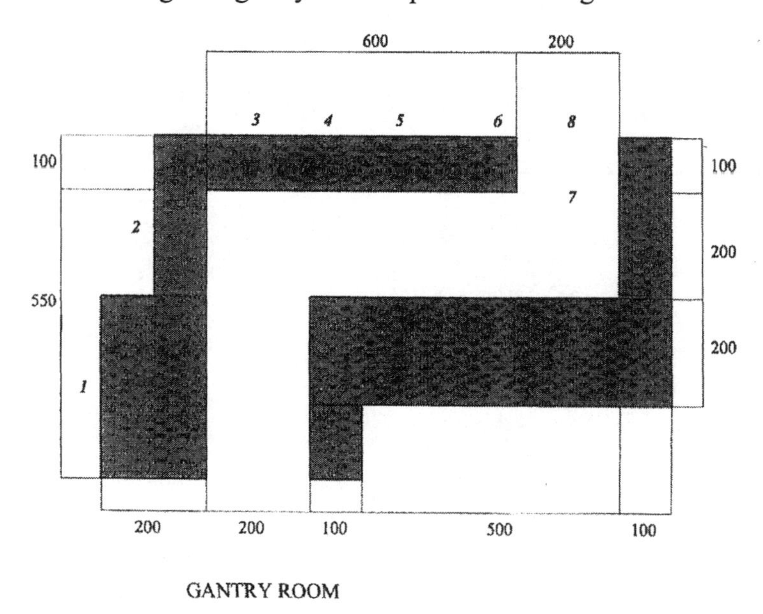

Figure 12.4. Maze geometry (dimensions in cm). The maze height is 3 m; after S. Agosteo et al. [12.2].

Chapter 13

GLOBAL COSTS AND FINANCIAL ANALYSIS OF THE ACTIVITIES OF THE PROTON CENTER

The introduction of every new treatment modality into the clinical routine is difficult, especially if it is much more expensive in setting up and running than conventional ones. The clinical benefit has to be proven in relation to the increased costs of treatment. Proton- and heavy-ion units are expensive and therefore many scientific projects have been stopped or never come to realization. It is necessary to survey the complete socioeconomic environment of this new treatment modality starting with indications and patient numbers, continuing with cost estimates, and finally discussing financing. The clinical relevance of particle therapy was already presented in Secs. 2.3 and Chapter 3. The discussion and the estimate of costs and prizes will usually depend on the individual social system and the overall cost level in different countries [13.1]. In this chapter the estimated costs of construction and exploatation of Protontherapy center will be presented.

13.1 Cost Considerations of Hadrontherapy

In general, the costs of the installation of a proton unit are high in comparison with conventional installations; about 5- to 10-fold. The energy consumption depends on the technique of the accelerator. Usually the technical running costs are higher. According the comparison made for the operation of proton beam treatment facility of 3 - 4 treatment bays versus a 3 - 4 linear accelerator facility with each designed to perform three-dimensional conformal optimize treatments, the cost increment of protontherapy over that of photontherapy is judged to be a factor of 1.5. The clinical personnel is almost equivalent to conventional systems and the accelerator staff will also be small because of modern-machine technology [13.1, 13.2].

Scenario and cost estimate. In order to estimate the personnel, space, capital and daily cost of an institution dedicated to particle radiotherapy, a scenario was built up [13.3] as presented in Table 13.1. A total of 27,000 fractions should be treatable per year at this institute. This results in a yearly number of 1,500 patient treatments for protons with 25 fractions as full treatment, and 12 fractions for boost treatment. The complete financial situation of personal, space and equipment was constructed based on German standards. Personal costs, capital, and running costs are given in Table 13.2. The machine costs of about 43 mln DM will concentrate to a proton facility with two gantries and one fixed beam, including the building. The yearly cost of the described unit will be somewhere between 14 and 15 mln DM, including amortization costs and running costs [13.1].

This scenario demonstrates a median size radiotherapy department in Germany according to the yearly patient numbers. Patient numbers are crucial for financing and must be estimated to a level as high as possible. Considering 1 500 patients and 27,000 fractions per year, those numbers would be around 540 DM / fraction, 13 500 DM per full treatment and 6 500 DM per boost treatment.

Setup time	15 min.
Irradiation time	5 min.
Fraction / hour / cave	2.5
Working hours/day	12
Working days/week	6
Working weeks/year	50
Fractions/day	100
Fractions/year	27,000
Full: boost treatment	1:1
Patients/year	1,500
Fraction size/full treatment	25
Fraction size/boost treatment	12
New setups/day	5.6

Table 13.1. Scenario for a clinical proton unit, one accelerator and three treatment rooms; after G. Gademann [13.1].

The other estimations were made by LASA (Milan) / CAL (Nice) studies and described by P. Mandrillon [6.1]. Assuming an amortization over 20 years and without the costs of the medical staff (they are the same as in case of complex conformal therapy), an approximate costs per patient can be derived as 2 400 Euros. Proton facility the estimations are made for, is equipped in 200 MeV superconducting cyclotron, and can treat 500 patients per year using two treatment rooms. The estimated costs are presented in Table 13.3 [6.1].

	mln. DM	
Long-term capital costs (18 years)		
accelerator, beam-line, gantry one	30	
leasing rate		3.4
building	13	
loan		1.3
Short- and intermediate-term capital costs (8 years)		
equipment (computer. CT, NMR)	8.65	
leasing rate		1.4
Running costs		
personnel, energy, maintenance		9
Total	51.65	15.1

Table 13.2. Cost per year for a running proton facility; after G. Gademann [13.1].

	Investment costs (mln. Euros)
Equipment	
Accelerator	3
Beam delivery	5
Building	4
Running costs	
Maintenance	0.30
Electricity	0.10
Salaries	0.20

Table 13.3. Tentative costs estimates for 200 MeV proton facility based on superconducting cyclotron; after P. Mandrillon [6.1].

13.2 Comparison of Treatment Costs

The above estimated costs for proton radiotherapy must be compared with the costs of conventional radiotherapy. However, there are no data available that are precise enough.

According to a number of publications, mentioned by G. G. Gademann [13.1], dealing with hadron and conventional radiotherapy of chordomas and chondrosarcomas, the 5-years control is only 62 vs. 20%. Every recurrent tumor needs second therapy consisting of new neurosurgery (15 500 DM), post-surgery rehabilitation (12 000 DM), other rehabilitation procedures (5 000 DM), and nursing for at least 5 years (48 500 DM). The total rehabilitation is estimated to about 80 000 DM.

Calculation of the overall cost for the German incidence of the above-mentioned tumor (72 patients per year) for hadron as well conventional radiotherapy provides the yearly profit for the insurance company. Primary treatment of all chordomas and chondrosarcomas with hadrontherapy will bring a pecuniary benefit of 1 670 400 DM yearly for the insurance companies (Table 13.4).

	Cost of conventional radiotherapy (DM)	Costs of- hadrontherapy (DM)
Single treatment costs	9 800	20 200
Costs for 72 patients	705 600	1 454 400
Rehabilitation cost	4 608 000	2 188 800
Overall yearly costs	5 313 600	3 643 200

Table 13.4. Estimated cost comparison of proton and conventional radiotherapy in case of chordomas and chondrosarcomas; after G. G. Gademann [13.1].

13.3 Global Costs of Existing the Proton Facilities

13.3.1 Costs of the Equipment

The estimated fabrication costs for replicating the LLUMC facility are shown in Table 13.5. They are compared with the estimated costs of the NPTC equipment subsystems. The budget of the first generation LLUMC facility was about 55.4 mln US$, and the costs of the second generation NPTC facility was 46.1 mln US$ [4.5, 6.12].

In general, the expected fee charged to the patients to cover the running costs are in the range of 10 000 to 15 000 US$ [1.3, 5.5].

The global costs of constructing of the other hospital – based hadron facilities are as follow:

- HIMAC, Chiba, Japan – 350 mln US$,
- Kashiwa Proton Treatment Facility – 55 mln US$,
- Hyogo Charged Particle Facility, Japan – 275 mln US$,
- PRMC, Japan – 70 mln US$ [13.4].

Subsystem	LLUMC - mln $ (%)	NPTC – mln $ (%)
Accelerator	10.4 (21)	12 (26)
Energy selection system		1.8 (4)
Beam transport system	4 (1)	4.6 (10)
Gantries	three gantries 19.7 (37)	two gantries 10.2 (22)
Fixed-beam room	3.8 (8)	6.9 (15)
Patient positioners (2)		3.7 (8)
Safety and control systems		6.9 (15)
Building and facilities	17.5 (33)	
TOTAL	55.4	46.1

Table 13.5. Comparison of estimated fabrication costs for replicating the LLUMC facility and approximate costs of the NPTC equipment subsystems; after J.M. Slater et al. [6.12] and M. Goitein [4.5].

13.3.2 Operational Expenses of Existing Proton Facilities

Minimizing operational expenses is essential for maintaining a successful program. This issue in a medical program increases in importance as the initial costs of the program rise because the expenses are directly proportional to the patient charges. On the other hand, the cost to patients is inversely proportional the number of patients treated per day. For example, costs of supplies for approximately 50 patients per day, maintenance, utilities and depreciation are listed in Table 13.6. It is the data based on the experience of LLUMC.

Supplies (moulds, boluses. patient needs)	$ 350 000
Maintenance	1 600 000
Utilities 24 h / day	300 000
Depreciation equipment	3 100 000
Depreciation building	700 000
Total annual costs	6 050 000

Table 13.6. Annual operation costs; after J.M. Slater et al. [6.12].

13.3.3 Staff Requirements and Costs

In estimating the needed staff, only the requirements relating to the specific activities of the facility have to be considered because the center will be inserted in the framework of the hospital complex, which should provide the general services. The available estimations for the Italian Hadrontherapy Center assume that the center will treat 1000 patients per year. Table 13.7 gives the general estimation of costs of the Italian Hadrontherapy Center; the special attention should be called to amount of staff foreseen and its distribution in the different departments of the center.

		Number	Costs in 1994 kLit	
Yearly cost of staff	Administration	8	8 820 000	
	Physicists and engineers	38		
	Medical doctors	38		
Buildings and conventional plants			34 500 000	
Management of the project			3 552 829	
Installation			26 982 922	
Technical com-ponents	Injector and main synchrotron		168 841 962	50 754 712
	Beam transport		8 517 200	
	Beam delivery		6 163 200	
	Patient positioning		10 277 350	
	Control and safety systems		7 955 000	
	RITA network		1 000 000	
Upgrading	Equipment for accelerating, transport and de-livering of ion beams		11 850 000	
	Radionuclide production		5 900 000	
	BNCT		410 000	

Table 13.7. Cost estimate for Italian Hadrontherapy Center; the exchange rate is: 1 DM = 1 kLit, 2.5 kLit = 1 UK£ (1994); after U. Amaldi et al. [1.4].

Chapter 14

PROPOSAL OF A DEDICATED PROTONTHERAPY FACILITY

14.1 Clinical Requirements for a Dedicated Protontherapy Center

Regardless of how the project is to be run, and in particular how the equipment is to be procured, it is extremely important to begin with a clear-cut, explicit set of specifications for the facility. In putting together a set of specifications there is a great temptation to include issues regarding implementation. However, it is a very good practice to begin with top-level functional specifications, which devolve explicitly from a clinical requirements [4.6]. Only when these have been written out and agreed upon by the potential users as well as sponsors, should specific implementation issues, with their often quite technical ramifications, be addressed. According M. Goitein [4.5], it is a signal that a clinical project is poorly conceived if the first initiative is the promotion of a particular technology say, a type of proton accelerator or style of gantry.

The specifications for a Protontherapy facility are, in line with the above comments, purely clinical specifications with minimal guidance as to the technical solutions. In contrast to the typical specifications for a high-energy physics accelerator, which might address the issues of beam energy, beam intensity, spill time structure, beam emittance, etc, the top four Protontherapy facility specifications concern: (1) safety, (2) reliability, (3) operability and (4) maintainability. The facility should be extremely safe. The minimal requirement would be that the probability of a fatality be < 0.01% over the anticipated 30 year life of the facility. The facility should be extremely reliable. The availability should be at least 95% of the scheduled operating time (98% is typical of electron linear accelerators used in radiotherapy, and would be preferred). The facility should be extremely easy to operate and maintain. There should be no operators in the main control room; in the daily practice radiation therapists should normally call for and get beam under computer control. Only a small engineering and technical support should be needed. The facility should be characterized by short mean time to repair.

A few additional specifications are provided in Table 14.1 to give a sense of the clinical requirements that are imposed.

Clinical specifications	**Preferred specifications**
Beam dose shape	
Range in patient	maximum 32 g/cm^2
	minimum 3,5 g/cm^2
Range modulation	steps 0,5 g/cm^2 or less over full depth; 0,2 g/cm^2 or less for ranges ≤ 5 g/cm^2
Range adjustment	steps 0,1 g/cm^2; 0,05 g/cm^2 for ranges < 5 g/cm^2
Distal dose fall-off (80% - 20%)	mot more than 0,1 g/cm^2 above physical limit from range straggling
Field size	fixed 40 × 40 cm^2; gantry 40 × 30 cm^2
Lateral penumbra (80% - 20%)	not more then 2 mm over the penumbra due to multiple scattering in the patient
Dose uniformity	± 2,5% over treatment field
Beam dose	
Average dose rate	a beam intensity sufficient to treat a 25 × 25 cm^2 at depth of 32 g/cm^2 to dose of 2 Gy in 1 min. or less
Dose accuracy	Daily accuracy and reproducibility ± 1.5%
Beam trajectory	
Beam position	Direct a beam to the target to better then ± 1 mm. (Results from several components including gantry mechanical isocenter accuracy, and PPS (patient positioner) accuracy.)
Other specifications	
Spill structure	"scanning upgradable"
Effective SAD (for gantry beam)	≥ 3 m

Table 14.1. Main clinical specifications for Protontherapy facility.

As it was suggested by NPTC designers, the beam range in the patient's body has been specified rather then the beam energy. The development of protontherapy is obviously linked to beam ranges (beam energies) required for a particular treatment. In general, the clinical uses of proton beams can be divided into two main groups: (1) using protons with energies less than 80 MeV, (2) using energies in the ranges of 150 - 250 MeV (Chapter 5). The minimum dose rate of a typical beam is specified rather than the beam intensity. Clinical specifications, if properly written, are essentially neutral with respect to technology [4.5]. Following this philosophy the designers are given the maximum flexibility to come up with innovative solutions, which are optimally matched to the clinical and other requirements.

14.1.1 Equipment for Protontherapy Facility

The core of a modular design is the division of the system into functionally distinct subsystems, as indicated in Sec. 5.3. Each subsystem performs well-defined functions, and they have straightforward and explicitly established relationships between them. Requirements, as well as the existing and proposed solutions of the above mentioned elements are to be widely discussed in the following Sections.

14.1.1.1 Technical Parameters and the Choice of a Proton Accelerator

Modern accelerator technology exists today to meet all of the clinical requirements within the reasonable budget for hospital-based hadrontherapy facilities. There are currently many debates going on concerning the optimum specifications and choice of accelerator for use in high-energy protontherapy. The accelerator should run in a hospital environment without technical support services found in nuclear physics laboratories. Therefore, technological choices are particularly relevant. The summary of the main proton accelerator specifications was presented in Sec. 6.1.

Choice of accelerator. The three principal options for the proton accelerator proposed for a dedicated Protontherapy center, are: cyclotron, synchrotron and the linear accelerator. A comparison of parameters of accelerators used and proposed for protontherapy were summarized in Table 6.3.

An accelerator for a dedicated Protontherapy facility should deliver a beam of accelerated protons with the maximum energy of at least 250 MeV (Table 14.2). What is particularly important is that this energy should be easily controlled over a wide range of values because of the need to have the depth distribution quickly modified. For the dose rate of several Gy min^{-1} to be produced in the volume of 1000 cm^3 the beam intensity should be about 10^{10} p s^{-1}. Most of the accelerator parameters should be fixed, because this facilitates the control of the machine. The accelerator should be safe and reliable.

Parameter		Specifications
Beam energy		
Energy range	maximum	250 MeV
	minimum	70 MeV
Energy precision		± 0.4 MeV
Energy variability		0.1 g/cm^2, or 0.05 g/cm^2 for ranges less than 5 g/cm^2
Energy variation of the extraction		± 0.1%
Beam intensity		
Average beam intensity		10^{11} protons / s
Time structure		cw beam or quasi-cw (with duty factor close to 100%)
Undesired beam intensity		< 20% in any time window
Beam abort time		< 10 μs after receipt of a trigger signal
Quality of the extracted beam		
Position stability of extracted be-am		± 1 mm
Angle stability of extracted beam		± 1 mrad

Table 14.2. Performance specifications for the beam produced by accelerator for Protontherapy facility.

At this stage, it is very difficult to decide which of dedicated proton accelerators is the best solution for radiotherapeutical purposes.

The synchrotron has a variable energy. This eliminates the need for a degrader with its associated scattering and momentum spread. Hence the target volume is discretized into slices of equal range. Each slice is scanned independently and then the beam energy is changed. Theoretically, with a synchrotron, this change could be made at the accelerator. In practice, in the existing solutions of dedicated proton synchrotrons the extraction energy can be chosen only out of some fixed values. The operation of a synchrotrons is complex because rapid magnetic field variations are required. Moreover, undesired intensity variations require careful controls. Nevertheless, is should be noted that many of the new designs of Japanese and European proton facilities are based on the use of synchrotrons.

At present, the isochronous cyclotron (parameters collected in the Table 6.4, Sec. 6.3.1) seems to be the best solution of a dedicated proton accelerator. Its technology is well-known and widespread as evidenced by the large number of similar and smaller machines installed. The cyclotron extracted beam is characterized by a fixed energy, a low energy dispersion and a fixed, low emittance. The accelerator is unable to produce an extracted beam with incorrect energy, energy dispersion or emittance pattern. But then, the cyclotron is a fixed energy accelerator, therefore it requires an additional energy selection system.

In choosing how to approach the design of the accelerator, a contractor should be found to provide help in designing and constructing the machine, e.g. a company should be selected with a possible largest experience on the field of proton facilities projects, such as a Belgium company IBA. NPTC and NCC designers have already chosen a similar solution. They have installed CYCLON 235 accelerator (Fig. 6.4). It should also be noted, that IBA company proposes a whole system for proton radiotherapy (Sec. 4.5).

14.1.1.2 Beam Transport and Delivery Subsystem

General performance specifications for a beam transport system were presented in Sec. 7.2. The choice of the beam transport system elements depends on the beam modification system used in the facility (Secs. 7.2 and 7.4). The set of elements depends on the configuration of the center. Therefore, at this stage it is not possible to give detailed information about the beam transport system elements and construction.

14.1.1.3 Gantries and Fixed Beams

The highest degree of conformity, while using proton radiotherapy, can be achieved by a combination of a rotating gantry with an advanced scanning beam delivery technique. This trend can be recognized in recently built or currently proposed proton cancer therapy facilities, especially in dedicated hospital-based facilities where a high level of flexibility of the beam delivery system is required to treat a large spectrum of tumor sizes and sites (Secs. 4.2.3 and 4.3). There are presently several high-energy gantries in operation or under construction. Consideration of gantry geometry possibilities leads to several paths. It is interesting to note that practically all these paths have been or are being pursued. No consensus has yet been reached about a preferred scheme, which would results in the least expensive overall system and yet meet all the clinical requirements. The use of passive scattering vs. active scanning may affect the gantry geometry. In particular, the location of the scattering or scanning devices may be an important factor. In the proposed dedicated Protontherapy facility, gantries performance must meet the specifications listed in Table 8.1.

For a proton gantry it is recommended to find a contractor to provide help in designing, constructing and installing the whole gantry. A good solution seems to be a large-throw gantry. This conventional single plane gantry was built for MGH / NPTC by IBA and GA (Fig. 4.2(4)). The diameter of the gantry is 9 m. The estimated costs are about U$ 5.2 mln.

14.1.1.4 Beam Shaping Subsystem

Scattering techniques have been predominantly used for patient treatment with protons by all centers, with the exception of NIRS and new PSI facility. The dedicated proton facility LLUMC is equipped with a scattering beam delivery system, which at a later stage may be upgraded to provide raster scanning or conformation therapy. A beam delivery system at NTCP includes a passive system and a wobbling system. Such a solution should also be employed in a dedicated facility.

Variable Range Shifters. The best way of spreading out the Bragg peaks at different depths is by changing the energy of the extracted proton beams from the accelerator. The entire beam transport system must also be changed in such a way that the tune and the positioning accuracy of the beam in the target volume remain constant.

When we are not able to extract the beam of variable energy from accelerator, the best way to change the range seems to be a double-wedge variable absorber (Sec. 7.5.1). As a variation, the wedge may be circularly shaped; each radius has a constant thickness, and its magnitude varies linearly with the angular displacement.

In the proton facility, what will be also need is a patient specific device for modifying the beam range according to the patient anatomy that is compensator or bolus. Its function is to adjust the range of the beam across the target to conform the distal edge of the Bragg peak to the distal geometry of the target volume. The entire process, from obtaining the compensator contours based on the treatment plan to cutting the material on a numerically controlled milling machine, should be automated.

Range Modulation. The function of range modulation may also be accomplished by designing a propeller or fan-shaped stepped absorber, which is made to rotate rapidly in the beam so that the appropriate thickness of the propeller blades intercept the beam. The blades are made of Lucite, or Plexiglass, which are water-like in their absorbing and scattering characteristics. For balancing of the rapidly rotating blades, the propeller is usually made of two or four blades. This technique is used at HCL and LLUMC and works very well.

The actual construction of the propellers is done by making a set of layers cut to the desired shape and then glued together. A water equivalent thickness of 4 mm of water is a practical Lucite thickness. Again Lucite is chosen because it is a low *Z* material thereby it reduces multiple scattering and the associated increase in beam divergence. At LLUMC, the propellers are machined out of plastic blocks. The useful radial size of the blades must be approximately three times the beam diameter. A small propeller for beams less than 2 cm in diameter can also be produced and used for ocular treatments.

Lateral Spreading of Particle Beams. The aim is to produce a large field (up to 30×30 cm^2) that covers the target area with a uniform dose with a variation of less than ±2% (Sec. 5.1.9). Other important considerations are: optimization of such beam characteristics as the sharpness of the lateral dose falloff, the sharp falloff of the distal-peak dose (Sec. 5.1.5), the beam utilization efficiency, dose rate (Sec. 5.1.8), neutron production, beam fragmentation in case of heavier ions, the ease of beam tuning, repeatability, stability of the delivered dose distributions, and patient safety.

Passive Beam Delivery Systems. To obtain a broader uniform dose distribution at the isocenter, some of the excess particles near the central ray must be removed. The best method seems to be a double-scattering method. Such scattering systems were developed for proton beams at HCL, and at LBL. This method should be employed at the first stage of the facility activity. It should make it possible to reduce the investment costs at this stage.

Active Beam Delivery Systems. An active beam delivery system should be an option in the Protontherapy facility. It should be possible to upgrade the facility with a spot scanning system.

A large field of a flat dose can be obtained by moving a beam spot across the field in discrete steps. To achieve efficient pixel scanning, very fast magnets and fast monitoring systems are required. Considerations of multiple scattering of protons in the treatment volume and the edge matching of neighboring pixels make the smallest acceptable size of the pixel bigger than 1×1 cm^2. The treatment volume may be divided into many layers and scanned layer by layer by changing the residual range of the beam. This method is used in PSI proton facility (Sec. 7.5.3.2) in order to reduce the number of layers.

Collimators. Fixed aperture collimators have an opening designed by clinical personnel based on the projected target shape. By automating the design and milling processes, these types of collimators may be fabricated routinely in the center.

It should also be possible to use automated multileaf collimators (Sec. 7.5.4 and Fig. 7.12) whose leaves are moved by actuators under computer control.

14.1.1.5 Dosimetry

In general, dosimetry systems for radiotherapy and radiosurgery have three main functions: (1) measurement of the dose being delivered to the patient in real time in order to terminate the irradiation at the prescribed dose, (2) measurement of the lateral distribution of the radiation delivered in order to insure that the patient prescription is satisfied, and (3) measurement of radiation field parameters for controlling the beam delivery system (Sec. 9.2.1). The components of the monitoring chain strongly depend on the choice of the beam delivery systems. For a passive system the techniques used in conventional radiotherapy can be the basis of the monitoring chain, but for an active system the most often used system of the monitoring should have to be of a novel design.

The proposed dosimetry systems for a dedicated Protontherapy facility are presented in Table 14.3.

Measurement type	Dosimeter type	Applications	Comments
Monitoring chain	Wire chambers	• On line monitoring of beam profile and centroid position • Monitoring of the beam profile of the dose delivered by spot-scanning methods	Used at NIRS, PARMS and LLUMC
	Plastic scintillators	• In the with CCD cameras as a range detector and for field uniformity control • With photocathodes for beam control	The systems containing scintillator and CCD camera is used at LLUMC as a range detector
Reference dosimetry systems	Calorimeters Bragg-Gray cavity ion chambers	• Calibration of dosimetry systems	Should be used according the appropriate dosimetry protocol
Relative dosimetry systems	Diodes	• Arrays of diodes for measuring profiles of radiation fields and Bragg ionization curves	
	Films	• For measuring spatial dose distributions • In the systems for patient position verification	

Table 14.3. Dosimetry systems for dedicated Protontherapy facility.

14.1.1.6 Proton Treatment Ancillary Facilities

A computer facility must be networked to nationwide protocol control, picture archiving and computer systems, therapy planning computers and treatment control computers.

A typical patient flow in the Department of Radiation Oncology would be that a patient is referred to the Department for consultation by a radiation oncologist. When the patient is determined to be a proton beam therapy candidate, the patient information will be entered into a database. A protocol coordinator will be notified to determine whether or not the patient can be entered into the existing or the immediate future protocols. The patient protocols as well as protocol accrual must be connected by PACS (Picture Archiving and Computer Systems) network for nationwide information transmittal. The computer, which will be used for these purposes must be able to communicate with any computers within the proton medical facility. When all these data collection is done or in place, the patient will go to a diagnostic work-up.

Therapy Planning System. The therapy planning system must provide the capabilities described in Sec. 10.1.2. At present, commercial therapy planning systems allowing dose distributions calculation for proton beams do not exist. The Swedish company Helax is improving an already existing system, which can deal with proton radiation. It is based on the pencil beam model. The program is used successfully at GWI in Uppsala. This system was also implemented in Polish oncological centers for conventional radiotherapy planning. Therefore, it seems reasonable to use it also for Protontherapy facility.

Systems for Patient Immobilization. There is a margin of error all around the target as seen from the beam's-eye-view, and the immobilization must hold the patient so the target remains inside that margin of error. The requirements for the immobilization were presented in Sec. 10.1.3.2. For many cases a contoured mask or cast may be produced using radiolucent thermoplastic material (Acquaplast^{TM}) with or without multiple perforations. In some cases, the thermoplastic masks should used in combination with other devices (bite block) to obtain a complete fixation of the head and neck in both the seated and supine position.

Treatment Couch and Chair. The main performance specifications for treatment couch and chair were described in Sec. 10.1.3.3, some examples of the existing solutions were presented in Sec. 10.3.2. The design and construction of the treatment couch can be accomplished in two ways: (1) by use of a commercially available couch like those that are provided with linear accelerators (this solution was adopted at LLUMC), and (2) by specially designed couch (this solution was chosen for the NPTC - Fig. 10.4). The second solution guarantees achieving more degrees of freedom and better precision of table positioning.

Different types of chairs are used to treat patients in the seated position and in particular to treat ocular tumors (Sec. 10.3.2). At this stage, it is difficult to give the precise specifications concerning the treatment chair. However, it has to emphasized that the construction of the chair must meet the requirements described in Secs. 10.1.3.3 and 10.3.2.

Patient Handling Systems in the Proton Facility. The patient should be prepared for the treatment in the preparation room. All operations should be performed with the patient lying immobilized in the supine position in his or her individually moulded couch. It should be possible to transport the patient couch using a special carriage system (patient transporter), as it is done at PSI facility (Sec. 10.3.5.1).

The patient couch should coupled to the CT table using the same clamping mechanism as for the proton gantry. In this way, a well-defined mechanical reproducibility of the position of the couch on the gantry and in the CT will be guaranteed for all phases of treatment. In order to achieve the required mechanical precision, the entire patient support system of the commercial CT is usually replaced by a new table designed specifically for this task. The positioning of the patient should be checked before each fraction by taking the same scout view images and comparing them on a computer workstation with the reference images. By checking the position of the patient using the same CT as that used for treatment planning, we obtain the most direct validation of the treatment planning data before each fraction. After the completion of the patient positioning in the CT, the patient is moved into the proton treatment room.

In the gantry room the couch should be coupled to the table of the proton gantry. If necessary, the position of the patient with respect to the couch is checked again with x-rays.

14.1.1.7 Control and Safety Subsystem

The control system automates and centralizes all the activities needed for a correct management of the Protontherapy center. The control system must provide the tools for the integration of the medical diagnostic facilities with the treatment planning and beam delivery equipment. A wide-band local area network is a good solution that makes possible the rapid distribution of large amounts of information related to the medical digital images.

The safety system should contain systems for: (1) patient monitoring, (2) emergency buttons, (3) door interlocks, (4) radiation monitors, (5) mechanical safety systems. The safety and control systems should be integrated with the parallel system installed at the oncological hospital complex. The choice of the company solution of Protontherapy facility, the control and safety systems are the integral part of the whole project.

14.1.1.8 Estimated Costs of Protontherapy Facility

The introduction of every new treatment modality into clinical routine is difficult, especially if it is much more expensive in setting up and running than the conventional one. The clinical benefit has to be proven in relation to the increased costs of treatment (Secs. 2.3 and 3). The discussion and the estimate of costs and prices will usually depend on a given social system and the overall cost level in different countries (Chapter 13).

The costs of existing dedicated Protontherapy centers (Tables 13.5 and 13.7) can be the basis for estimating the costs of constructing the proposed dedicated proton facility. According to the preliminary estimations, the costs of design, equipment and construction of the center will be about U$ 40 – 50 mln. The estimated relative Protontherapy facility equipment costs are presented in Table 14.4 (on the basis of M. Goitein publication [4.6]).

Subsystem	Relative costs (%)
Accelerator	26
Energy selection system	4
Beam transport system	10
Two gantries	22
Fixed-beam room	15
Patient positioners (2)	8
Safety and control systems	15

Table 14.4. Relative costs of Protontherapy equipment; after M. Goitein [4.6].

14.1.2 Building for Protontherapy Facility

In choosing how to approach the building construction M. Goitein [4.6] considered two approaches: (1) designing the building, and then bidding out its construction based on the established design, and (2) so-called design / build approach in which a single contractor designs and builds the building. For NPTC they have chosen the design / build approach and they have been very satisfied with it. It had the consequence that the contractor, the architect, the constructor and all the other related disciplines worked together as a team to solve problems and most importantly, they took on board the project's four goals of quality, preserving (and, indeed, expanding) the project scope, keeping within budget and on schedule.

The advantage of the design / built approach is that one avoids the rather common situation in which, when a problem arises, it is unclear whether the designer or builder is responsible for it with the result that, in any event, the customer is likely to have to pay for the neccessary correction.

One aspect of building construction comes under the title of "tight vs. loose fit". The traditional way that high-energy accelerators and some particle medical facilities (i.e. Chiba facility) are built is to locate the accelerator in a more or less rectangular, somewhat oversized, building so that there is likely to be adequate room to accommodate evolutions of the design. This approach has the enormous advantage of flexibility for future modifications and even expandability However, it is not uncommon to have severe site restrictions. This was the case with the NPTC. Then, it may be necessary to have a very tight fit of the building to the equipment. In that case, it is vital to have a clearly established building / equipment interface.

14.1.2.1 Space Solution in Protontherapy Facility

The proton facility building will house the protontherapy equipment and the related program space. But at the same time it is important to keep the clinical and technical areas separated (see Secs. 5.2.1 and 5.4).

The clinical area should be located at the different levels than the accelerator and services. The advantages in terms of an optimized patient flow and decoupled clinical and research / service activities must be balanced against the possible additional complexity of the building, shielding requirements, a more complex beam transport and overall costs.

The treatment floor of the building should be underground, containing the accelerator, treatment rooms and some equipment support space, and pits to accommodate gantry rotation extending below that floor. The facility should be designed so as to be capable of subsequent expansion to additional treatment rooms. It should also include clinical space for patient examination and care. The ground level should contain administrative, staff and miscellaneous support space as well as aspects of equipment support and clinical space.

The space of the center should provide approximately 40 - 60 000 net feet (14 – 18 000 m^2) of program space including three clinical treatment rooms, accelerator equipment, and offices (on the basis of the existing solutions of dedicated Protontherapy facilities, Table 5.2).

14.1.2.2 Treatment Rooms

A dedicated proton medical facility should have two or three rotating gantry treatment rooms and at least one horizontal fixed-beam treatment room (Sec. 5.2.1). If the construction proceeds in a phased mode, at least one gantry room and one fixed-beam room must be provided in the initial phase. The initial facility must be constructed in such a way that it will be compatible with future expansion in a cost-effective manner. Excellent solutions of treatment levels of the operating Protontherapy facilities are presented in Figs. 5.5 and 5.8.

The horizontal fixed-beam room must be designed so that small-field treatments (e.g. for eye treatment) and the large-field (up to 40 × 40 cm^2) treatments can be performed. The equipment switch time between the eye treatment and large-field, must be less then 10 min.

The rotating gantry room must be designed to provide at least 40 × 30 cm^2 fields. Together with the couch specification, the facility must be able to irradiate any part of the patient from any direction.

14.2 Conclusions

A proposed dedicated full-energy Protontherapy facility is characterized by the specifications listed in Table 14.5.

<table>
<tr><td rowspan="7">Accelerator</td><td rowspan="2">Energy range</td><td>maximum</td><td>250 MeV</td></tr>
<tr><td>minimum</td><td>70 MeV</td></tr>
<tr><td colspan="2">Energy precision</td><td>± 0.4 MeV</td></tr>
<tr><td colspan="2">Average beam intensity</td><td>10^{11} protons s^{-1}</td></tr>
<tr><td colspan="2">Beam structure</td><td>continuous or quasi-continuous, scanning ready</td></tr>
<tr><td colspan="2">Exploitation</td><td>most of the accelerator parameters fixed</td></tr>
<tr><td colspan="2">Accelerator type</td><td>cyclotron or synchrotron</td></tr>
<tr><td colspan="3">Beam transport system</td><td>configuration and parameters depending on the facility configuration and types of the equipment</td></tr>
<tr><td rowspan="5">Proton gantries</td><td colspan="2">Number of gantries</td><td>2 - 3</td></tr>
<tr><td colspan="2">Dimensions (diameter)</td><td>< 13 m</td></tr>
<tr><td colspan="2">Rotation angle</td><td>370°</td></tr>
<tr><td colspan="3">Passive beam modification system installed predominantly; active system – as an option</td></tr>
<tr><td colspan="2">Gantry type</td><td>„large-throw"</td></tr>
<tr><td rowspan="6">Beam modification systems</td><td colspan="2" rowspan="2">Variable range shifting</td><td>Extraction the beam of variable energy from accelerator</td></tr>
<tr><td>Circularly shaped double-wedge absorber</td></tr>
<tr><td colspan="2">Range modulation</td><td>Plexiglass propeller or rotating fan-shaped stepped absorber</td></tr>
<tr><td rowspan="2">Lateral beam spreading</td><td>passive delivery systems</td><td>Double scattering foils</td></tr>
<tr><td>active delivery systems</td><td>Pixel scanning method</td></tr>
<tr><td colspan="2">Collimators</td><td>Aperture and multileaf</td></tr>
<tr><td rowspan="4">Dosimetry systems</td><td colspan="2">Monitoring chain</td><td>Wire chambers, plastic scintillators</td></tr>
<tr><td colspan="2">Reference dosimetry systems</td><td>Calorimeters, Bragg-Gray cavity ion chambers</td></tr>
<tr><td colspan="2">Relative dosimetry systems</td><td>Diodes, dosimetric films</td></tr>
<tr><td colspan="3">For scanning systems– dedicated monitoring chain</td></tr>
<tr><td rowspan="3">Treatment ancillary facilities</td><td colspan="2">Treatment planning system</td><td>TMS</td></tr>
<tr><td colspan="2">Treatment couch and chair</td><td>Dedicated</td></tr>
<tr><td colspan="2">Patient handling system</td><td>Dedicated system, including treatment couch, patient positioning and immobilization devices</td></tr>
<tr><td rowspan="3">Facility building</td><td colspan="2">Estimated space of the facility</td><td>14 – 18 000 m^2</td></tr>
<tr><td colspan="2">Architecture</td><td>Two floors: treatment and administrative</td></tr>
<tr><td colspan="2">Treatment rooms</td><td>2 – 3 gantry rooms, 1 fixed beam room, 1 experimental room</td></tr>
</table>

Table 14.5. Proposed parameters and equipment for a dedicated full-energy Protontherapy facility.

REFERENCES

[1.1.] N. Ivanov et al.: Kompaktni protonove synchrotrony pro hadronovou terapii in [1.2]

[1.2.] Cekoslovensky casopis pro fyziku, No. 47, Fyzikalni Ustav, Akademie Ved Ceske Republiky, Praha, 1997

[1.3.] W. H. Scharf, O. A. Chomicki: Medical Accelerators in Radiotherapy: Past, Present and Future, Phys. Med., No. 4, Oct.-Dec., 1996

[1.4.] U. Amaldi, M. Silari: The TERA Project and the Centre for Oncological Hadrontherapy, Rome, July 1994

[1.5.] J. Debus et al.: Proposal for a Dedicated Ion Beam Facility for Cancer Therapy, Eds.: K. D. Groß, M. Pavlovic (GSI), September 1998

[1.6.] W. T Chu: Ion Beams for cancer treatment □ a perspective in [1.7]

[1.7.] Proceeding of CAARI96, Eds. J. L. Duggan and I. L. Morgan, AIP Press, New York, May 1997

[1.8.] Anonymous: Heavy Ion Medical Accelerator in Chiba, NIRS, 1993

[1.9.] Vermoken AJM and Schermer FATM (Eds.): Towards Coordination of Cancer Research in Europe, IOS PRESSS, Amsterdam-Oxford-Washington, D.C.-Tokyo, 1994

[1.10.] M. Regler and Th. Auberger: The Austron project: in-depth study of the clinical aspect of AUSTRON, in [1.11]

[1.11.] U. Amaldi, B. Larsson, Y. Lemoigne (Eds.): Advances in Hadrontherapy, Excerpta Medica, Int. Congress Series 1144, Elsevier, Amsterdam, 1997

[1.12.] U. Amaldi: TERA Programme: medical applications of protons and ions, private communication

[1.13.] G, Kraft, G. Gademann: Einrichtung einer experimentellen Strachlentherapie bei der GSI, Darmstadt, GSI-Report 93-23, Darmstadt 1993

[1.14.] Anonymous: Tumor Therapy with Heavy Ions, GSI-Nachrichten, 4-96, Darmstadt

[1.15.] U. Amaldi and S. Rossi: The TERA Foundation and its projects in oncological hadrontherapy in 1.11

[1.16.] E. Nahum: Prospects for proton-beam radiotherapy, Europ. Journ. of Cancer, Vol. 30A, No. 10, 1994

[1.17.] K. Prokes: Radioterapie nadorovych onemocneni na prelomu 20. stoleti in [1.2]

[1.18.] O. Chomicki and W. Scharf: Akceleratory medyczne w radioterapii: historia, stan obecny i perspektywy, Postępy Fizyki, No. 45 (5), 1994
[1.19.] Pietraszek et al.: Obyśmy byli zdrowi, Gazeta Stołeczna, 8.11.1998
[1.20.] H. D.Suit et al.: Increased efficacy of radiation therapy by use of proton beam, Strahlenther. Onkol. 166(No.1), 1990
[1.21.] U. Amaldi: Overview of the world landscape of hadrontherapy and projects of TERA foundation, Phys. Med., Vol.XIV, Suppl. 1 July 1998
[1.22.] W. Scharf Biomedical Particle Accelerators, AIP Press, New York, 1994
[1.23.] Proton beam therapy for better quality of life, PRMC, Univ. of Tsukuba, 1994
[1.24.] Proposal for a dedicated ion beam facility for cancer therapy
[1.25.] D.T.L. Jones Present status and future trends of heavy particle radiotherapy in [4.52]
[1.26.] W. Wieszczycka Preliminary project of Polish dedicated Protontherapy facility, Warsaw, 1999
[1.27.] R.A. Gough The Light Ion in Biomedical Research Accelerator (LIBRA), LBL-23413, March 1987

[2.1.] B. Schaffner Range Precision of Therapeutic Proton Beams, Diss. ETH No. 12474, PSI, 1997
[2.2.] G. Kraft Radiobiology of heavy charged particles, GSI-Preprint-96-60, Nov. 1996
[2.3.] G. Kraft The radiobiological and physical basis for radiotherapy with protons and heavier ions, Strahlenther. Oncol. 166 (1990), 10-13 (Nr. 1)
[2.4.] G. Kraft What kind of radiobiology should be done at hadrontherapy centre? in [1.11]
[2.5.] R. Oozeer et al. A model for the lateral penumbra in water of 200 MeV proton beam devoted to clinical applications, Med. Phys. 24(10), Oct. 1997
[2.6.] L. D. Skarsgard Radiobiology with heavy charged particles, Phys. Med. Vol 14, Suppl. July 1998
[2.7.] Wainson et al. The RBE of accelerated protons in different parts of the Bragg curve, Brit. J. Rad. 45, 1972
[2.8.] J. Robertson et al. Radiobiological studies of high energy modulated proton beam utilizing cultured mammalian cells, Cancer 35, 1975
[2.9.] R. Cox et al. Inactivation and mutations of cultured mammalian cells by aluminium characteristic ultrasoft x-rays. Dose responses of Chinese hamster and human diploid cells to aluminium x-rays and radiations of different LET, Int. Rad. Biol. 36, 1979
[2.10.] D. Bettega et al. Relative biological effectiveness for protons of energies up to 31 MeV, Rad. Res. 77, 1979

[2.11.] E. Blumquist et al. Relative biological effectiveness of intermediate energy protons. Comparison with ^{60}Co gamma-radiation using two cell lines, Rad. Oncol. 28, 1993

[2.12.] M. R. Raju Radiotherapy: heavy ions, mesons, neutrons and protons in Physics in Medicine and Biology, Encyclopedia, ed. T.F. McAinsh, Pergamon Press, Oxford

[3.1.] R. Miralbell Treatment precision with proton beams: clinical experience and potential new indications in [1.11]

[3.2.] B. Schaffner Range uncertainties of protons and their consequences for proton radiotherapy planning in PSI Annual Report 1997 / Annex 2, Villigen, 1997

[3.3.] K.M. Hanson et al. Computed tomography using proton energy loss, Phys. Med. Biol. 26(6), 1981

[3.4.] R. Wilson Radiological use of fast protons, Radiology 46, 1946

[3.5.] C. Tobias et al. Pituitary irradiation with high energy proton beams: a preliminary report, Cancer Res. 18, 1958

[3.6.] J. Sisterson Particles 26, July 2000

[3.7.] J. Sisterson Protontherapy, in 1996 in CAARI'96 Proceeding, AIP Press, New York, 1997

[3.8.] M.R. Raju Proton radiobiology, radiosurgery and radiotherapy, Int. J. Radiat. Biol. 67(3), 1995

[3.9.] K.N. Johnson 35 years of patient treatments at the HCL: a geografical perspective in Abstracts of the PTCOG, Boston, April-May 1997

[3.10.] J. Munzendrider Protontherapy with the Harvard Cyclotron in U. Amaldi, D. Larsson (eds.) Hadrontherapy in Oncology, Amsterdam, Elsevier Science B.V., 1994

[3.11.] R.N. Kjellberg et al. Life time effectiveness a system of therapy for pituitary adenomas, emphasizing Bragg peak proton hypophysectomy, in Recent Advances in the Diagnosis and Treatment of Pituitary Tumors, Edited by J.A. Linfoot (New York, Raven)

[3.12.] B. Kliman Proton beam therapy of acromegaly: a 20-year experience; in Progress in Endocrine Research and Therapy, eds. P. McL. Black (New York, Raven)

[3.13.] Y. Minakova Proton beam therapy in neurosurgical clinical practice, in Medical Radiology 32 (in Russian), Moscow, 1987

[3.14.] B.A. Konnov Protontherapy at Lenningrad Synchrocyclotron, in Proceeding of PTCOG, Loma Linda, Ca, Oct. 1987

[3.15.] M. Krengli et al. Clinical experience with proton beam therapy at MGH, MEEI and HCL: skull base sarcoma and uveal melanoma in [1.11]

[3.16.] G. Coutrakon et al. Design considerations for medical proton accelerators in [3.17]

[3.17.] Proceedings of the 1999 Accelerator Conference, New York, 1999
[3.18.] Good Experience with Protontherapy in PSI Annual Report 1999, General Report, 1999

[4.1.] W.T. Chu et al. Instrumentation for treatment of cancer using proton and light-ion beams, Rev. Sci. Instrum. 64 (8), August 1993.
[4.2.] T. Chu et al. Performance specifications for proton medical facility, LBL-33749, UC-000, Lawrence Berkeley Laboratory, March 1993
[4.3.] J. Flanz et al. Overview of the MGH-Northeast Protontherapy Center plans and progress, Nucl. Instr. Meth. B 99, 1995
[4.4.] D.E. Bonnett Current developments in Protontherapy: a review, Phys. Med. Biol. 38, 1993
[4.5.] IBA Protontherapy System, IBA, 1994
[4.6.] M. Gotein The technology of hadrontherapy: the context within which technical choices are made in [1.11]
[4.7.] F.J.M. Farley Advanced European Medical Facilities in EPAC'94 Proc.
[4.8.] J.R. Alonso Magnetically scanned ion beams for radiation therapy, Nucl. Instrum. Methods Phys. B40/41, 1987
[4.9.] J.M. Slater Protontherapy: an advanced form of radiation therapy, http://www.llu.edu/proton/physician/technical.html
[4.10.] S. Graffman et al. Proton radiotherapy with the Uppsala cyclotron. Experience and plans, Strahlenterapie 161(12), 1985
[4.11.] Machine development, Development of radiotherapy with proton beam in TSL Progress Report 1996 – 97
[4.12.] J.E. Munzenrider et al. Protontherapy at Harvard, Strahlenterapie 161(12), 1985
[4.13.] G. Dutto Recent performances of the TRIUMF cyclotron and status of the facility, TRI-PP-95-79, Sept. 1995
[4.14.] I.V. Chuvilo ITEP synchrotron proton beam in radiotherapy Int. J. Radiat. Oncol. Biol. Phys. 10, 1983
[4.15.] H. Tsujii Planning for coordinate program for expansion of proton and ion beam therapy, in Japan in CAARI'96 Proceeding, AIP Press, New York, 1997
[4.16.] D. Jones et al. The NAC particle therapy facilities, private communication, Feb. 1994
[4.17.] History, The Center of Protontherapy of Orsay, http://www.inria.fr/epidaure/bondiau/bondiau.html
[4.18.] P. Chauvel Status report on the installation of proton and neutrontherapy in Centre Antoine-Lacassagne, EPAC'90 Proc.
[4.19.] Proton Medical Research Center, in KEK Annual Report 1990
[4.20.] The OPTIS project at PSI, http://www.psi.ch
[4.21.] The new Protontherapy facility of PSI, http://www.psi.ch

[4.22.] J. M. Slater et al. The proton treatment center at Loma Linda University Medical Center: rationale for and description of its development, Int. J. Radiat. Oncol. Biol. Phys. Vol. 22, 1991
[4.23.] J. M. Slater et al. Applying charged particles physics technology for cancer control at Loma Linda University Medical Center, USA, EPAC'90 Proc.
[4.24.] G. Coutracon et al. A prototype beam delivery system for the proton medical accelerator at Loma Linda, Med. Phys. 18, 1991
[4.25.] Massachusetts General Hospital, Department of Radiation Oncology, http://www.mgh
[4.26.] J.B. Flanz et al. Recent performance of the NPTC equipment compared with clinical specifications, January 1998, personal communication
[4.27.] J.M. Sisterson Update on the world wide Protontherapy experience through 1997 in Abstracts of the PTCOG Meeting, Sept. 14-16 1998
[4.28.] U. Amaldi et al. A hospital based accelerator complex, EPAC'94 Proc.
[4.29.] S. Frullani The TOP-LINAC for the Protontherapy Project of the Italian National Institute of Health (ISS), http://www.tera.it
[4.30.] M.P.R. Waligórski et al. The Hadron Radiotherapy Center in Kraków: a project based on 20 years clinical experience with fast neutrons, Brachytherapy 13(1), Jan. 1997
[4.31.] G. Cesari et al. Feasibility Study of a Synchrotron for the European Light Ion Medical Accelerator
[4.32.] P.J. Bryant Progress of the Proton – Ion Medical Machine Study (PIMMS), GSI 97-07
[4.33.] P.J. Bryant Progress of the Proton – Ion Medical Machine Study (PIMMS), Phys. Med. Vol. XIV, Suppl. 1, July 1997
[4.34.] S. Yamada et al. HIMAC and medical accelerator projects in Japan in CAAA'98
[4.35.] T. Ogino et al. Proton treatment facility at NCC, Kashiwa, Japan: a progress report, in Abstracts of XXVI PTCOG Meeting, April 30-May 2 1997
[4.36.] T. Tachikawa et al. Protontherapy system for National Cancer Center in Abstracts of XXVI PTCOG Meeting, April 30-May 2 1997
[4.37.] Operation of treatment facilities in NAC Annual Reoprt 1998
[4.38.] B. Gottschalk et al. Synchrotron survivor to bow out after 50 years, CERN Courier Feb. 1999
[4.39.] M. Inoue Status of accelerators in Japan in [4.40]
[4.40.] Proceeding of the First Asian Particle Accelerators Conference, March 23-27, 1998, KEK, Tsukuba, Japan
[4.41.] W.P. Jones et al. Design of a beam transport system for a proton radiation therapy facility in [3.17]
[4.42.] B. Glimelis et al. Project of bio-medical hall, in TSL Progress Report 1998 – 1999, TSL, Uppsala, Sweden

[4.43.] V.M. Abazov et al. Medical facility for the radiation therapy with JINR proton phasotron beams, JINR, 1994
[4.44.] Phasotron at the Laboratory of Nuclear Problems, JINR, and its beams, Dubna 1992
[4.45.] Abstracts of the XXXI PTCOG Meeting, Bloomington, Indiana, USA, 11 - 13 Oct. 1999
[4.46.] Y. Takada Gantry design and beam dellivery system of the new Protontherapy facility, in PMRC, Tsukuba in [4.45]
[4.47.] S. Reimoser et al. Practical design of a 'Reisenrad' ion gantry in [4.45]
[4.48.] P.G.Seiler et al. First position dependent beam gating of the PSI gantry during a phantom irradiation: a step towards conformal Protontherapy with real time tumor tracking, in [4.45]
[4.49.] S. Reimoser et al. Overview of the proton-ion medical machine study (PIMMS) in [4.45]
[4.50.] S. Fukuda Project of proton facility at WERC (Wakasa Energy Research Center), private correspondence, 2000
[4.51.] J.C. Cornell et al. New beamlines for Protontherapy at NAC in [4.52]
[4.52.] Proceedings of Conference on Cyclotrons and Their Applications 98, Caen, 14-19 June 1998
[4.53.] P.J. Bryant Progress of the Proton – Ion Medical Machine Study (PIMMS), Physica Medica, Vol. XIX, Supplement, 1 July 1998, in [1.56]
[4.54.] Proton Ion Medical Machine Study (PIMMS), CERN Report of Activities, Annnual Report, 1997, Vol. II
[4.55.] M. Blom Development of a scanning system for Protontherapy in Uppsala, in [4.56]
[4.56.] Proceedings of EPAC, 1998
[4.57.] P.J. Bryant Developments in the design of proton and ion accelerators for medical use, in [4.56]
[4.58.] S. Lorin The proton radiation therapy project, Uppsala Accelerator news, January 2000
[4.59.] M. Blom et al. Development of a scanning system for Protontherapy in Uppsala, in [4.56]
[4.60.] L.W. Funk A medical facility proposal to use the SSC linac, Nuclear Instruments and Methods in Physics Research B99 (1995)
[4.61.] M. Goitein The Northeast Protontherapy Center, Boston, USA: A status report in [3.6]
[4.62.] J. Sisterson Particles 25, January 2000
[4.63.] A.E. Morrison FDA approval and Medicare coverage and payment issues in Protontherapy, in [4.63]
[4.64.] LIBO Project, in CERN Report of Activities in the Divisions, Annual Report 1999, Vol.II

[4.65.] Sonia Ternier A Protontherapy system for in-hospital operation, IBA, private correspondence, 2000
[4.66.] U. Amaldi The importance of particle accelerators, EPAC 2000, Wien, 26-30 June 2000
[4.67.] H. Tsujii et al. Preliminary results of phase I/II carbon-ion therapy at the National Institute of Radiological Sciences, in Int. J. of Brachytherapy, Vol. 13 no 1, Jan. 1997
[4.68.] S. Yamada Commissioning and performance of the HIMAC medical accelerator, in IEEE 1996
[4.69.] A. Maruhashi PMRC Design - figures, private correspondence, 2000

[5.1.] W.T. Chu et al. Performance specifications for proton medical facility, LBL-33749, UC-000, Lawrence Berkeley Laboratory, March 1993
[5.2.] K. P. Gall et al. State of Art.? A New Proton Medical Facilities for MGH and UCDMC, Nuclear Instruments and Methods in Physics Research B79 (1993)
[5.3.] Proton Cancer therapy equipment, http://www.triumf.ca
[5.4.] The proton gantry, http://www.psi.ch
[5.5.] M. Siliari Basic design considerations for a national laboratory to support advanced clinical applications of ion beams, Nucl. Instr. Meth. B 99, 1995

[6.1.] P. Mandrillon Accelerators for Hadrontherapy, in [1.11]
[6.2.] T. Stammbach Cyclotrons, JUAS'97
[6.3.] G. Coutracon et al. Spill uniformity measurements for a raster scanned proton beam, in CAARI'96 Proceeding, AIP Press, New York, 1997
[6.4.] R.W. Hamm et al. Preliminary Design of a Dedicated Protontherapy Linac, in PAC'91 Proceedings
[6.5.] L. Picardi et al. The TOP – ISS Linear Accelerator: a High Freaquency Proton Linac for Therapy, in EPAC'96 Proceedings
[6.6.] L. Picardi et al. The ISS protontherapy linac, in CAARI'96 Proceeding, AIP Press, New York, 1997
[6.7.] D. Vandeplassche et. al. 235 MeV cyclotron for MGH's Northeast Protontherapy Center (NPTC): present ststus in EPAC'96 Proceeding
[6.8.] W.T. Chu Ion beams for cancer treatment – a perspective, Nucl. Instrum. Meth. B(99), 1995
[6.9.] S.I Kukarnikov et al. Design of dedicated proton synchrotron for Prague Radiation Oncology Centre in Proceeding of the EPAC'98
[6.10.] G. Arduini et al. An H^- / light ion synchrotron for radiation therapy, Nucl. Instr. Meth. A(365), 1995
[6.11.] Song Shipeng et al. Conceptual design of Bejing Protontherapy synchrotron, Journal of Brachytherapy 13(1), Jan. 1997

[6.12.] J.M. Slater et al. Overview of technical and operational essentials for a hadrontherapy facility, in [1.11]
[6.13.] P.Mandrillon A compact facility for high energy Protontherapy based on superconducting cyclotron, in EPAC'94 Proceedings
[6.14.] C. Bieth et al. A very compact protontherapy facility based on an extensive use of high-temperature superconductors (HTS), paper presented at Cyclotrons and Their Applications 98, Caen, 14-19 June 1998
[6.15.] Protontherapy accelerator and gantry – conceptual design, Pantechnik, CAL, INP3, Institut Curie, Jan. 1998, private corresspondence A. Laisne, April 2000
[6.16.] A. Laisne Detailed pre-study of the compact isochronous cyclotron PK210, private corressspondence, April 2000
[6.17.] U. Amaldi et al. A 3 GHz proton linac booster of 200 MeV for cancer treatment

[7.1.] H. Blattmann Beam delivery systems for charged particles, Radiation and Environmental Biophysics, 31, 1992, 219-231
[7.2.] J.B. Flanz et al. Design considerations for a Protontherapy beamline with an energy degrader, in Proceedings of CAARi'96, 1257-60
[7.3.] Th. Haberer et al. Magnetic scanning system for heavy ion therapy, Nucl. Instr. Meth. A330, 1993, 296-305
[7.4.] Y. Futami Development of 3-dimentional irradiation system for heavy-ion radiation therapy in JAERI-Conf 95-021, 442-444
[7.5.] U. Weber, G. Kraft Bragg-peak-widening using a ripple filter, GSI-97-09, Darmstadt, 1997
[7.6.] H. Blattman i in. Spot scanning for 250 MeV protons, Strahlenther. Oncol. (1), 1990, 45-48
[7.7.] Spot Scanning: gantry equipment, properties, html://www.psi.ch\scan_depth.html

[8.1.] M. Pavlovic Gantries for light-ion cancer therapy, GSI 97-09, Darmstadt, 1997
[8.2.] J. Flanz Large medical gantries in the PAC'95 Proceeding
[8.3.] M.E. Schulze Commissioning results of the LLUMC beam switchyard and gantry, in the PAC'91 Proceedings
[8.4.] E. Pedroni et al. The 200 MeV Protontherapy at Paul Scherrer Institute: conceptual design and practical realization, Med. Phys 22(1), Jan. 1995
[8.5.] E. Pedroni, H. Enge Beam optics design of compact gantry for Protontherapy, Med. Biol. Engin. Comput., May 1995
[8.6.] B. Astrakhan AntiGANTRY (AG) – a new system of engineering and technology of rotational-scanning proton / heavy ion - therapy without gantry, in [3.6]

[8.7.] Abstracts of the XXXII PTCOG MEETING and the Workshop on Computational Methods for Proton Beam Treatment Planning, Uppsala, Sweden, April 15-20 2000
[8.8.] B. Astrakhan AntiGANTRY (AG) – a new system of engineering and technology of rotational-scanning proton / heavy ion - therapy without GANTRY, in [8.7]
[8.9.] M.F. Lomanov Remark on the report of Dr. Astrakhan about antigantry system, in [8.7]
[8.10.] Report on the Utilisation of Particle Accelerators for Protontherapy, 7-10 July, 1998, F1-AG-1010, Vienna
[8.11.] Teiji Nishio Protontherapy facilities, private corresspondence, June 2000
[8.12.] Present status and planning of facilities for proton and heavy ion cancer treatment in Japan, J. At Energy Soc. Japan, Vol. 41, No. 11, 1999

[9.1.] J. Mijnheer Protocol for the determination of absorbed dose in the patient irradiated by beams of nuclear particles in radiotherapy procedures, in Nuclear and atomic data for radiotherapy and related radiobiology IAEA, Vienna, 1987, 141-54
[9.2.] P.J. Kliauga et al. The relative biological effectiveness of 160 MeV protons, Int. J. Radiat. Oncol. Biol. Phys. 4, 1978, 1001-1018
[9.3.] ICRU Report No. 49. Stopping Powers and Ranges for Protons and Alpha Particles. ICRU, Bethesda MD, 1993
[9.4.] Mazal et al. La protontherapie: bases physiques et technologiques, Bull. Cancer / Radiother. (83), 1996, 230-246, Elsevier, Paris
[9.5.] E. Grussel et al. Faraday cup dosimetry in a Protontherapy beam without collimation, Phys Med. Biol. 40, 1995, 1831-40
[9.6.] S. Vynckier, D.E. Bonnett, D.T.L. Jones Supplement to the code of practice for clinical proton dosimetry, Radiotherapy and Oncology (32), 1994, 174-79
[9.7.] S. Vynckier, D.E. Bonnett, D.T.L. Jones Code of practice for clinical proton dosimetry, Radiotherapy and Oncology (20), 1991, 53-63
[9.8.] A.R. Smith Dosimetry intercomparisons and protocol for heavy charged particle therapy beams, Int. J. Radiat. Biol. Phys. (8), 1982, 2061-63
[9.9.] Vatnisky et al. Proton dosimetry intercomparison, Radiotherapy and Oncology 41, 1996, 169-77.
[9.10.] Brusasco et al. A detection system for the verification of three dimentional dose distributions, in [1.11]

[10.1.] Lomax and M. Goitein Patient assessment, treatment planning and dose delivery in [1.11], 233-250

[10.2.] R. Ragona et al. Treatment planning of proton beams using the GEANT Monte Carlo in [1.11], 304-308
[10.3.] S. Garelli et al. Real and simulated data comparitive studies for proton energy in [1.11], 331-334
[10.4.] Treatment planning; http://www.psi.ch/tr_plan.html
[10.5.] M. Goitein et al. Planning treatment with heavy charged particles, Int. J. Radiat. Biol. Phys. Vol. 8, 1982, 2065-70
[10.6.] A.N. Schreuder et al. The PROXELPLAN system used at NAC for proton treatment planning
[10.7.] Patient handling; http://www.psi.ch/handl.html
[10.8.] J.B. Flanz et al. Design approach for a highly accurate patient positioning system for NPTC in [1.11]
[10.9.] K. Noda et al. A treatment beam control system for irradiation gated by respiration in JAERI-Conf 95-021, 439-441
[10.10.] K. Noda et al. Performance of a respiration-gated beam control system for patient treatment in EPAC'96 Proceeding, 2656-58
[10.11.] A.N. Schreuder et al. Three years' experience with NAC prothontherapy patient positioning system in [1.11], 251-259
[10.12.] E. Grusell Protontherapy at the The Svedberg Laboratory in Uppsala Accelerator News (14), Automn 1996

[12.1.] S. Agosteo et al. Shielding for proton medical accelerator facility, private communication
[12.2.] S. Agosteo et al. Shielding for proton medical accelerator facility, private communication
[12.3.] H.B. Knowles et al. Shielding and activation study for proton medical accelerators
[12.4.] J.V. Siebers Shielding measurements for a 230 MeV proton beam, PhD thesis, Univ. of Wisconsin, Madison, 1990
[12.5.] International Commission on Radiological Protection. ICRP Publication 60, Pergamon Press, 1990
[12.6.] S. Agosteo et al. Monte Carlo calculations as shielding design tools for heavy charged particle accelerators, Journal of Brachytherapy 13(1), Jan. 1997

[13.1.] G. Gademann Cost-benefit considerations of hadrontherapy, in [1.11]
[13.2.] H.D. Suit, M. Krengli Basis for interest in proton beam radiation therapy, in [1.11]
[13.3.] Einhei tlicher Bewertungsmasstab 1.6.1996, Deutschher Arzteverlag 1996
[13.4.] U. Amaldi Overview of the world landscape of hadrontherapy and the projects of the TERA Foundation, GSI 97-09, Darmstadt, 1997

[13.5.] J.M. Slater et al. Accelerators for cancer therapy, in Proc. PAC'99

[14.1.] M.P.R. Waligórski et al. The hadron radiotherapy Centre in Krakow – a project on 20 years of clinical experience with fast neutrons

INDEX

Quality factor, 27